D1531305

Progress in
Medicinal Chemistry 48

Progress in Medicinal Chemistry 48

Editors:

G. LAWTON, B.SC., PH.D., C.CHEM.

Garden Fields
Stevenage Road
St. Ippolyts
Herts SG4 7PE
United Kingdom

and

D. R. WITTY, B.A., M.A., D.PHIL., C.CHEM., F.R.S.C.

GlaxoSmithKline
New Frontiers Science Park (North)
Third Avenue
Harlow, Essex CM19 5AW
United Kingdom

ELSEVIER

AMSTERDAM · BOSTON · HEIDELBERG · LONDON · NEW YORK · OXFORD
PARIS · SAN DIEGO · SAN FRANCISCO · SINGAPORE · SYDNEY · TOKYO

Elsevier
Radarweg 29, PO Box 211, 1000 AE Amsterdam, The Netherlands
Linacre House, Jordan Hill, Oxford OX2 8DP, UK

First edition 2009

British Library Cataloguing-in-Publication Data
A catalogue record for this book is available from the British Library

Library of Congress Cataloging-in-Publication Data
A catalog record for this book is available from the Library of Congress

ISBN: 978-0-444-53358-6
ISSN: 0079-6468

For information on all Elsevier publications
visit our website at elsevierdirect.com

Printed and bound in United Kingdom

09 10 11 12 13   10 9 8 7 6 5 4 3 2 1

Working together to grow
libraries in developing countries

www.elsevier.com | www.bookaid.org | www.sabre.org

ELSEVIER   BOOK AID International   Sabre Foundation

Preface

More than a decade has passed since Chris Lipinski's seminal paper describing the 'rule of 5' which represented the beginning of the 'numbers' era of medicinal chemistry. Potency is clearly by no means the only important number in determining the effectiveness of a drug. Various structure/property correlations are now an integral part of every medicinal chemistry programme. The binding affinity is made up of enthalpic and entropic terms; understanding which of these factors is most important for a particular lead series can considerably aid design decisions. It is also increasingly recognised that the equilibrium binding affinity of a drug for its receptor and its pharmacokinetic profile are not the only important parameters in determining clinical efficacy. The kinetics of drug–receptor interaction can make a significant contribution. In Chapter 1 of this year's volume, Graham F. Smith reviews the various numbers involved in 'drug-likeness' and in the binding equilibrium and kinetics of drug–receptor interactions and how these determine the observed pharmacodynamic effects of the drug.

Research into the mechanisms which control gastric motility offers the promise of treatments for gastrointestinal disorders, including diabetic gastroparesis and conditions induced by chemotherapeutic agents. The motilin receptor, a G-protein coupled receptor which responds to the eponymous peptidic hormone, was also known to be agonised by Erythromycin analogues, but with a limited therapeutic window and suboptimal DMPK characteristics. In Chapter 2, Susan M. Westaway and Gareth Sanger describes the discovery of a range of small molecule drug-like ligands that can modulate this receptor, offering potentially safer treatment options either for enhancing or reducing gastric motility. Examples which have reached clinical assessment are included.

There has been a burgeoning of targets under evaluation as routes to novel oncology therapeutics. These include the phosphoinositide 3-kinase family of lipid kinases which regulate a network of signal transduction pathways controlling cellular processes. A number of ATP-competitive, small molecule inhibitors with distinct subtype selectivity profiles have recently entered clinical trials. Progress in the design, development and clinical evaluation of these compounds is reviewed by Stephen Shuttleworth and colleagues in Chapter 3. This is likely to be a rich vein of drug discovery in the future as more understanding is achieved of the required selectivity profiles for effective personalised therapy.

Dementia is a large and growing problem for society and many medicinal chemists are occupied in developing medicines in this disease area. A great deal of work has focused on the secretase enzymes which are involved in the processing of the amyloid proteins and are thought to play an important role in Alzheimer's disease. HIV protease and renin are examples of aspartyl proteases that have provided successes for medicinal chemists. B-secretase has proven difficult; high-throughput screening has largely been unproductive and the shallow and polar nature of the active site leads to difficulty in achieving 'drug-like', brain-penetrating lead compounds by substrate-based design. The fragment-based approach to ligand-efficient molecules using X-ray, NMR and surface plasmon resonance screening methods has been more productive. Recent progress in all of these approaches is reviewed by Jeffery S. Albert in Chapter 4.

The discovery of ligands for the 5-HT$_6$ receptor has opened up the possibility to treat a number of disease areas believed to be modulated through this GPCR, including cognitive deficits associated with Alzheimer's disease, and metabolic syndromes associated with excessive eating. Recent research has also uncovered a number of compound classes which show dual action at 5-HT$_6$ and other specific receptors. These may offer a wider range of opportunities for therapeutic interventions in conditions such as anxiety, schizophrenia or pain. In Chapter 5, David Witty and colleagues cover current work in this field and review several compounds which have shown efficacy in clinical studies.

September 2009   G. Lawton
                                                         D. R. Witty

Contents

List of Contributors

Mahmood Ahmed
GlaxoSmithKline R&D China, Singapore Research Centre, Biopolis at One-North, 11 Biopolis Way, The Helios, #06-03/04, Singapore 138667

Jeffrey S. Albert
CNS Discovery Research, AstraZeneca Pharmaceuticals, 1800 Concord Pike, PO Box 15437, Wilmington, DE 19850-5437, USA

Alexander Cecil
Karus Therapeutics Ltd., 2 Venture Road, Southampton Science Park, Southampton, SO16 7NP, UK

Tsu Tshen Chuang
GlaxoSmithKline, Sirtris, a GSK Company, 200 Technology Square-Suite 300, Cambridge, MA 01139, USA

Thomas Hill
Karus Therapeutics Ltd., 2 Venture Road, Southampton Science Park, Southampton, SO16 7NP, UK

Helen Rogers
Karus Therapeutics Ltd., 2 Venture Road, Southampton Science Park, Southampton, SO16 7NP, UK

Gareth J. Sanger
Neurogastroenterology Group, Wingate Institute of Neurogastroenterology, Barts and The London School of Medicine and Dentistry, Queen Mary College, University of London, London E1 2AT, UK

Stephen Shuttleworth
Karus Therapeutics Ltd., 2 Venture Road, Southampton Science Park, Southampton, SO16 7NP, UK

Franck Silva
Karus Therapeutics Ltd., 2 Venture Road, Southampton Science Park, Southampton, SO16 7NP, UK

Graham F. Smith
Merck Research Laboratories, 33 Avenue Louis Pasteur, Boston, MA 02115, USA

Cyrille Tomassi
Karus Therapeutics Ltd., 2 Venture Road, Southampton Science Park, Southampton, SO16 7NP, UK

Paul Townsend
School of Medicine, Southampton General Hospital, University of Southampton, SO16 6YD, UK

Susan M. Westaway
Immuno-Inflammation CEDD, GlaxoSmithKline, Medicines Research Centre, Gunnels Wood Road, Stevenage, Herts SG1 2NY, UK

David Witty
Neurosciences Centre of Excellence for Drug Discovery, GlaxoSmithKline, New Frontiers Science Park, Coldharbour Road, Harlow, Essex, CM19 5AW, UK

1 Medicinal Chemistry by the Numbers: The Physicochemistry, Thermodynamics and Kinetics of Modern Drug Design

GRAHAM F. SMITH

Merck Research Laboratories, 33 Avenue Louis Pasteur, Boston, MA 02115, USA

Progress in Medicinal Chemistry – Vol. 48   1
Edited by G. Lawton and D.R. Witty
DOI: 10.1016/S0079-6468(09)04801-2

INTRODUCTION

At the beginning of my career as a medicinal chemist my job seemed to me more like a mysterious dark art than a science. It appeared then that the experienced medicinal chemists had a more intuitive approach rather than a rigorous scientific approach to finding active molecules. My training in synthetic chemistry taught what *could* be made, but it was hard to see what *should* be made, in the absence of knowledge of the target structure to enable structure-based drug design. More recently, a number of rules and equations based on scientific fact and precedent have been developed, and these promote good compound design and a statistically greater chance of successful drug development. In this chapter are outlined key aspects of modern physicochemistry, thermodynamics and kinetics which lead to good drug design. For an aspiring chemist, the ability to articulate the specific reasons for designing the features of compounds demonstrates a true mastery of the medicinal chemistry art.

It takes around 10 years to train a synthetic chemist to be a medicinal chemist and *en route* it is necessary to learn and understand many of these guidelines and why they are important. The description of the principles in this chapter, and their limitations, is intended to accelerate this training and lead to even better drug design.

Medicinal chemists have sometimes seemed to be sceptical of the predictions and rules. It is the view of this author that these guidelines are backed by scientific fact and evidence and therefore ignoring them can be folly.

PHYSICOCHEMISTRY

The landmark papers in the area of physicochemistry came over 10 years ago from Pfizer's Chris Lipinski [1, 2]. His analysis of 2,243 Phase II drug-like compounds from the USAN and INN (United States Adopted Name, International Non-proprietary Name) named drugs databases yielded a guideline, called the 'rule of 5', which has been widely adopted by those designing orally bioavailable drugs. Put simply, it states that 90% of all bioavailable drugs have molecular weight less than 500 or Log$P$ less than 5 or fewer than 5 H-bond donors and fewer than 10 H-bond acceptors. Lipinski stated that 'poor absorption is more likely if' these physicochemical limits are surpassed, however many people reacted strongly initially focusing largely on exceptions to the rule rather than embracing the spirit of the analysis and conclusions. In the years that have followed though, physicochemistry has become a mainstay of modern drug discovery, as evidenced by the fact that Lipinski's original paper has been cited over 2,204 times [3] to date.

DRUG-LIKE MOLECULAR PROPERTIES

Drug-like – rule of 5

In the decade that has followed Lipinski's first publication, a deeper understanding of probabilities and causes of the physicochemical contribution to drug-likeness has led to the refinement of the rule of 5, most notably by workers at Pfizer, AstraZeneca and GSK. The rule of 5 has become deeply intuitive to medicinal chemists and they now routinely use either cLogP versus activity, or molecular weight versus activity, plots as principle components of compound design. In fact Gleeson [4] has shown that molecular weight and cLogP are the two principle components in discriminating drug-like physicochemical properties. Other researchers have added to and refined the rule of 5 still further. Veber *et al.* [5] and Pickett *et al.* [6] from GSK showed that it was possible to conclude that fewer than 10 rotatable bonds and polar surface area (PSA) of less than 140 Å^2 were both related to better oral bioavailability. Kelder [7] and others showed that for CNS penetration the physicochemistry ranges were more tightly defined than for general oral bioavailability and that a lower value of PSA (less than 70 Å^2) was important. Norinder and Haeberlein [8] have also shown that a more subtle balance of physicochemical properties, derived from $(N + O) - \text{Log}P$, must be positive to achieve a high probability of blood–brain barrier penetration.

Lead-like – rule of 4

Teague [9], Oprea [10] and Hann [11, 12] proposed that the physicochemistry of drug leads should be more tightly defined than the original rule of 5 to allow for the seemingly inevitable increase in the values of physicochemical parameters in the progression from leads to development candidates. Thus a rule of 4 for project leads from sources such as HTS can be defined (Table 1.1).

Table 1.1   DRUG-LIKE, LEAD-LIKE AND FRAGMENT-LIKE PROPERTIES

	MWT	LogP	H-bond donors	H-bond acceptors (N+O)	PSA (Å^2)	References
Rule of 5 drugs	<500	<5	<5	<10	<140	[1,5,6]
Rule of 4 leads	<400	<4	<4	<8	<120	[9–11]
Rule of 3 fragments	<300	<3	<3	<3	<60	[13]

Fragment-like – rule of 3

In recent years, the pursuit of lead-like properties by screening very small molecules to find weakly active hits, but having high ligand efficiency, has become widely adopted. This has popularly become known as fragment-based drug discovery (FBDD) and is the subject of several recent high-quality reviews [14–17]. Congreve *et al.* [13] have defined a still tighter rule of 3 for molecules used as screening tools for FBDD which defines the smallest hit molecules that can be detected by specialist methods due to their low binding affinities.

These rules or guidelines now combine to signpost a sensible range of molecular weight and lipophilicity which should be tolerated at early stages of drug discovery. To go beyond these ranges at early stages of research when absorption, selectivity, toxicity and pharmacokinetics have yet to be addressed can easily lead to molecules that are non-optimisable at the later and more costly stages of pre-clinical research. Scheme 1.1 shows the well-known example of AT7519 (1) developed by Astex [18, 19] from a low molecular weight fragment (2). At each stage of its optimisation as

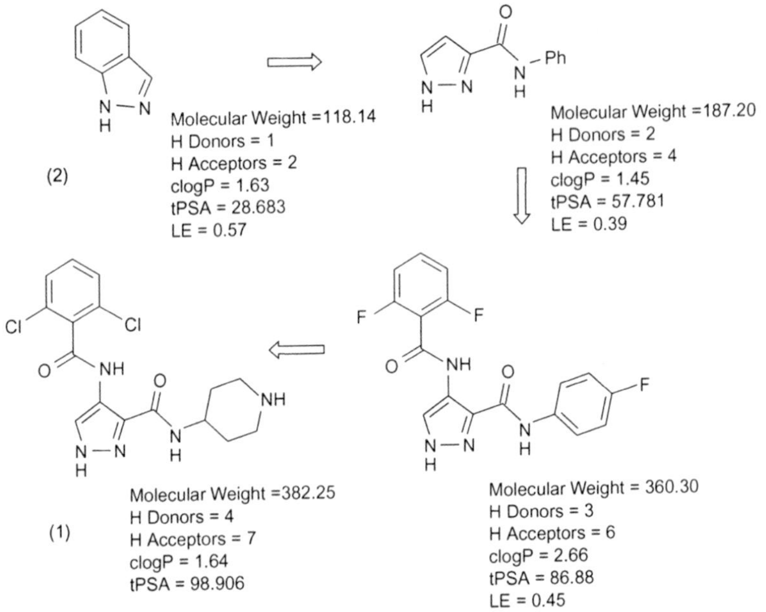

Molecular Weight = 118.14
H Donors = 1
H Acceptors = 2
clogP = 1.63
tPSA = 28.683
LE = 0.57

(2)

Molecular Weight = 187.20
H Donors = 2
H Acceptors = 4
clogP = 1.45
tPSA = 57.781
LE = 0.39

Molecular Weight = 382.25
H Donors = 4
H Acceptors = 7
clogP = 1.64
tPSA = 98.906

(1)

Molecular Weight = 360.30
H Donors = 3
H Acceptors = 6
clogP = 2.66
tPSA = 86.88
LE = 0.45

Scheme 1.1   The evolution of Astex PhII clinical candidate AT7519 (1) from fragment hit.

a development candidate its properties obey the relevant rules from Table 1.1.

SHAPE

The classical view of medicinal chemistry is to work with the lock and key approach first postulated in 1894 by Emil Fischer. That is to say, design the small molecule key to fit the lock, which is the target enzyme or receptor. However, it is an oversimplification of both the lock and the key (target and receptor) to trust this analogy so far as to make important design decisions without confirmation by empirical means. René Magritte famously produced the picture 'this is not a pipe' (Figure 1.1). In the view of the author, medicinal chemists should hold this philosophical thought in mind when they use target structure to design small molecule inhibitors. Both protein and ligand are flexible objects that change shape dynamically under physiological conditions, and perhaps even more so when they interact with each other. So, any static picture of a protein does not robustly represent reality, as it does not usually show either the potential for movement or the multitude of water molecules that surround the protein or drug. Thus, very small differences in binding energy components, and flexibility in both the shape of the protein and the binding mode of the ligand to the protein can invalidate design assumptions based on the static picture of a protein.

To a large extent, 3D molecular conformation is calculated reasonably accurately and automatically by software, at least for small molecules in

Fig. 1.1   René Magritte. 'This is not a pipe'.

vacuum. It is therefore recommended to design molecules of similar shape and dipoles to a known active compound rather than to design a key to fit a lock, although structure knowledge does give insights into opportunities for binding to other parts of the lock (target) which might not yet be fully exploited. This is supported by the analysis of Martin *et al.* [20] which showed that increasing chemical distance from an active lead results in lower probability of activity. Medicinal chemists typically use Tanimoto similarity of greater than 0.85 to select molecules similar to one another, but this can be highly dependent on the chemical descriptors or fingerprints used in the measure of structure similarity. Recently, Hajduk *et al.* [21] have elegantly shown that belief theory can help by using the strengths of different similarity methods to give statistical support to the chances of finding activity in any two pairs of molecules. The theory sets guidance for values typical of close analogues (20–50% joint belief) and those typical of alternate series or lead-hopping approaches (1–5% joint belief). These low percentages are essentially the chance of successfully predicting activity, and show that accurately predicting activity even from a known active compound is not as easy as one might be led to believe.

Typical small changes that, if used with consideration, will give rise to a deeper understanding of SAR and binding mode include the following:

- hetero atom insertions into rings and chains,
- optimisation of existing dipoles in the lead,
- addition or deletion of methylene in rings and chains,
- addition of small substituents to rings and chains.

All of the above strategies are well summarised in a recent paper by Barriero [22] on the use of bioisosterism. Additional strategies exist, but usually have even lower chance of success. These are listed below:

- looking to fill unoccupied spaces in the target binding site,
- utilising non-binding regions to modulate molecule physicochemistry without adverse effects on binding,
- forming or breaking rings to affect flexibility,
- targeting selected hydrogen bonds within easy reach of the ligand,
- trying to displace ordered water from the enzyme.

BINDING

Perhaps the area of physicochemistry still least understood in the mainstream medicinal chemistry literature is that of the biophysical forces of drug-like compounds associated with binding to their biological targets. These are non-covalent interactions governed by the physical chemistry of two molecules, a drug and its target. To demystify structure activity relationships one must understand this physical interaction in full, knowing what gains, losses and compromises are being made with every change in ligand structure. In the 1990s, this was exemplified by the Andrews binding energy approach [23], which had its uses but focussed too much on the positive forces of polar interactions. Only in the past few years have researchers, including Freire [24] and Chaires [25], used modern micro-calorimetric methods, such as isothermal scanning calorimetry (ITC) and differential scanning calorimetry (DSC), to take a detailed look at binding energy with some very interesting results.

ENERGY

Attempts in the 1990s to measure binding energy using structure-based drug discovery (SBDD), focussing mainly on the formation of polar interactions such as hydrogen bonds, have largely failed to assist medicinal chemists in predicting activity, despite their widespread use. Medicinal chemists find themselves relying more on empirical evidence and then use SBDD to provide hypothesis support. Kunz *et al.* [26] initially highlighted that the binding energy of ligands is in general largely governed by van der Waals (intermolecular attractive) forces and the mostly entropic forces of the hydrophobic effect. Only gradually do medicinal chemists now begin to understand the importance of these terms (Equations (1.1) and (1.2)) and the impact of their values on the potency of drugs. Equally, only recently has the biophysical screening technology been available to separate enthalpy and entropy contributions to binding affinity. In fact, the calculated ΔG is often a small number calculated as the difference of two much larger terms, ΔH and ΔS, and thus understanding the effect on binding affinity of even small structural changes is not trivial.

$$\Delta G_{\text{binding}} = \Delta G_{\text{hydrogen bonds}} + \Delta G_{\text{electrostatic}} + \Delta G_{\text{hydrophobic}} + \Delta G_{\text{vdw}} \quad (1.1)$$

$$\Delta G_{\text{binding}} = \Delta H - T\Delta S = -RT\ln K_{\text{eq}} \quad (1.2)$$

Equations (1.1) and (1.2). *Gibbs free energy of binding.*

A molecule with a 10 nM binding constant is showing an overall Gibbs free binding energy made up of all contributions totalling 46 kJ mol^{-1}. At 300 K a 10-fold improvement in potency is seen for every ~ 5.7 kJ mol^{-1} gain in binding energy. Owing to the ease of dominance of the van der Waals and hydrophobic terms, a 10-fold increase in potency is expected from the surface area and van der Waals binding of a well-positioned methyl group of molecular surface area 48 Å^2. In fact, a potency of 10 nM can be achieved from the binding energy achieved through the burial of 440 $Å^2$ of non-PSA alone [27]. The molecular surface area of drug-like molecules increases by 95 $Å^2$ for every 100 unit increase in molecular weight. It has been the tendency of medicinal chemists to focus on the SAR and potency achieved through the hydrophobic effect and the resulting trend in Log$P$, rather than any specific binding interactions, that has made the area of physicochemistry and binding energy more important over the past 10 years. In this era of high throughput medicinal chemistry and high throughput screening, many analogues could be made and screened from commercially available starting materials, but these did not fully reflect the diversity of polar interactions available. Neither was the synthesis and purification of these more polar analogues as easy. This caused many of the most potent hits and leads to be driven by higher molecular weight and lipophilicity [28] rather than specific polar interactions.

ENTHALPY

The enthalpy of binding term is dominated by the values associated with bonding of the polar groups in molecules through dipoles. Since these polar interactions have exacting distances and geometries they also confer stereo- and regiospecificity, unlike some of the less-specific driving forces associated with entropy. These are electrostatic forces arising due to charge and dipole interactions from polar atoms of up to 0.42 kJ mol^{-1} $Å^{-2}$, but they can also include smaller contributions from the weaker van der Waals–London forces, due to the snugness of fit of two non-polar surfaces and the induction of small temporary dipoles. Because of the presence of water in the system, the solvation and desolvation energy of water from polar groups on the target or the drug can often totally undermine any binding enthalpy gained by the direct interaction of drug with the receptor. The desolvation energy of polar groups in water is some 10-fold greater than that of typical lipophilic groups [29].

Hydrogen bonds

It is important to consider that specific water molecules, and networks of associated waters in and around binding sites, are often an extension of the drug target site and can be bound to, or displaced, to give different SAR and binding affinities. Hydrogen bonds are the most important examples of polar interaction, and due to the peptidic nature of most drug targets these can dominate the landscape of a drug receptor site. The complementarity of polarity and dipole is usually enthalpically favourable even if it is entropically disfavoured. The maximal benefit from a single hydrogen bond is estimated by Freire to be between 16 and 21 kJ mol^{-1}. This equates to a 1,000–5,000-fold increase in potency if fully achieved, which from the experience of the author happens very rarely. Ligands, however, often form nearly ideal but weaker hydrogen bonds in water and thus hydrogen bonds to protein have to be almost ideal to compete with this water interaction. Hydrogen bonds are strongly directional (C$=$O$\cdots$H angle: $180\pm30°$), very sensitive to distance (2.5–3.2 Å) and difficult to optimise in terms of angles and distance from the drug molecule with current structure-based design methods [24]. Hunter *et al.* [30] have shown in model systems that unless the newly formed hydrogen bond is above an optimum strength, it cannot overcome the energy penalties needed to form the interaction. Not all hydrogen bonds are equal, and tables exist showing the relative strengths or abilities of drugs to form hydrogen bonds, such as the ones proposed by Abraham [31] and reviewed recently by Lawrence [32].

There are many other diverse examples of enthalpically positive polar interactions for drug-like molecules, such as halogen bonds [33] and various forms of π interactions with other π systems and polar groups. These others are generally weaker than hydrogen bonds but nonetheless just as hard to optimise in terms of geometry using structure-based approaches.

Van der Waals

Gaps between the target and drug in the binding pocket are essentially a missed van der Waals binding opportunity and therefore binding energy lost. Achieving a good fit, for instance with the ideal chirality for the biological target, is of course the most atom-efficient strategy to enhance ligand efficiency and achieve target selectivity, since the opposite enantiomer has the same connectivity and physicochemistry but lower affinity. This, in fact, is a long-known principle known as Pfeiffer's rule [34] where the relative potency difference of enantiomers is proportional to the potency of the more active enantiomer.

ENTROPY

Entropy is of course the measure of disorder, but a term which is sometimes hard to visualise in the same concrete way as enthalpy. In most cases, the largest contribution to the binding energy of drugs is entropy and so it is important to grasp its principles for good drug design. Entropy is made up of two major energy terms. The first is due to the desolvation of more ordered but weakly bound water molecules, in the binding site and

surrounding the ligand, releasing them into the chaos of the surrounding
free solution. The second term is derived from differences in conformational
energy of the drug or the target. The entropy of desolvation contribution
dominates the binding for lipophilic molecules and substituents, adding
around $0.1\,kJ\,mol^{-1}\,\text{Å}^{-2}$ of non-PSA [35]. Due to the nature of the van der
Waals force and the hydrophobic effect for many non-polar fragments, the
gain in activity for optimally achieving binding through shape and non-PSA
should actually be much greater than is usually seen and accepted. Thus, it
is important to know when medicinal chemists are underachieving
lipophilicity with their substituents so that more efficient alternatives can
be tested.

The entropy of conformation and translation change upon binding is
always unfavourable, due to loss of tumbling and torsional degrees of
freedom and any induced fit occurring in the target. This energy equates to
around $-105\,kJ\,mol^{-1}$ for a drug-like molecule [36] and is a price which
must be paid by all drugs upon binding to their target. In order to minimise
this entropy penalty it is very important to minimise conformational
flexibility in drug molecules. The chemistry strategy of freezing conformations is hard to get exactly right, since the ligand may need to flex to some
degree to get into the binding site. These design strategies can lead to the
making and testing of many inactive molecules before finding the best fit.
Medicinal chemists are, though, very adept at visualising and optimising
binding conformations and locking in those, perhaps higher energy,
conformations during drug synthesis. The penalty for binding with a
suboptimal ligand or protein conformation is a loss of binding enthalpy
equal to the difference in conformational energy. This can also be seen as
the strain on either molecule to achieve binding. The enthalpy cost of
freezing a single rotatable bond to form an active conformation is
$2.1–5.0\,kJ\,mol^{-1}$ which is equivalent to a 3–10-fold decrease in activity.

Hydrophobic effect

Removing weakly bound water from a hydrophobic surface has a low
enthalpy penalty that is largely outweighed by the entropy gains from
displaced water in a chaotic solution state leading to the so-called
'hydrophobic effect'. In total, this can be worth up to $6.3\,kJ\,mol^{-1}$ for
each methylene group added to a molecule (from $0.1\,kJ\,mol^{-1}\,\text{Å}^{-2}$ of non-
PSA). This translation to an approximately 10-fold increase in binding
energy has nothing to do with the shape of a drug specifically binding to the
target or receptor. Increasing lipophilicity can be a cheap way of buying
potency, driven only by the increased entropy (disorder) of released water.

THERMODYNAMIC SIGNATURES

Freire has proposed the concept of a thermodynamic signature which is the balance of specific enthalpy and entropy contributions for a molecule which makes up the Gibbs free energy of binding. This is a development of the enthalpy–entropy compensation principle [37]. Analyses of the specific enthalpy and entropy data for HIV protease inhibitors [38, 39] (Figure 1.2) and statins [40] seem to suggest that the best drugs optimally bind when the enthalpic contribution makes a greater contribution to the total free energy of binding. It is suggested that optimal leads have around a third of their Gibbs free energy made up from polarity-driven enthalpic contributions. As can be seen in Figure 1.3, a so-called enthalpy funnel [41, 42] is formed when optimising potency. This is a subtly different way of driving SAR since it requires that polar groups are used in driving specific binding enthalpy and not just used to moderate solubility and pharmacokinetic related properties.

These still largely retrospective analyses show that early leads often rely more or less heavily on entropic contributions to binding and this can be less target-specific than the enthalpic polar interactions. The optimisation of hits and leads with an eye on balancing their enthalpic contributions should more rapidly lead to more robust drug-like clinical candidates, with increased probability of being more target-specific and selective. Equally, the

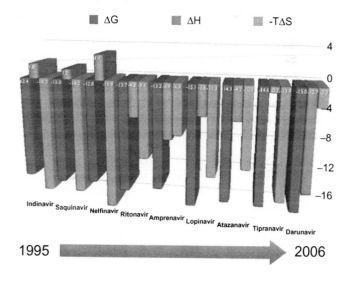

Fig. 1.2   The balance of enthalpy and entropy contributions for HIV protease inhibitors. (Reproduced with permission from 'Drug Discovery Today')

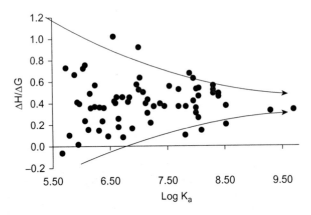

Fig. 1.3   Enthalpic contribution to the Gibbs energy of binding [ΔH/ΔG] versus the logarithm of the binding affinity [Log k_a] for inhibitors of plasmepsin II (adapted with permission from Wiley).

polar groups needed to form these enthalpic interactions should provide more favourable physicochemistry than groups which give binding energy from non-PSA entropic contributions.

LIGAND EFFICIENCY

When many molecules and chemical series are involved in a discovery project, it is sometimes hard to empirically select the future winners as so many parameters are involved. It was not until Kunz [26] proposed there was an optimal efficiency of ligands (related to the Gibbs free energy of binding divided by the heavy atom (HA) count) that a window to optimal ligand efficiency was widely accepted by medicinal chemists. Hopkins and Groom [43] noted that this free energy per atom term was particularly useful for series selection post high throughput screening (Equation (1.3)).

$$LE[\Delta g] = \frac{\Delta G}{N} = -\frac{RT \, log[K_d]}{N \, \text{non-hydrogen atoms}} \qquad (1.3)$$

Equation (1.3). *Kunz's Gibbs free energy per atom.*

$$\text{Maximal LE} = \frac{0.0715 + 7.5328}{HA} + \frac{25.7079}{[HA]^2} + \frac{361.4722}{[HA]^3}$$

$$FQ = \frac{LE}{\text{Maximal LE}} \qquad (1.4)$$

Equation (1.4). *Fit quality score normalised for ligand size.*

In general, a ligand efficiency of greater than 0.3 is seen as being an effective lead, although the maximal values attainable are both target and chemotype specific. An analysis of historical target classes by workers at Pfizer [44] gave these values as typical for certain common gene families (Table 1.2). Reynolds *et al.* [45] recently showed from the analysis of a broad range of SAR that ligand efficiency is dependent on ligand size and that smaller ligands can often show higher ligand efficiencies (Figure 1.4). As molecular size increases, ligand efficiency decreases to a maximal obtainable value of around 0.2. For this reason they proposed a new efficiency term called fit quality score (FQ) (Equation (1.4)), which compensates for the number of HAs in the molecule and allows a better comparison between molecules from different sources such as fragment screening and high throughput screening.

Table 1.2   AVERAGE LIGAND EFFICIENCY VALUES FOR KNOWN LIGANDS OF SPECIFIC GENE FAMILIES

Gene family	Mean ligand efficiency
Aminergic GPCRs	0.4
Ion channels	0.4
Metalloproteases	0.4
Nuclear hormone receptors	0.3
Peptide GPCRs	0.3
Phosphodiesterases	0.3
Protein kinases	0.3
Serine proteases	0.3

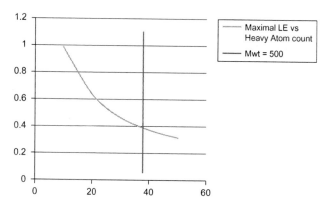

Fig. 1.4   Maximal ligand efficiency per heavy atom.

Recently, there has been some discussion in the medicinal chemistry literature of various ligand efficiency indices as useful tools in the hit to lead phase of drug discovery. Each variation has its strengths and weaknesses (Table 1.3). Workers from Abbott [46] have proposed three new quick ways to calculate ligand efficiency indices as being useful in the high throughput screening triage process. These are: percentage efficiency index (PEI), binding efficiency index (BEI) and surface binding efficiency index (SEI).

Recently, the concept of lipophilic ligand efficiency (LLE) as defined by Leason [28], or LipE as defined by Edwards [47, 48], (Equation (1.5)) is a useful check on the tendency to use lipophilicity-driven binding to increase potency. LLE ensures that the maximum amount of specific binding for a given LogD is being given by a molecule and thus enables the design of more efficient ligands. It is yet to be shown whether this also relates to a higher ratio of enthalpic binding for these ligands.

$$LLE = -\text{Log } K_i - \text{Log } D \qquad (1.5)$$

Equation (1.5). *Calculation of lipophilic ligand efficiency.*

Historical analysis shows that medicinal chemists alter physicochemistry in a fairly predictable way when progressing a compound from hit to clinical candidate. Teague [10] and Hann [11] have shown by analysis of lead–drug pairs from different databases that molecular weight increases, usually by between 42 and 80 atomic mass units, during the drug discovery process.

Keserü and Makara [49] recently published a further refinement of ligand efficiency known as LELP which is defined in Equation (1.6). Using a

Table 1.3   AN OVERVIEW OF LIGAND EFFICIENCY CALCULATIONS
AND VARIATIONS

Index	Calculation	Comments	Ideal values	References
LE (Δg)	$\Delta G/N$	Sometimes over compensates for CF_3 and under compensates for I	>0.3	[26, 39]
FQ	LE/LE_{max}	Allows proper comparison of different size ligands	0–1	[41]
PEI	% inhibition (conc)/MWT	Quick to calculate for raw HTS hits	2–4	[42]
BEI	pK_i/MWT (kDa)	Compensates for size and weight of different row elements versus LE	15–35	[42]
SEI	pK_i/100 Å^2 PSA	Ensures best potency for exposed surface area and optimises absorption	5–20	[42]
LipE [LLE]	$-\text{Log } K_i - \text{Log } D$	Allows optimisation of activity wrt lipophilicity	5–10	[42–44]
LELP	$\text{Log } P/LE$	Allows optimisation of efficiency wrt lipophilicity	0–7.5	[46]

database of previously described leads and drugs, they have shown that using LELP ensures the optimum use of lipophilicity.

$$\text{LELP} = \frac{\log P}{\text{LE}} \qquad (1.6)$$

Equation (1.6). *Calculation of LELP.*

Fragment-based leads

FBDD has become a popular alternative route to high quality, novel, ligand-efficient leads. This starts with highly ligand-efficient hits that are low in molecular weight but have very low binding energies and therefore binding affinities. These low affinities usually require biophysical methods such as NMR, PSR or X-ray to measure them. FBDD has been widely hailed as a complementary approach to high throughput screening and as an approach that favours the production of highly efficient and lower molecular weight leads. Keserü and Makara [49] recently analysed the literature to find that in the optimisation of fragment-based hits, the resultant drug development leads were in fact no more efficient or lead-like than those from HTS. Recent reviews [15, 17] of fragment-based discovery also suggest that medicinal chemists are yet to fully understand how best and quickest to optimise such weak hits into leads for clinical drug development without compromising their initial appealing atom-binding efficiency or drug-like properties.

BINDING KINETICS

Binding kinetics is also referred to as: slow offset, slow off-rate, slow dissociation, insurmountable antagonism, ultimate physiological inhibition, tight binding and non-equilibrium blockade.

The binding constant (K_d) calculated at a single time point assumed to be at system equilibrium is used routinely as a measure of potency in medicinal chemistry, and for almost all compound selection criteria in the lead optimisation phase. It had been easily forgotten until recently that binding is actually a more complicated and time-sensitive equilibrium. A thermodynamic binding constant shows us nothing about what might be happening over time as the drug and biological target interact. In fact, by using a standard high throughput assay to determine apparent K_d, the subtleties of binding kinetics have always been lost, or sometimes put down to 'strange phenomena' and assay artefacts. Compounds that exhibit interesting slow on or slow off kinetics, with say $t_{1/2}$ of the order of hours, would have been poor performers in most binding assays, as they have insufficient time to exert their full equilibrium IC_{50} or K_i during the time in which they are incubated with the target. With this rate of binding or dissociation they would need about five half lives to pass before showing 95% of their full equilibrium effect. Such assays would need incubation periods of around 24 h instead of the minutes which a typical high throughput IC_{50} assay is allotted [50].

Since the pharmacodynamic behaviour of drug molecules is a key parameter used in selection of compounds in clinical development, many medicinal chemists have been perhaps missing an opportunity for the exploration of more detailed SAR, in which pharmacodynamics is influenced by binding kinetics as well as adsorption, distribution, metabolism and excretion (ADME) physicochemistry. Much of the design effort undertaken between lead generation and drug candidate is currently focussed on optimising potency and minimising dose to avoid toxicity. In using only the thermodynamic equilibrium constant, medicinal chemists have largely been missing any opportunity to perturb duration of action through modification of binding kinetics. The holy grail of a 24 h half life can be, and is, much more easily attained by a balance of reasonable plasma half life and moderately slow offset. In fact slow offset can also give a greater therapeutic index and even assist target selectivity (*vide infra*).

We are now beginning to enter an era where k_{on}, k_{off} and K_d can all be used in compound selection, instead of using only the last to determine the relative utility of lead compounds. In this era, medicinal chemists will design for certain binding kinetic parameters rather than just for potency and ADME physicochemical properties alone. Using this approach will depend on the desired target profile and the nature of tolerable toxicity risks. The pharmacology literature does contain many, mainly retrospective, analyses of drugs selected for development with unusual pharmacodynamics which have been linked later to the kinetics of their binding. Table 1.4 shows some

Table 1.4   HISTORICAL COMPOUNDS WHOSE CLINICAL EFFICACY IS ASSOCIATED WITH BINDING KINETICS

Drug	Disease area	Dissociative $t_{1/2}$ (hours)	References
Granisetron (5-HT$_3$)	Emesis	0.25	[51,52]
Candesartan (angiotensin II)	Hypertension	2	[53]
Tiotropium (M3)	COPD	34.7	[54]
Desloratidine (H1)	Allergy	>6	[55]
Saquinavir (HIV protease)	AIDS virus	50.2	[56]

historical examples in which slow offset binding kinetics has been retrospectively proposed to give a better clinical profile due to enhanced pharmacodynamics, improved selectivity or lower dose. It is likely that these compounds were initially selected for development based on their absolute potency and only later did the chance finding of advantageous slow offset kinetics *in vivo* give rise to a better drug.

THE THEORY OF SLOW OFFSET

The overall Gibbs free energy of binding (ΔG) determines the thermodynamic equilibrium of the drug (L) and its biological target (R) and the binding potency. However, any equilibrium is established through a balance of on-rates (k_{on}) and off-rates (k_{off}) of the drug and receptor (Equation (1.7)). Constants k_{on} and k_{off} provide a dynamic or kinetic view of the equilibrium between a drug and its target, how fast the states are attained and the dynamics of their subsequent interconversion. Copeland [57] postulates that most druggable targets are in fact best described by Equation (1.8) rather than Equation (1.7), where a post-binding event, such as a protein conformational change to a higher energy conformation of the receptor complex, gives rise to a more potent bound complex (RL^*), which is no longer in such rapid equilibrium with the system (plasma, agonist or other ligands) and thus becomes apparently insurmountable. Copeland [58], Vauquelin [59, 60], Kenekin [61] and Swinney [62, 63] have all postulated in recent reviews that increased pharmacodynamic half life, receptor occupancy, selectivity and therapeutic index can all be driven by slow offset kinetics, often associated with ligands that have these more complex RL^* equilibria. If this is true then understanding of what is going on in our assays probably requires more complicated interpretation than we thought.

$$[R] + [L] \underset{k_{off}}{\overset{k_{on}}{\rightleftharpoons}} [RL] \quad K_d = \frac{k_{off}}{k_{on}} \tag{1.7}$$

$$[R] + [L] \underset{k_2}{\overset{k_1}{\rightleftharpoons}} [RL] \underset{k_4}{\overset{k_3}{\rightleftharpoons}} [RL]^* \quad K_d^* = \frac{k_2}{k_1 + (k_1 k_3 / k_4)} \tag{1.8}$$

Equations (1.7) and (1.8).

Recognition rate – k_{on} $M^{-1} s^{-1}$ (association rate)

There are two components to k_{on}; the first of these is collision between drug target (enzyme or receptor) and drug. The second is the interaction where the two participants organise themselves, or reorganise themselves, for release of optimum binding energy. Collision theory sets the fastest rate for k_{on} as the diffusion limit for a target and a drug in proximity which is about 10^8–$10^9 \, M^{-1} \, s^{-1}$. Since the collision of drug and target may not always result in binding to the active site there are a significant number of random diffusion-based events happening before a drug is successful in finding its target. If we estimate that approximately 3% [64] of the biological target's surface is responsible for binding, this then gives us something that comes close to a typical observed k_{on} value which is often in the order of $10^7 \, M^{-1} \, s^{-1}$. If significant conformational changes of inhibitor or binding site need to occur after binding and/or major desolvation of either partner is required, then this will reduce k_{on} still further and drive slow association, giving k_{on} values typically in the region of around 10^4–$10^5 \, M^{-1} \, s^{-1}$. This can be seen as increasing the energy barrier in the transition state between a drug and its complex with biological target (E_{act} for Equation (1.7)). However, if the protein preorganises before drug binding collision occurs, then k_{on} can theoretically be closer to the maximal values of $10^7 \, M^{-1} \, s^{-1}$.

It is noteworthy that the rate of receptor/ligand complex formation is dependant on the local concentration of the ligand, which is controlled by pharmacokinetics; the adsorption, distribution metabolism and excretion parameters. A very rapid k_{on} implies that a lower free drug concentration would be needed to bind to the target, and that perhaps subsequently these compounds will tolerate greater plasma protein binding, and perhaps lower bioavailability, for equivalent receptor occupancy and subsequent efficacy.

Slow k_{on} compounds will need increased doses and higher free drug levels to achieve acceptable receptor occupancy/efficacy and this may erode therapeutic index, unless special target-specific dosing regimens such as use of the inhaled route are available. If potency is driven by slow k_{on} it is necessary to maintain high free drug levels through good pharmacokinetics to maintain occupancy if the drug is also fast off. Targeting increased potency via structure activity relationships which increases k_{on} will always be limited by the upper limits of diffusion control ($10^9 \, M^{-1} \, s^{-1}$) and so is not seen as a robust strategy for lead optimisation. However, having a fast k_{on} may be important for rapid onset of activity in some indications and the need for high drug concentrations may be removed. It has also been observed that the range of variation in k_{on} is more limited than for k_{off} and therefore k_{off} probably offers more opportunity for structure kinetic relationships (SKR).

Dissociation rate $k_{off} \, s^{-1}$ (diffusion rate)

Dissociation rate of the simplest ligand–receptor binary complex (RL) is dependant only on the lifetime of the bound state and shows typical first-order exponential decay to predict half life (Equation (1.9)). For those complexes with post-binding conformational changes, or so-called induced fit, then the prediction of half life is much more complicated (Equation (1.10)).

$$\text{Dissociation } t_{1/2} = \frac{\text{Log}_n 2}{k_{\text{off}}} = \frac{0.693}{k_{\text{off}}} \qquad (1.9)$$

$$t_{1/2} = \frac{0.693(k_2 + k_3 + k_4)}{k_2 k_4} \qquad (1.10)$$

Equations (1.9) and (1.10). *Calculation of half life for ligand receptor complexes.*

The apparent pharmacodynamic half life of the drug can be significantly increased by its slow offset, such that the predicted half life calculated from plasma half life is less than that observed *in vivo*. Actual drug clearance from the system *in vivo* is therefore lower than predicted from metabolic half life since there is a significant delay in drug dissociation from the receptor before subsequent clearance (Table 1.5).

KINETIC MAPS

SKR or kinetic maps show that compounds with the same K_d lie on the 45° lines but can achieve the K_d with different k_{off} and k_{on} profiles. The kinetic map shown in Figure 1.5 indicates how a set of well-known compounds, some of which are discussed below, gain their advantage by using their kinetic profiles. Some best-in-class agents exhibit their potency in part due to an unusually slow off-rate and lie to the left of the centre of the graph. Typically a slow off-rate would be categorised at being less than $10^{-3}\,\text{M}\,\text{s}^{-1}$ and a slow on-rate would be categorised as less than $10^5\,\text{M}\,\text{s}^{-1}$.

Table 1.5   TABLE SHOWING HOW K_{OFF} VALUES CAN GIVE SIGNIFICANT HALF LIFE INCREASES OVER PK PREDICTED VALUES

$k_{off}\ (s^{-1})$	$t_{1/2}$
1	0.69 s
10^{-1}	6.9 s
10^{-2}	69 s (1.1 min)
10^{-3}	690 s (11 min)
10^{-4}	6,930 s (1.9 h)
10^{-5}	69,300 s (19.2 h)

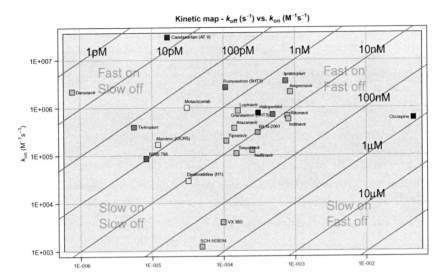

Fig. 1.5   Kinetic map showing how k_{on} and k_{off} contribute to K_d for various drugs.

IMPROVED PHARMACODYNAMICS

Compounds exhibiting slow offset kinetics do provide a superior clinical profile compared to fast off compounds, as their increased half life leads to better dosing regimens and improved pharmacology. This is achieved through longer receptor occupancy and slow washout rather than via the pharmacokinetic profile. Indeed for intracellular targets, the pharmacodynamic half life is often enhanced beyond the value modelled by using k_{off} and plasma half life. This is due to unbound free drug in cells being compartmentalised at a high local concentration and not subject to normal clearance from plasma before a further binding event takes place. In cases of slow offset and reasonable pharmacokinetics it is possible to achieve such a high half life that the receptor essentially remains inhibited until resynthesis by the organism. These have been termed by Schramm et al. [65] as 'ultimate physiological inhibitors' and are exemplified by DADMe-ImmH (3); they behave in the same way as would mechanism-based inactivators which are covalent binders. Just as with covalent mechanism-based inactivators, the off-target toxicity profile of slow offset inhibitors is driven by high specificity for the target or by SKR where the half life for other receptors in the system is not slow offset.

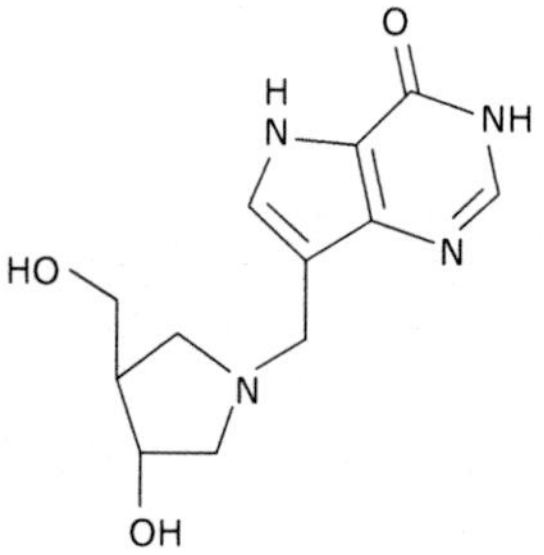

DADMe-ImmH
(3)

Ultimate physiological inhibitor of Purine Nucleoside Phosphorylase

CASE STUDIES

Inhaled anti-muscarinic agent, Tiotropium (4) [54, 66, 67] is the first once-daily bronchodilator used for the treatment of chronic obstructive pulmonary disorder. It has a half life at the M3 receptor of 34.7 h. Its trough plasma levels are in the low pM range after daily dosing and cannot explain the extended efficacy observed in the clinic. Tiotropium also shows kinetic receptor subtype selectivity for M3 over M2 and M1 due to its very slow dissociation rate from the M3 receptor. The measured K_d for each receptor is within threefold of the M3 value.

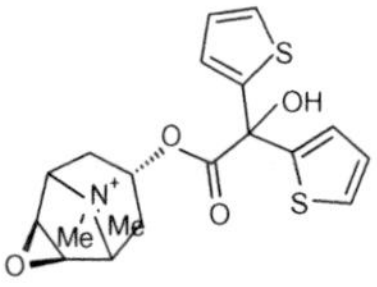

Tiotropium
(4)

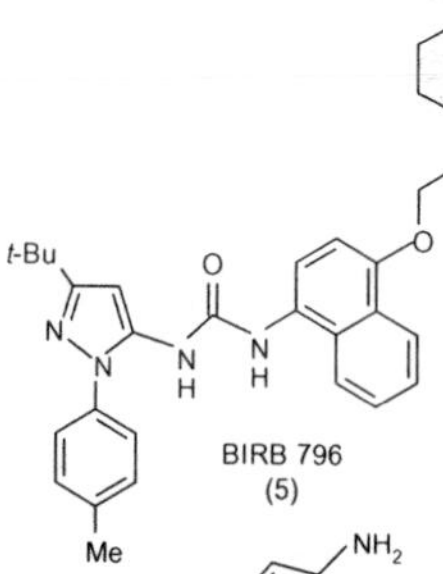

BIRB 796
(5)

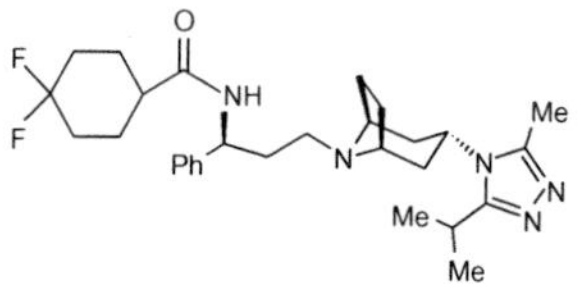

Maraviroc
(6)

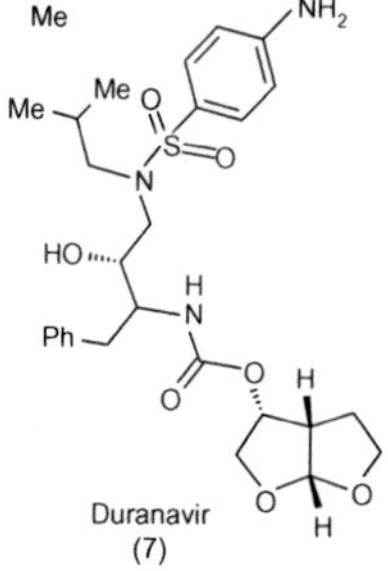

Duranavir
(7)

BIRB 796 (doramapimod) (5) [68, 69] shows a very slow off-rate from p38 kinase and a half life of 24 h. This slow off-rate is achieved by stabilising the 'inactive' 'DFG out' kinase conformation after initial binding, making an RL* type complex. This 24 h half life is probably equal to the cellular turnover of p38 which makes BIRB 796 essentially an irreversible p38 inhibitor. Interestingly BIRB 796 was discovered by optimisation of a classical ATP site hinge-binding kinase inhibitor.

Maraviroc (6) is a CCR5 type chemokine receptor antagonist approved as an HIV entry inhibitor. This compound shows a dissociative half life from human CCR5 receptors of 16 h [70], which results in a longer duration of action of the CCR5 blockade than predicted from its pharmacokinetic profile. A more robust protection of cells from HIV infection is produced.

Darunavir [71, 72] (7) is a second-generation HIV protease inhibitor approved in 2006 for the treatment of AIDS. It shows practically irreversible inhibition of wild type HIV protease with a dissociation half life of more than 10 days, thought to be the lifespan of the target in the cell. In contrast, its plasma half life is measured at 15 h. As discussed in an early section, this drug also achieves a large proportion of its binding energy from enthalpy change. Darunavir shows excellent resistance profiles against wild type and protease resistant strains.

Clozapine (8) and haloperidol (9) are both D_2 receptor antagonists but show differentiated pharmacology [73] in vivo. Clozapine shows fast dissociation from the D_2 receptor with a half life of 15 s. Haloperidol in comparison shows slow dissociation kinetics with a half life of 58 min. In the case of D_2 receptors 60–75% occupancy is needed for clinical efficacy, however higher inhibition of D_2 function gives rise to a mechanism-based side effect of tremor. Inhibitors such as clozapine are more readily displaced from the receptor by the natural agonist dopamine and so give a cleaner profile as anti-psychotics, with side effects controlled by pharmacokinetics and dose and not by slow offset.

Clozapine
(8)

Haloperidol
(9)

Amlodipine
(10)

Verapamil
(11)

The L-type calcium channel blocker amlodipine (10) is pharmacologically differentiated from verapamil (11) by its dissociation kinetics from the ion channel. The dihydropyridine, amlodipine, has a dissociative half life of 38 min whereas verapamil has a bound half life of 0.25 s [62, 63]. Verapamil is used for the treatment of arrhythmias whereas amlodipine is used for the treatment of hypertension.

THE APPLICATION TO KINASES

The human genome encodes 518 kinases [74] however they remain one of the largest classes of druggable targets not yet fully exploited by pharmaceutical research [75]. Researchers have often found it difficult to achieve good potency and pharmacokinetics in inhibitors without severe toxicity, which has largely been ascribed to lack of selectivity within the kinase family. This has limited their commercial utility to areas of acute care such as oncology [76]. Much work in recent years has been focused on the formation of ligand–kinase complexes in the so-called 'DFG-out' or 'αC-Glu' out formations (Type II, inactivated forms) [77, 78], as opposed to initial strategies that principally focussed on the classical ATP site hinge-binding region (Type I, active forms). These DFG loop out complexes should all show the typical binding kinetics associated with RL^* complexes (Equation (1.8)) since there is conformational change and induced fit after initial ligand binding. In the case of BIRB-796 (5) and its interactions with the kinases p38 (also known as MAPK14) and Abl these complexes have dissociation half lives of $\sim$24 h and associated pharmacodynamic benefits as discussed above. Five of the eight kinase inhibitors marketed so far target inactive kinase conformations (e.g. lapatinib (12) and imatinib (13)). In these two cases, there is significant post-binding reorganisation which is able to kinetically protect these complexes from the competition of the high intracellular ATP concentrations. Additionally, the typical physicochemistry of the kinase inhibitors in development is significantly out of line with that of historical orally available drugs [76]. On average the kinase inhibitors in development have a molecular weight 110 mass units higher than oral drugs and a $cLogP$ 1.5 units higher than oral drugs making them potentially more challenging for traditional pharmacokinetic controlled pharmacology which is largely driven by physicochemistry [79].

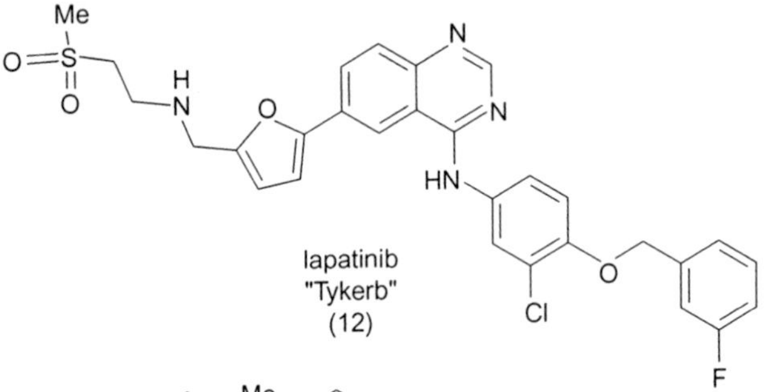

lapatinib
"Tykerb"
(12)

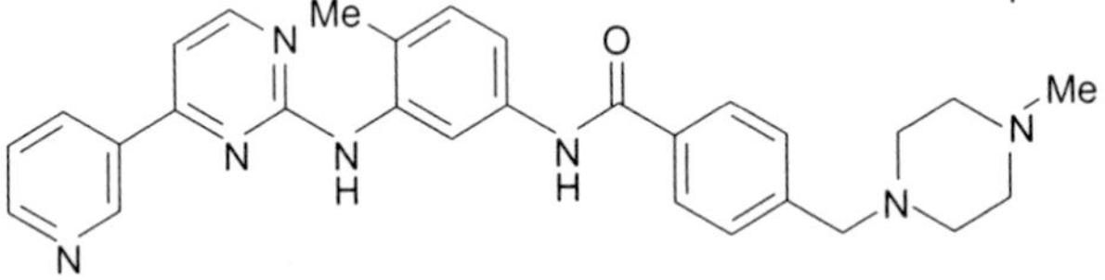

Imatinib
"Gleevec"
(13)

Marketed kinase inhibitors with slow k_{off}

It has already been shown above that for several drug classes, including GPCRs, proteases and ion channels, that therapeutic index can be significantly increased by slow offset kinetics giving rise to increased selectivity, due to better pharmacodynamics over other targets. These findings have now been successfully applied in the area of kinase drug discovery, and this is a current area of significant research focus [80], although the associated SKRs are only now beginning to appear in the medicinal chemistry literature. In a recent landmark paper, workers at Cephalon [81] have categorised structure data for 82 kinases and their ligand-binding topologies. They have been able to classify 25 pyrimidine-based hinge-binding kinase ligand classes, 27 other ligand structural motifs that bind the hinge region, and 5 classes of ligand which show DFG-out binding capabilities. This comprehensive analysis gives medicinal chemists empirical guidance to structure activity relationships for kinases and their ligands using a simple 2D pharmacophore model (Figure 1.6). It is not yet clear what fraction of kinases can be inhibited in the DFG-out

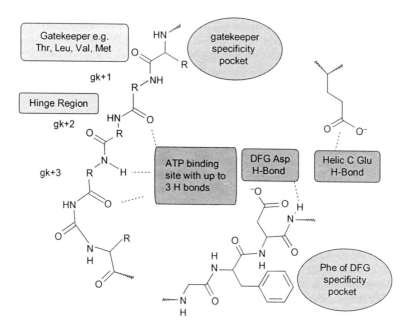

Fig. 1.6 2D pharmacophore model for kinases and their ligands.

conformation. The analysis focuses on opportunities of DGF-out ligands to access two additional hydrogen bonds and a lipophilic pocket to enhance selectivity over hinge-binding ligands, although it does not highlight the implications of the DFG-out conformation for binding kinetics which must also be considered a key to success in the area of kinase drug discovery. As discussed above, this level of understanding requires more detailed screening information to be gathered in the discovery phase, determining kinetic and thermodynamic parameters in addition to the equilibrium Gibbs free binding energy obtained from a K_d.

CONCLUSIONS

The physicochemistry of molecules can often indicate which molecules not to make, but is less prescriptive of what should be synthesised. Additionally, sometimes the numbers just do not add up and the various parameters required to deliver the drug concept are not available within the chemotype under evaluation or perhaps not even in the vastness of chemical space.

In the author's opinion, following the known facts (numbers) will result in improved 'luck' in drug discovery and development, over close pursuit of *in vitro* structure activity relationships without holding the long-term clinical view. Indeed, it is already a difficult task to get selectivity and a high therapeutic safety index by designing molecules for activity, but since drugs are a clinical commercial product, they need to be designed for use in that setting and not solely for the high throughput research environment.

There is still ample space for creative synthetic chemistry to achieve the right shapes and numbers in the vastness of the available chemistry we know which is estimated at around 10^{60} molecules. However, one of the most important numbers the industry is dealing with at present is the cost of making compounds and testing hypotheses. The approximate cost to provide a single sample for screening is around \$4,000. Scientists are often shielded from such numbers lest their creativity be inhibited, however a poorly designed compound may also be a waste of that precious investment which medicinal chemists are privileged to wield.

Many questions remain unanswered in these areas of medicinal chemistry and although the price of obtaining each answer is a high one, the understanding achieved could transform the way we pursue this science.

Physicochemistry based on the rule of 5 is now used in all medicinal chemistry settings. In this chapter we have seen that physical chemistry, in the form of the balance of binding energy contributions and target on- and off-rates, also plays a key part in delivering a good drug from a lead series. The importance of the enthalpic contributions to binding constant and to increased ligand efficiency and selectivity seem now ever clearer. Equally, consideration of dissociation constants and the SKR of different leads is a tool perhaps almost as important as the understanding of ADME parameters. It is clear that we now have readily available tools for the selection of leads based on both of these new parameters. Rather than this being an additional burden of structure activity relationships, I would propose that this is in fact now a tool for clarification of what were, until recently, seen as inexplicable series behaviours, which caused some lead chemical series to fall mysteriously by the wayside *en route* to clinical development.

REFERENCES

[1] Lipinski, C.A., Lombardo, F., Dominy, B.W. and Feeney, P.J. (1997) *Adv. Drug. Deliv. Rev.* **23**, 3–25.
[2] Lipinski, C.A., Lombardo, F., Dominy, B.W. and Feeney, P.J. (2001) *Adv. Drug Deliv. Rev.* **46**, 3.

[3]   According to SciFinder there are >2204 papers which reference Lipinski's 1997 paper [Ref. 1].
[4]   Gleeson, M.P. (2008) *J. Med. Chem.* **51**, 817–834.
[5]   Veber, D.F., Johnson, S.R., Cheng, H.-Y., Smith, B.R., Ward, K.W. and Kopple, K.D. (2002) *J. Med. Chem.* **45**, 2615–2623.
[6]   Clark, D.E. and Pickett, S.D. (2000) *Drug Disc. Today* **5**, 49–58.
[7]   Kelder, J., Grootenhuis, P.D.J., Bayada, D.M., Delbressine, L.P.C. and Ploemen, J.P. (1999) *Pharm. Res.* **16**, 1514–1519.
[8]   Norinder, U. and Haeberlein, M. (2002) *Adv. Drug Deliv. Rev.* **54**, 291–313.
[9]   Teague, S.J., Davis, A.M., Leeson, P.D. and Oprea, T.I. (1999) *Angew. Chem. Int. Ed. Engl.* **38**, 3743–3748.
[10] Oprea, T.I., Davis, A.M., Teague, S.J. and Leeson, P.D. (2001) *J. Chem. Inf. Comput. Sci.* **41**, 1308–1315.
[11] Hann, M.M., Leach, A.R. and Harper, G. (2001) *J. Chem. Inf. Comput. Sci.* **41**, 856–864.
[12] Hann, M.M. and Oprea, T.I. (2004) *Curr. Opin. Chem. Biol.* **8**, 255–263.
[13] Congreve, M., Carr, R., Murray, C. and Jhoti, H. (2003) *Drug Disc. Today* **8**, 876–877.
[14] Verdonk, M.L. and Hartshorn, M.J. (2004) *Curr. Opin. Drug Disc. Dev.* **7**, 404–410.
[15] Hajduk, P.J. and Greer, J. (2007) *Nat. Rev. Drug Disc.* **6**, 211–219.
[16] Sutherland, J.J., Higgs, R.E., Watson, I. and Vieth, M. (2008) *J. Med. Chem.* **51**, 2689–2700.
[17] Murray, C.W. and Rees, D.C. (2009) *Nat. Chem.* **1**, 187–192.
[18] Wyatt, P.G., Woodhead, A.J., Berdini, V., Boulstridge, J.A., Carr, M.G., Cross, D.M., Davis, D.J., Devine, L.A., Early, T.R., Feltell, R.E., Lewis, E.J., McMenamin, R.L., Navarro, E.F., O'Brien, M.A., O'Reilly, M., Reule, M., Saxty, G., Seavers, L.C.A., Smith, D.-M., Squires, M.S., Trewartha, G., Walker, M.T. and Woolford, A.J.-A. (2008) *J. Med. Chem.* **51**, 4986–4999.
[19] Squires, M.S., Feltell, R.E., Wallis, N.G., Lewis, E.J., Smith, D.-M., Cross, D.M., Lyons, J.F. and Thompson, N.T. (2009) *Mol. Cancer Ther.* **8**, 324–332.
[20] Martin, Y., Kofron, J.L. and Traphagen, L.M. (2002) *J. Med. Chem.* **45**, 4350–4358.
[21] Muchmore, S.W., Debe, D.A., Metz, J.T., Brown, S.P., Martin, Y.C. and Hajduk, P.C. (2008) *J. Chem. Inf. Model.* **48**, 941–948.
[22] Lima, L.M. and Barriero, E.J. (2005) *Curr. Med. Chem.* **12**, 23–49.
[23] Andrews, P.R., Craik, D.J. and Martin, J.L. (1984) *J. Med. Chem.* **27**, 1648–1657.
[24] Freire, E. (2008) *Drug Disc. Today* **13**, 869–874.
[25] Chaires, J.B. (2008) *Annu. Rev. Biophys.* **37**, 125–151.
[26] Kuntz, I.D., Chen, K., Sharp, K.A. and Kollman, P.A. (1999) *Proc. Natl. Acad. Sci.* **96**, 9997–10002.
[27] Al-Lazikani, B., Gaulton, A., Paolini, G., Lanfear, J., Overington, J. and Hopkins, A. (2007) *In* "Chemical Biology: From Small Molecules to Systems Biology and Drug Design". Schreiber, S.L., Kapoor, T.M. and Wess, G. (eds), Vol. 3, pp. 806–808. Wiley-VCH, Weinheim.
[28] Leeson, P.D. and Springthorpe, B. (2007) *Nat. Rev. Drug Disc.* **6**, 881–890.
[29] Cabani, S., Gianni, P., Mollica, V. and Lepori, L. (1981) *J. Solution Chem.* **10**, 563–595.
[30] Chekmeneva, E., Hunter, C.A., Packer, M.J. and Turega, S.M. (2008) *J. Am. Chem. Soc.* **130**, 17718–17725.
[31] Abraham, M.H., Duce, P.P., Prior, D.V., Barratt, D.G., Morris, J.J. and Taylor, P.J. (1989) *J. Chem. Soc. Perkin II* 1355–1375.

[32] Laurence, C., Brameld, K.A., Graton, J., Le Questel, J.-Y. and Renault, E. (2009) *J. Med. Chem.* **52**, 4073–4086.
[33] Auffinger, P., Hays, F.A., Westhof, E. and Ho, P.S. (2004) *Proc. Natl. Acad. Sci.* **101** (48), 16789–16794.
[34] Pfeiffer, C.C. (1956) *Science (Washington D.C.)* **124**, 29–31.
[35] Sharp, K.A., Nicholls, A., Friedman, R. and Honig, B. (1991) *Biochemistry* **30**, 9686–9697.
[36] Chang, C.A., Chen, W. and Gilson, M.K. (2007) *Proc. Natl. Acad. Sci.* **105**(5), 1537–1539.
[37] Starikov, E.B. and Norden, B. (2007) *J. Phys. Chem. B* **111**, 14431–14435.
[38] Ohtaka, H. and Freire, E. (2005) *Prog. Biophys. Mol. Biol.* **88**, 193–208.
[39] Ohtaka, H., Muzammil, S., Schön, A., Velazquez-Campoy, A., Vega, S. and Freire, E. (2004) *Int. J. Biochem. Cell Bio.* **36**, 1787–1799.
[40] Carbonell, T. and Freire, E. (2005) *Biochemistry* **44**, 11741–11748.
[41] Ruben, A.J., Kiso, Y. and Freire, E. (2006) *Chem. Biol. Drug Des.* **67**, 2–4.
[42] Gilson, M.K. and Zhou, H.-X. (2007) *Annu. Rev. Biophys. Biomolec. Struct.* **36**, 21–42.
[43] Hopkins, A.L., Groom, C.R. and Alex, A. (2004) *Drug Disc. Today* **9**, 430–431.
[44] Paolini, G.V., Shapland, R.H.B., van Hoorn, W.P., Mason, J.S. and Hopkins, A.L. (2006) *Nat. Biotechnol.* **24**, 805–815.
[45] Reynolds, C.H., Tounge, B.A. and Bembenek, S.D. (2008) *J. Med. Chem.* **51**, 2432–2438.
[46] Abad-Zapetero, C. and Metz, J.T. (2005) *Drug Disc. Today* **10**, 464–469.
[47] Denonne, F. (2008) *iDrugs* **11**, 870–875.
[48] Ryckmans, T., Edwards, M.P., Horne, V.A., Correia, A.M., Owen, D.R., Thompson, L.R., Tran, I., Tutt, M.F. and Young, T. (2009) *Bioorg. Med. Chem. Lett.* **19**, 4406–4409.
[49] Keserü, G.M. and Makara, G.M. (2009) *Nat. Rev. Drug Disc.* **8**, 203–212.
[50] Zang, R. and Monsma, F. (2009) *Curr. Opin. Drug Disc. Dev.* **12**, 488–496.
[51] Akuzawa, S., Ito, H. and Yamaguchi, T. (1998) *Japan J. Pharmacol.* **78**, 381–384.
[52] Blower, P.R. (2003) *Support Care Cancer* **11**, 93–100.
[53] Fierens, F.L.P., Vanderheyden, P.M.L., Roggeman, C., De Backer, J.-P., Thekkumkar, T.J. and Vauquelin, G. (2001) *Biochem. Pharmacol.* **61**, 1227–1235.
[54] Disse, B., Speck, G.A., Rominger, K.L., Witek, T.J. and Hammer, R. (1999) *Life Sci.* **64**, 457–464.
[55] Anthes, J.C., Gilchrest, H., Richard, C., Eckel, S., Hesk, D., West, R.E., Jr.., Williams, S.M., Greenfeder, S., Billah, M., Kreutner, W. and Egan, R.W. (2002) *Eur. J. Pharm.* **449**, 229–237.
[56] Markgren, P.-O., Schaal, W., Hamalainen, M., Karlen, A., Hallberg, A., Samuelsson, B. and Danielson, U.H. (2002) *J. Med. Chem.* **45**, 5430–5439.
[57] Tummino, P.J. and Copeland, R.A. (2008) *Biochemistry* **47**, 5481–5492.
[58] Copeland, R.A., Pompliano, D.L. and Meek, T.D. (2006) *Nat. Rev. Drug Disc.* **5**, 730–739.
[59] Vauquelin, G. and Van Liefde, I. (2006) *Trends Pharmacol. Sci.* **27**, 356–359.
[60] Vauquelin, G. and Szczuka, A. (2007) *Neurochem. Int.* **51**, 254–260.
[61] Kenakin, T. (2009) *ACS Chem. Biol.* **4**, 249–260.
[62] Swinney, D.C. (2004) *Nat. Rev. Drug Discov.* **3**, 801–808.
[63] Swinney, D.C. (2009) *Curr. Opin. Drug Disc. Dev.* **12**, 31–39.
[64] Veith, M., Siegel, M.G., Higgs, R.E., Watson, I.A., Robertson, D.H., Savin, K.A., Durst, G.L. and Hipskind, P.A. (2004) *J. Med. Chem.* **47**, 224–232.

[65] Lewandowicz, A., Tyler, P.C., Evans, G.B., Furneaux, R.H. and Schramm, V.L. (2003) *J. Biol. Chem.* **278**, 31465–31468.

[66] Gradman, A.H. (2002) *J. Hum. Hypertens.* **16**, S9–S16.

[67] Van Noord, J.A., Bantje, T.A., Eland, M.E., Korducki, L. and Cornelissen, P.J.G. (2000) *Thorax* **55**, 289–294.

[68] Regan, J., Pargellis, C.A., Cirillo, P.F., Gilmore, T., Hickey, E.R., Peet, G.W., Proto, A., Swinamer, A. and Moss, N. (2003) *Bioorg. Med. Chem. Lett.* **13**, 3101–3104.

[69] Sullivan, J.E., Holdgate, G.A., Campbell, D., Timms, D., Gerhardt, S., Breed, J., Breeze, A.L., Bermingham, A., Pauptit, R.A., Norman, R.A., Embrey, K.J., Read, J., Van Scyoc, W.S. and Ward, W.H.J. (2005) *Biochemistry* **44**, 16475–16490.

[70] Napier, C., Sale, H., Mosley, M., Rickett, G., Dorr, P., Mansfield, R. and Holbrook, M. (2005) *Biochem. Pharmacol.* **71**, 163–172.

[71] Ghosh, A.K., Dawsona, Z.L. and Mitsuya, H. (2007) *Bioorg. Med. Chem.* **15**, 7576–7580.

[72] Dierynck, I., De Wit, M., Gustin, E., Keuleers, I., Vandersmissen, J., Hallenberger, S. and Hertogs, K. (2007) *J. Virol.* **81**, 13845–13851.

[73] Seeman, P. (2006) *Exp. Opin. Ther. Targets* **10**, 515–531.

[74] Manning, G., Whyte, D.B., Martinez, R., Hunter, T. and Sudersanam, S. (2002) *Science (Washington D.C.)* **298**, 1912–1934.

[75] Hopkins, A. and Groom, C. (2002) *Nat. Rev. Drug Disc.* **1**, 727–730.

[76] Gill, A.L., Verdonk, M., Boyle, R.G. and Taylor, R. (2007) *Curr. Top. Med. Chem.* **7**, 1408–1422.

[77] Alton, G.R. and Lunney, E.A. (2008) *Exp. Opin. Drug Disc.* **3**, 595–605.

[78] Liu, Y. and Gray, N.S. (2006) *Nat. Chem. Biol.* **2**, 358–364.

[79] Smith, D.A., van de Waterbeemd, H., Walker, D.K. (2006) *In* "Pharmacokinetics and Metabolism in Drug Design, Methods and Principles in Medicinal Chemistry". Mannhold, R., Kubinyi, H., Folkers, G. (eds), Vol. 31. Wiley-VCH, Weinheim.

[80] Anderson, K., Lai, Z., McDonald, O.B., Darren, J.S., Nartney, E.N., Hardwicke, M.A., Newlander, K., Dhanak, D., Adams, J., Patrick, D., Copeland, R.A., Tummino, P.J. and Yang, J. (2009) *Biochem. J.* **420**, 259–265.

[81] Ghose, A.K., Herbertz, T., Pippin, D.A., Salvino, J.M. and Mallamo, J.P. (2008) *J. Med. Chem.* **51**, 5149–5155.

2 The Identification of and Rationale for Drugs Which Act at The Motilin Receptor

SUSAN M. WESTAWAY[1] and GARETH J. SANGER[2]

[1]*Immuno-Inflammation CEDD, GlaxoSmithKline, Medicines Research Centre, Gunnels Wood Road, Stevenage, Herts SG1 2NY, UK*

[2]*Neurogastroenterology Group, Wingate Institute of Neurogastroenterology, Barts and The London School of Medicine and Dentistry, Queen Mary College, University of London, London E1 2AT, UK*

Progress in Medicinal Chemistry – Vol. 48   31
Edited by G. Lawton and D.R. Witty
DOI: 10.1016/S0079-6468(09)04802-4

INTRODUCTION

The gastrointestinal (GI) tract is central to the process of eating. It is no longer thought of as a simple receptacle of food from which nutrients are absorbed and the waste excreted. Instead, this organ plays a vital role in promoting the sensation of hunger, whilst helping other parts of the body cope with the intake of nutrients ingested during a meal, and simultaneously providing defence against accidental ingestion of toxins (emesis, diarrhoea) and playing a key role in immune protection. As part of its role in the process of digestion, the gut is also equipped with systems that can signal a need to reduce eating, while it deals with the ingested food. These two basic roles of promoting and reducing appetite operate via the intrinsic and extrinsic nervous systems of the gut, acting together with a variety of hormones that are released from the gut to signal its different needs. In general, the hormones released during fasting tend to promote appetite and gastric motility, and/or exert different metabolic actions; these include orexin, ghrelin and motilin (1). Other hormones released at different times after a meal in response to the volume and physical/chemical composition of ingested material, tend to reduce appetite and/or promote digestion by slowing the transit of the ingested material. This group includes gastrin, cholecystokinin, leptin, enterostatin, peptide YY, apolipoprotein A-IV, glucagon-like peptide-1 (GLP-1), glicentin, pancreatic polypeptide, oxyntomodulin and amylin [1].

Considerable effort is ongoing to identify drugs that mimic or prevent the actions of some of these hormones. GLP-1 receptor agonists, for example, are now available for treatment of type II diabetes (acting to increase glucose-dependent insulin secretion, suppress glucagon secretion and slow gastric emptying [2]). In terms of identifying drugs that stimulate GI motility, perhaps the best understood and most advanced approach has focused on the motilin receptor, a G-protein coupled receptor previously known as GPR38 [3]. A major contributor to understanding the functions of this receptor was the discovery that the 14-membered macrolide and antibiotic drug erythromycin-A (2) is also a motilin receptor agonist [4].

H₂N–Phe–Val–Pro–Ile–Phe
Gln–Leu–Glu–Gly–Tyr–Thr
Arg–Met–Gln–Glu–Lys–Glu
HO₂C–Gln–Gly–Lys–Asn–Arg

(1) Motilin

(2) Erythromycin-A

The powerful ability of erythromycin-A to stimulate gastric motility then helped define a role for endogenous motilin in the regulation of gastric motility during fasting and in particular, with Phase III of the migrating motor complex (MMC). MMCs are a pattern of GI movements that occur at regular intervals in the stomach and small intestine during fasting, which culminate in high-amplitude waves of propulsive contractions (this is Phase III of the MMC). These Phase III contractions are thought to be at least partly mediated by the release of motilin from the gut, and are linked to the sensation of hunger. MMCs are also thought to prevent bacterial overgrowth and clear the upper gut of any undigested material [1, 5–7]. MMCs are abolished by eating a meal and to date a role for endogenous motilin in the pattern of gastric motility which follows eating has not been clearly identified (see Section, Motilin Receptor Antagonists for a discussion on a possible role for motilin in certain forms of diarrhoea).

Despite the fact that a clear association between the functions of endogenous motilin and GI motility has only been demonstrated during fasting, the application of a drug which activates the motilin receptor has important implications in the treatment of patients who require increased gastric emptying of meals. Thus, erythromycin-A can increase gastric emptying of meals and such activity provides a therapeutic use for this drug in treating disorders associated with delayed gastric emptying [8]. More recently, a number of other studies have helped to further open up motilin receptor research. In particular, these include the discovery of the motilin receptor, the discovery of non-peptide, non-macrolide, small molecule structures that selectively act at the motilin receptor and significantly, a change in preclinical research focus which has helped to clarify the role of motilin (by placing more emphasis on the neuronal rather than the myogenic actions of motilin receptor agonists).

This review describes these advances in motilin biology, which have important implications in terms of providing feedback to medicinal chemistry with translational value. The early and more recent developments in the medicinal chemistry of motilides (erythromycin-A analogues) and peptidic agonists will then be summarized, before describing the new areas of small molecule motilin receptor agonists and motilin receptor anta-gonists. The potential therapeutic utility of new drugs acting as agonists or as antagonists at the motilin receptor is then discussed, with emphasis on the better understood use of the motilin receptor agonists.

ADVANCES IN MOTILIN RECEPTOR NEUROPHARMACOLOGY

Despite the antibiotic and other actions of erythromycin, the availability of this drug has been invaluable in exploring the potential clinical uses of a more selective motilin receptor agonist [9]. In this regard, it is important to understand that as a stimulant of gastric motility and as a treatment for GI disorders, the doses of erythromycin-A must be smaller (e.g. 4×40–70 mg/kg p.o.) than those normally given for antibiotic use. These lower doses minimize the occurrence of adverse events and provide optimal activation of the motilin receptor, avoiding the bell-shaped dose–response phenomena whereby higher doses lead to loss of GI efficacy (see below for explanation and discussion). As a stimulant of gastric motility, erythromycin-A has been used in patients needing rapid intubation or endoscopy and when removal of gastric contents is required prior to endoscopy or surgery. Erythromycin-A is also used to facilitate enteral feeding (e.g. in German Intensive Care Units, approximately 39% of patients may receive prokinetic therapy as a standard, including erythromycin-A [10]), help control blood glucose levels in diabetic patients and treat symptoms in patients with diabetic- or non-diabetic gastroparesis. Importantly, by carefully selecting the dose of erythromycin, it may be possible to administer the drug repeatedly over several weeks without substantial loss of efficacy. This has been reported in three different pilot studies. In the first [11], erythromycin-A was given by repeated intravenous administration to diabetic patients with severe gastroparesis, with the dose titrated to achieve both efficacy and tolerance in each patient. In this study, symptom improvements were maintained for up to 8 weeks. In the second [12], improved gastric emptying and symptom control was achieved over at least 4 weeks in patients with idiopathic or diabetic gastroparesis, when erythromycin-A was given orally in doses up to 500 mg four times daily, the improvements returning to baseline after washout. Finally, improved glycaemic control was achieved in patients with type II diabetes mellitus who were treated with erythromycin-A 200 mg orally three times a day for 4 weeks [13].

In the GI tract, it is now appreciated that the motilin receptor is expressed largely on nerves that are intrinsic to the stomach, as well as on the smooth muscle itself [14, 15]. This observation, described for rabbits and humans, is thought likely to be true for most other mammalian species that express the receptor. It is worth noting, however, that the previously confusing literature on the function of motilin in rodents (rats, mice) and guinea-pigs (which either suggested or refuted the idea that motilin may have function), seems to be clarifying as neither a motilin peptide nor functional orthologues of the motilin and motilin receptor genes can be detected in rodents. This suggests a loss of a functional motilin system at some point during the evolution of these animals [9].

Stimulation of gastric emptying in response to motilin receptor activation occurs because the neuronally located receptors are activated, facilitating excitatory cholinergic activity. This conclusion can be illustrated by a series of experiments which dissected the actions of motilin receptor agonists in rabbit isolated stomach, a species chosen because of the close homology of the rabbit motilin receptor to the human orthologue. The experiments demonstrated that low concentrations of erythromycin-A caused a long-lasting facilitation of cholinergic activity, whereas higher concentrations of erythromycin-A evoked very large increases in cholinergic activity which did not appear to be sustained over time [14, 16]. At these higher concentrations, erythromycin-A also directly contracted the smooth muscle, an activity that rapidly faded. It was argued [9, 16] that these different actions of erythromycin-A correlate with different actions of the drug in healthy volunteers. Thus, the facilitation of cholinergic activity by low concentrations correlates with an increase in gastric emptying which does not appear to easily fade with repeat dosing, whereas the effects of the higher concentrations may evoke non-cholinergic, non-propagating increases in gastric motility [17], induce nausea, stomach cramping [18] and increase meal-induced satiety [19]. The translational value of these experiments is also consistent with the ability of low doses of erythromycin-A to exert a maintained therapeutic benefit when given by repeat dosing (see above).

The long-lasting ability of erythromycin-A to increase gastric cholinergic activity was also observed with the non-peptide, non-macrolide selective motilin receptor agonist GSK962040 (60) [20]. However, in the rabbit isolated stomach, the maximum ability of GSK962040 to facilitate cholinergic activity appeared smaller than that of erythromycin-A or motilin. This observation was not seen when studying the effects of these molecules on the function of the recombinant receptor. Such studies raise the possibility of identifying partial agonists at the motilin receptor that is located on the cholinergic nerves of the gut. The use of such compounds might be expected to avoid overstimulating this system and thereby also avoid the side-effects described above.

Interestingly, the peptide motilin has a different kinetic profile from both erythromycin-A and GSK962040. Thus in the rabbit isolated stomach model, the facilitation of cholinergic activity caused by all concentrations of motilin appears not to be sustained in the manner that is observed with erythromycin-A and GSK962040 [16, 20]. The reason for this different profile of activity is unknown but it can be speculated that the occurrence of an additional binding site for motilin, when compared with those for erythromycin-A [21], may be somehow involved in the desensitisation process. Whatever the mechanism, the different kinetic responses to motilin, erythromycin-A and GSK962040, in the nerve or muscle cells of the stomach, have two important implications:

1. The rapidly desensitising activity of motilin is consistent with a role for endogenous motilin in the aetiology of the MMCs, short bursts of motility

that pass down the stomach and small intestine. Such short-lived but intense activity would require a rapid removal of the response to motilin. In contrast, the reduced ability of erythromycin-A to desensitise the neuronally mediated functions of the motilin receptor is consistent with the use of this non-peptide molecule as a prokinetic agent for disorders associated with delayed gastric emptying. This difference between the actions of motilin and erythromycin-A may also have implications for the design of peptide ligands as potential drugs (see later).

2. The marked variations in the kinetic responses to different motilin receptor agonists in different cell types, suggests that until more is understood, it is essential to profile new compounds in isolated tissues which express the native receptor in the therapeutically relevant cell type. Studies using recombinant motilin receptors expressed in host cells, or which are focused on the rapid desensitisation properties of the motilin receptor expressed by smooth muscle (both of which are common assays, as can be seen from the medicinal chemistry programmes described in the next sections) must therefore be treated with caution until more is known about the pharmacophores that are generated in each of the above models.

Finally, it needs to be recognized that in spite of the recent advances in motilin receptor research, gaps remain in understanding the full range of activity of motilin. In part, this may be attributed to the fact that a functional motilin system is not expressed by rodents (see above), limiting laboratory experimentation. However, studies in anaesthetized pigs, conscious dogs and human volunteers suggest that motilin may stimulate vagal nerve activity, thereby helping to facilitate gastric emptying [22] and possibly increase pancreatic polypeptide and insulin release [23–25]. Given the molecular and functional links between motilin and ghrelin [9], it will now be interesting to explore further the effects of low doses of motilin receptor agonists on other vagal nerve-mediated functions, such as those involved in the mechanisms of appetite, feeding and nausea.

MOTILIN RECEPTOR AGONISTS

MOTILIDES

Following the discovery of the prokinetic effect of erythromycin-A(2), the gastric prokinetic activity of other macrolide antibiotics in the dog was investigated [26]. Although somewhat limited, these early data suggested the importance of the 14-membered macrolide ring structure in combination with the basic desosamine sugar at C-5, since oleandomycin (3) showed a similar pattern of activity to erythromycin-A (although it was 20-fold less potent). The 16-membered macrolides examined (leucomycin, acetylspiromycin and tylosin) were inactive. This suggested that the gastric prokinetic and antibiotic activities are not inextricably linked for these compounds.

(2) Erythromycin-A   (3) Oleandomycin

Omura, Itoh and co-workers then proceeded to design and synthesise analogues of erythromycin-A combining improved gastric prokinetic activity with reduced antibiotic activity, targeting compounds suitable for clinical use as GI modulators. Their initial studies [27] showed that the cyclic enol ether (4) (produced by treatment of erythromycin-A with mild acid) showed a 10-fold increase in gastric prokinetic activity in the dog when dosed intravenously. Initial studies into the effects of modifying the amino group of the desosamine sugar at C-5 were also undertaken [27]. Conversion of the N,N-dimethyl group in (4) to a N-ethyl-N-methyl group resulted in a more potent compound (5), code-named EM523, with gastric prokinetic activity 18 times that of erythromycin-A in the dog. Quaternization of the N,N-dimethyl group in (4) resulted in compounds with gastric prokinetic activity up to 2,890 times that of erythromycin-A, e.g. compound (6). The 8,9-dihydro- analogues of these quaternary salts were also prepared, and although they exhibited significant gastric prokinetic activity, the levels were somewhat less than those for the parent enol ethers. All of these compounds showed very little, if any, antibacterial activity thus confirming that the gastric prokinetic and antibiotic properties of erythromycin-A derivatives could be separated.

Although these initial studies showed that the compounds exhibited strong GI prokinetic activity, evidence that at least part of this activity could result from interaction with the, as yet unidentified, motilin receptor came from radiolabel binding studies. Erythromycin-A and some of the above-mentioned derivatives were shown to inhibit specific binding of [125]iodinated motilin to rabbit duodenal muscle strips [28]. The most potent compound in *in vivo* dog studies, (6), gave a K_D value of 30 nM, approaching that of motilin itself (K_D 6 nM) (Table 2.1), whilst erythromycin-A was somewhat less potent

Table 2.1   SUMMARY OF BINDING AFFINITY DATA TO MOTILIN RECEPTORS IN
RABBIT (AND HUMAN) TISSUE OF MOTILIDES (4)–(30)

Cpd	Rabbit tissue type[a]	Cpd result	Motilin result	Erythromycin-A result	References
6	Duodenum	K_D 30 nM	K_D 6 nM	–	[28]
4	Gastric antrum	pIC_{50} 8.40	pIC_{50} ~9.0	pIC_{50} 7.24	[31]
6		pIC_{50} 8.36			
8		pIC_{50} 5.74			
5	Gastric antrum	pIC_{50} 8.40	pIC_{50} 9.17	–	[32]
		K_D 3.7 nM	K_D 0.53 nM		
5	Human gastric antrum	pIC_{50} 7.89	pIC_{50} 8.84	–	[32]
		K_D 13 nM	K_D 1.8 nM		
7	Gastric antrum	pIC_{50} 8.21	pIC_{50} 9.20	pIC_{50} 7.37	[39]
11	Small intestine	pIC_{50} 8.04	–	pIC_{50} 7.36	[44]
12		pIC_{50} 8.42			
13		pIC_{50} 8.22			
5		pIC_{50} 8.50			
11	Small intestine with acid	pIC_{50} 8.05	–	pIC_{50} 7.15	[44]
12		pIC_{50} 8.19			
13		pIC_{50} 8.10			
5		pIC_{50} 6.52			
14	Small intestine	pIC_{50} 8.60	pIC_{50} 9.31	–	[47]
	Small intestine with acid	pIC_{50} 8.75	–	–	
15	Small intestine	pIC_{50} 8.12	–	–	[48]
16		pIC_{50} 8.33			
17		pIC_{50} 8.16			
15	Small intestine with acid	pIC_{50} 7.40	–	–	[48]
16		pIC_{50} 8.08			
17		pIC_{50} 7.82			
29	Gastric antrum	pIC_{50} 7.86	pIC_{50} ~9.0	pIC_{50} 7.24	[31]
30	Gastric antrum	pIC_{50} 8.3	pIC_{50} 9.2	pIC_{50} 7.2	[63]

[a]Unless specified as human.

with a K_D value in the micromolar region. Work on this series was further expanded [29]; systematic derivatization of core and sugar hydroxyl groups did not give an increase in gastric prokinetic activity compared with parent enol ether (4). The substitution of the desosamine nitrogen was also assessed further and whilst compounds containing –NHMe, –NHEt, –NMe"Pr and –NEt$_2$ groups showed levels of gastric prokinetic activity similar to those of both compounds (4) and (5), the N-methyl-N-isopropyl analogue (7), code-named EM574, showed the largest increase in potency with a 248-fold increase in gastric prokinetic activity compared with erythromycin-A.

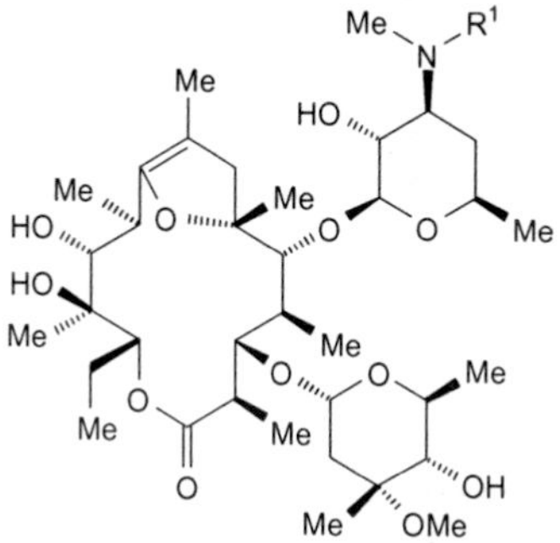

(4) R^1 = Me
(5) R^1 = Et (EM523)
(7) R^1 = ^{i}Pr (EM574)

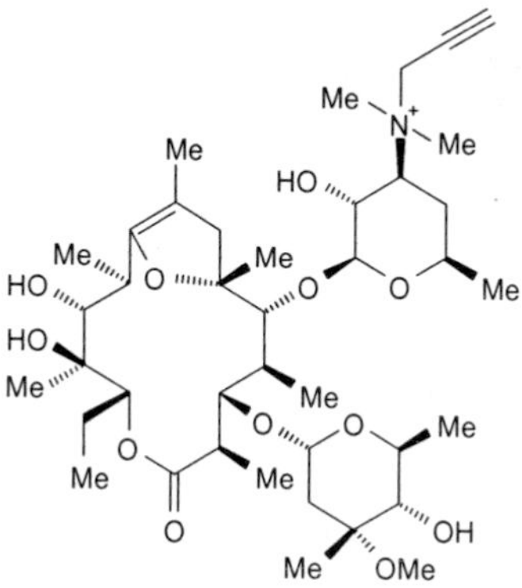

(6)

One of the disadvantages of these compounds is their instability to acid since the enol ether moiety reacts with the 12-hydroxyl to produce the spiroketal containing anhydroerythromycin-A (8) and this possesses much reduced gastric prokinetic activity. Reduction of the cyclic enol ether to the corresponding epoxide was envisaged as one solution to this problem and compounds such as (9) did show much improved levels of acid stability. However, the very large increases in gastric prokinetic activity were lost, such that (10) the dihydro- analogue of (7) showed a 14-fold increase compared with erythromycin-A as opposed to the 248-fold increase observed with (7). Again for all these compounds showing enhanced gastric prokinetic effects, antibacterial activity was minimal or non-existent. In an accompanying paper [30], the initial studies on the quaternary salts of (4) are extended and whilst a number of other active analogues were identified, none surpassed that of the propargyl compound (6).

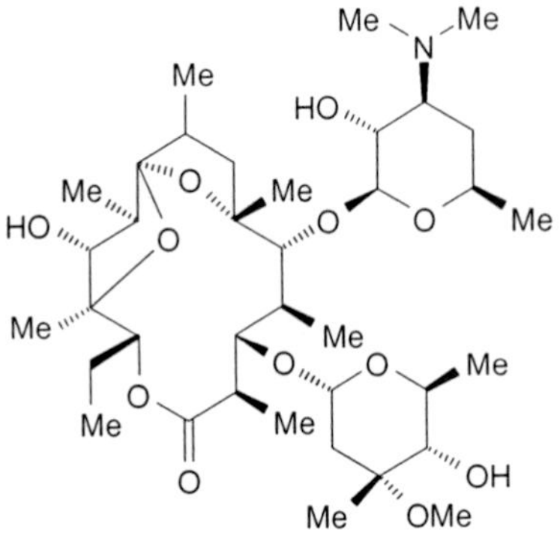

(8)

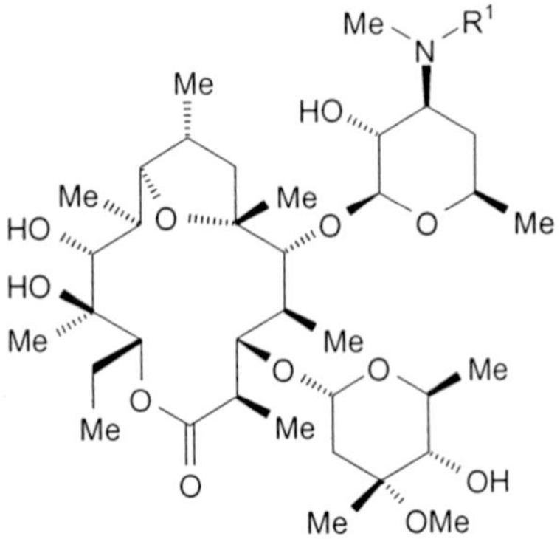

(9) R¹ = Me
(10) R¹ = ⁱPr

At the same time, Peeters and co-workers carried out an extensive SAR evaluation of some 60 erythromycins; the naturally occurring compounds, a selection of derivatives (including some of the early compounds described above), and seven fragments of the macrolide structure [31]. The compounds were investigated for their ability to competitively bind to the motilin receptor in a rabbit gastric antrum tissue homogenate (Table 2.1), and to induce contractions of rabbit duodenal smooth muscle tissue (Table 2.2). Their results also showed that the rank order of potency for the general ring analogues is anhydroerythromycin-A enol ether (4) > erythromycin-A (2) > anhydro-erythromycin-A (8). The importance of the combination of the macrolide ring and the sugar moieties was also demonstrated together with SAR around the amino group of the desosamine sugar at C-5.

As mentioned earlier, the groups of Omura and Itoh [28] showed that compound (5) displaced radiolabelled motilin from rabbit duodenal muscle strips in a concentration-dependent manner. Peeters and co-workers [32] showed that (5) also displaced radiolabelled motilin from rabbit and human gastric antrum preparations with competitive binding kinetics and K_D values of 3.7 and 13 nM respectively (Table 2.1). In addition, this study also demonstrated significant contractile activity of (5) in rabbit duodenal tissue (Table 2.2) and human duodenal tissue (pEC$_{50}$ 6.1 c.f. motilin pEC$_{50}$ 7.8). Further studies by Omura, Itoh and co-workers [33] demonstrated that (5) also caused contraction of rabbit small intestinal tissues (with the highest level of activity in the duodenum) (Table 2.2), but not those of rat or guinea pig (species where the motilin receptor is not thought to exist [9]).

This compound has been evaluated *in vivo* where it demonstrated motilin-like behaviour when dosed to fasted and fed dogs [34–36], with the former showing MMC-like contractions and the latter increased gastric

Table 2.2   SUMMARY OF RABBIT DUODENUM[a] SMOOTH MUSCLE
CONTRACTION ASSAY DATA FOR MOTILIDES (4)–(18)

Cpd	Cpd result	Motilin result	Erythromycin-A result	References
4	pEC_{50} 7.06	pEC_{50} ~8.7	pEC_{50} 5.16	[31]
8	pEC_{50} 3.40			
5	pEC_{50} 8.19	pEC_{50} 8.69	–	[32]
5	EC_{30} 5.8 nM (duo)	EC_{30} 1.3 nM (duo)	–	[33]
	EC_{30} 11 nM (jej)	EC_{30} 2.2 nM (jej)		
	EC_{30} 430 nM (ile)	EC_{30} 26 nM (ile)		
7	pEC_{50} 8.26	pEC_{50} 8.69	pEC_{50} 6.52	[39]
11	pEC_{50} 6.93	–	pEC_{50} 6.50	[44]
12	pEC_{50} 7.36			
13	pEC_{50} 7.41			
5	pEC_{50} 7.32			
14	pEC_{50} 8.21	pEC_{50} 8.34	–	[47]
15	pEC_{50} 7.71	–	–	[48]
16	pEC_{50} 7.64			
17	pEC_{50} 7.89			
18	pEC_{50} 8.5	pEC_{50} 8.5		[49]

[a]Except where specified: duo = duodenum, jej = jejunum, ile = ileum.

contractions with co-ordinated antral and duodenal motility. Due to the
acid instability of (5) all of these studies involved either intravenous or intra-
duodenal dosing. The results of these studies, and others (e.g. the effects of
the compound on the human GI tract [37]) resulted in development of (5)
(EM523) being undertaken by Takeda. A study has since been published
detailing positive effects on gastric emptying, in both healthy volunteers and
diabetic patients with signs of neuropathy, following i.v. dosing [38].
However, development of the compound has now been discontinued.

   The ability of the closely related analogue (7) to directly cause muscle
contraction has also been investigated in rabbit intestinal preparations [39]. In
the early dog studies, (7) was found to possess a much higher level of gastric
prokinetic activity when compared with (5), but in these rabbit studies its
activity (pEC_{50} 8.26) was more similar to that of (5) (pEC_{50} 7.95), (Table 2.2).
One possible reason for this discrepancy is that in the rabbit intestine assay, the
compounds caused contraction by direct activation of receptors located on the
muscle (a cell type where the motilin receptor is likely to desensitise rapidly; see
Introduction). However, the gastric prokinetic activities of these compounds in
the dog were inhibited by anticholinergic agents [40], implicating the

involvement of motilin receptors located on nerves (a cell type where the motilin receptor may be less readily desensitised by motilides; see Introduction). The *in vitro* contractile activity of (7) on human gastric antrum muscle strips has been evaluated [41] and, as for the rabbit experiments, the effects were found to be due to direct activation of the smooth muscle; the EC_{50} value for (7) was 1.4 µM compared with 0.14 µM for motilin. Together, these data illustrate the need for caution when selecting the right type of assay that will fully translate to a clinically meaningful endpoint. Further *in vivo* studies showed that (7) accelerated gastric muscle contractility and gastric emptying in fed dogs when dosed intra-duodenally [42] and that oral administration to healthy human volunteers also resulted in improved rates of gastric emptying [43]. This oral activity, in combination with its higher gastric prokinetic stimulating activity, gave (7) (EM574) a particular advantage over (5) and hence this compound was also progressed into development by Takeda. However, no studies in patients have appeared and development has been discontinued.

In order to address the issue of acid instability, researchers at Chugai Pharmaceuticals chose to adopt an approach whereby *O*-alkylation of the 12-hydroxyl group would prevent the acid-mediated conversion of the enol ether (4) to the ketal (8) [44]. Selective alkylation of the 12-hydroxyl group over the 11-hydroxyl group of a suitably protected enol ether was achieved by first oxidizing the 11-hydroxyl to the corresponding ketone and then selectively methylating the 12-hydroxyl. However, following deprotection of the sugar hydroxyl groups, it was not possible to reduce the 11-oxo group back to the hydroxyl. Hence the substitution of the desosamine nitrogen was investigated in this 11-oxo-12-methoxy series of compounds. In binding studies it was found that the parent *N,N*-dimethyl– (11), the *N*-ethyl-*N*-methyl– (12) and *N*-isopropyl-*N*-methyl– (13) compounds showed similar levels of specific binding to motilin receptors in rabbit small intestine homogenate to that of (5) (Table 2.1). To assess the acid stability of this new structural subtype, the same experiments were carried out in the presence of hydrochloric acid, and, while the binding affinities of (11)–(13) changed very little if at all, that of (5) dropped approximately 100-fold (Table 2.1). Contractile activity in rabbit duodenum muscle strips was also similar to that of (5) with the isopropyl analogue (13) being the most potent with pEC_{50} 7.41 (Table 2.2). When dosed intravenously to fasted dogs (13) was equipotent with respect to inducing gastric prokinetic activity (measured as motility index (MI); MI_{100} 1.0 µg/kg) when compared to (5) (MI_{100} 0.9 µg/kg), but on intra-gastric administration it was approximately 10-fold more active (MI_{100} 1.5 µg/kg c.f. MI_{100} 14.9 µg/kg for (5)), presumably due to better oral bioavailability resulting from increased acid stability. No antibiotic activity was observed for (13) and this compound, code-named GM-611 or mitemcinal, was progressed into development. Further results from a number of preclinical animal studies (rabbits, dogs, rhesus monkeys and mini-pigs) have appeared [45], demonstrating its prokinetic effects *in vivo*, and (13) has progressed to clinical trials. Its efficacy in increasing gastric emptying in healthy volunteers has been stated [46] but no details have been published. However, the results of a full Phase II study in patients with diabetic gastroparesis have been published and mitemcinal showed statistically significant symptomatic relief over a three-month period in a subset of these patients with a Body Mass Index of less than $35 \, kg/m^2$ and good glycaemic control [45, 46]. Further clinical trials are currently ongoing.

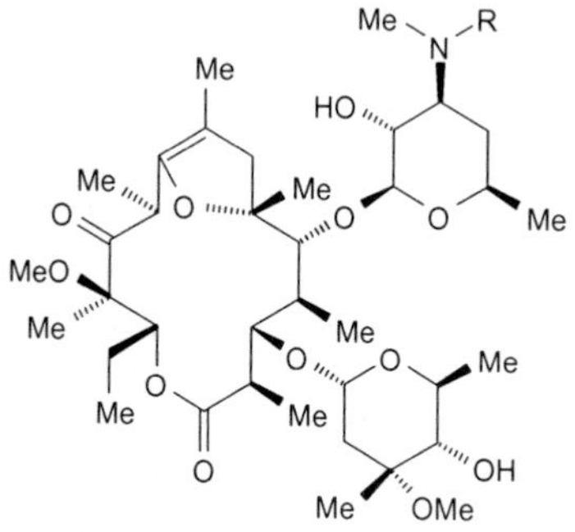

(11) R = Me
(12) R = Et
(13) R = ^{i}Pr (GM611, mitemcinal)

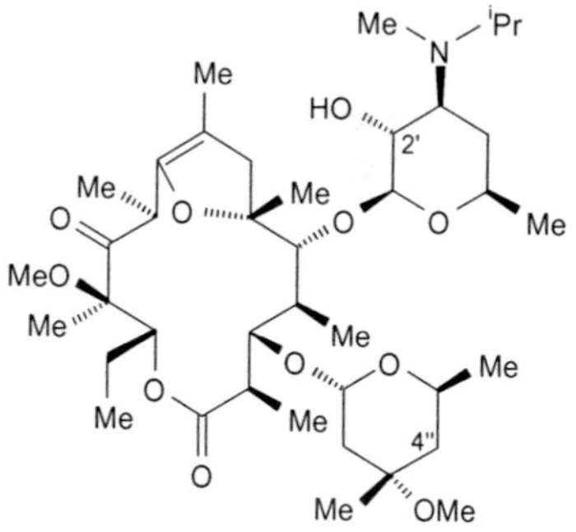

(14)

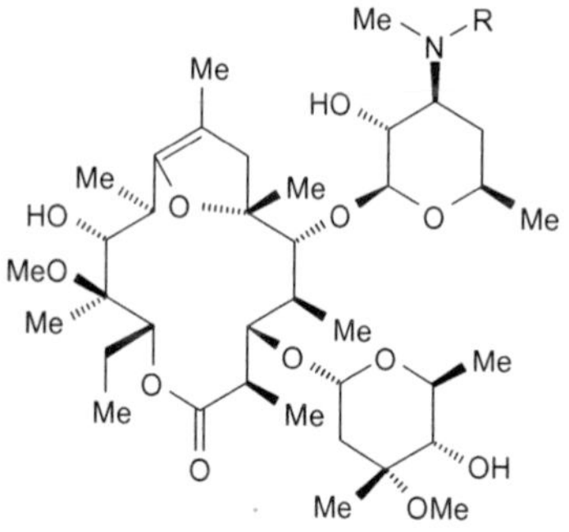

(15) R = Me
(16) R = Et
(17) R = ^{i}Pr

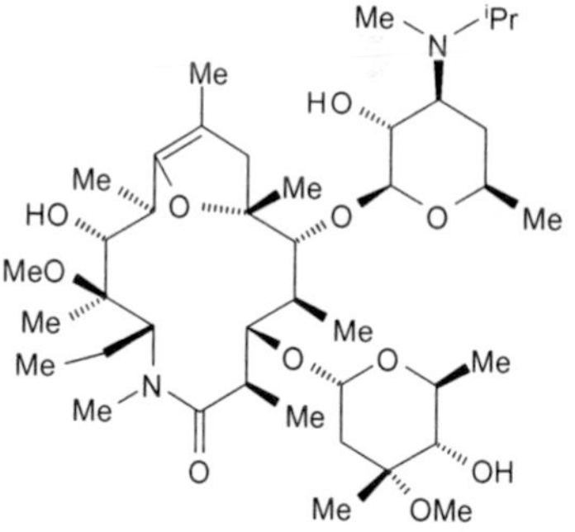

(18)

The importance of the cladinose 4″-hydroxyl and the desosamine 2′-hydroxyl groups for motilin agonist activity were also investigated [47]. Removal of the former, e.g. compound (14), gave a small increase in binding affinity (pIC_{50} 8.6) when compared with (13) (Table 2.1) which was not affected by acid, and a larger increase in contractile activity on rabbit duodenum (Table 2.2). In contrast, removal of both groups gave an increase in binding affinity (pIC_{50} 8.7) but a decrease in contractile activity which was mirrored *in vivo* in the dog. The increased activity of the 4''-deoxy compounds *in vitro* translated into more potent compounds *in vivo* with 4''-deoxy GM-611 (14) showing gastric prokinetic activity approaching the levels of motilin in these dog experiments. An alternative synthetic approach via a benzyl protection strategy enabled access to the 11-hydroxy-12-methoxy analogues (15)–(17) [48], and these compounds were also acid stable, potent motilin agonists devoid of antibacterial activity, with levels of activity *in vitro* (Table 2.1) and *in vivo* similar to those observed for (11)–(13). In addition, this group has applied for a patent on a series of macrolactams [49] where the *N*-Me lactam analogue (18) of 11-hydroxy compound (17) shows a level of activity the same as that of motilin in the rabbit duodenum contractility assay with a pEC_{50} of 8.5.

Researchers at Abbott Laboratories investigated the effects of removal of the cladinose 4''-hydroxyl group on prokinetic activity and at the same time addressed the issue of acid lability through complete removal of the 12-hydroxyl group, thus removing the potential for spiroketal formation [50]. It was found that the 4''-deoxy enol ether of erythromycin-A (19) showed the same level of activity in the rabbit duodenum contractility assay as enol ether (4), however, the corresponding erythromycin-B analogue (20), lacking the 12-hydroxyl group, was approximately 700-fold more potent in this assay (Table 2.3).

(19) R^1 = Me, R^2 = OH
(20) R^1 = Me, R^2 = H
(21) R^1 = Et, R^2 = H (ABT229)
(22) R^1 = Et, R^2 = OH

Changes to the desosamine amine group were then examined and it was found that the SAR tracked with the other series, since the *N*-methyl-*N*-ethyl analogues were more potent, with the erythromycin-B derivative (21) giving a pED_{50} value in excess of 12 (c.f. 5.85 for erythromycin-A). Furthermore, this compound showed a promising pharmacokinetic profile in fasted dogs with

Table 2.3 SUMMARY OF RABBIT DUODENUM SMOOTH MUSCLE
CONTRACTION ASSAY DATA FOR MOTILIDES (19)–(28)

Cpd	Cpd result	Erythromycin-A result	References
19	pEC_{50} 8.41	pEC_{50} 5.85	[50]
20	pEC_{50} 11.26		
21	pEC_{50} >12.0		
22	pEC_{50} 11.15		
21	pEC_{50} 7.22		[59]
23a	pEC_{50} 7.93		
23b	pEC_{50} 7.80		
24	pEC_{50} 8.54		
25	pEC_{50} 7.75		
26	pEC_{50} 8.11		
27	pEC_{50} 7.5	pEC_{50} 5.85	[60]
28	pEC_{50} 8.0		

oral bioavailability (F_{po}) of 39%. This was a significant improvement on that observed with EM523 (5), which had an oral bioavailability of only 1.4%, and with the corresponding erythromycin-A derived analogue (22), with an F_{po} of 2.7% in this study. This level of oral bioavailability led to evaluation of (21), code-named ABT-229 or alemcinal, for its gastric prokinetic activity in fasted conscious dogs in which it was found to be highly potent with an ED_{50} value of 0.5 μg/kg. Compound (21) was subsequently progressed into development. Initial studies in healthy human volunteers showed that (21) dosed at 4 or 16 mg p.o. significantly improved gastric emptying of a meal consumed 45 min later, but not of a second meal taken 4 h post dosing [51]. Interestingly, the plasma concentrations of the compound at this point were higher than for the lower dose at the initial meal. Furthermore, (21) failed in a number of studies to improve symptoms in patients with diabetic gastroparesis [52], functional dyspepsia (FD) [53] and gastroesophageal reflux disease [54]. This failure has been the focus of extensive discussion [9, 55–58] and a number of possible reasons have been put forward but tachyphylaxis (or desensitisation) caused by an inappropriately high dose of the compound is one of the key issues that will be discussed later.

In a later publication [59], the Abbott group disclosed their work on reduction of the cyclic enol ether in 4″-deoxy-erythromycin-A derivative (19) to the 6,9-epoxide (23) in an alternative approach aimed at improving acid stability. Molecular modelling based on known X-ray structures was used to predict which of the four possible diastereoisomers that result from hydrogenation of the cyclic enol ether would be most active. The reduction of (19) produced three of the four diastereoisomers and the results of rabbit duodenal contractility experiments agreed with the predictions from modelling, with the and (8S, 9R)- and (8R, 9R)- isomers (23a) and (23b) showing the highest level of activity in the rabbit duodenum contraction model (Table 2.3). The (8S, 9R)-diastereoisomer (23a) was used to investigate substitution of the desosamine amine group in this series. Unlike previous series, the N-ethyl-N-methyl analogue (24) (pEC$_{50}$ 8.54) was more potent than the N-isopropyl-N-methyl analogue (25) (pEC$_{50}$ 7.75) as was the N-(2-hydroxyethyl)-N-methyl compound (26) (pEC$_{50}$ 8.11). Compounds (24)–(26), together with (21), were assessed for stability to acid by exposure to pH 3.0 phosphate buffer. Somewhat surprisingly, (21) was very unstable with a half-life of only 0.15 h under these conditions whereas the half lives for (24)–(26) were 5.5 h, showing that the 6,9-epoxy motilides are considerably more stable. The authors propose them as potential back-up compounds to (21) although no development of these compounds has since been reported. This may be due to poor oral bioavailability as a later paper [60] discloses that the (8R, 9R)-isomer (23b) possessed a poor profile in the dog with F_{po} 5%. In this paper, a series of macrolactams or 'motilactides' is described, with the most potent compounds (27) and (28) (Table 2.3) showing good oral bioavailability profiles in the fasted dog, with F_{po} values of 40% and 61% respectively. Compound (28) is reported to possess an ED$_{50}$ of 0.036 mg/kg for contraction of GI smooth muscle when dosed orally to fasted dogs.

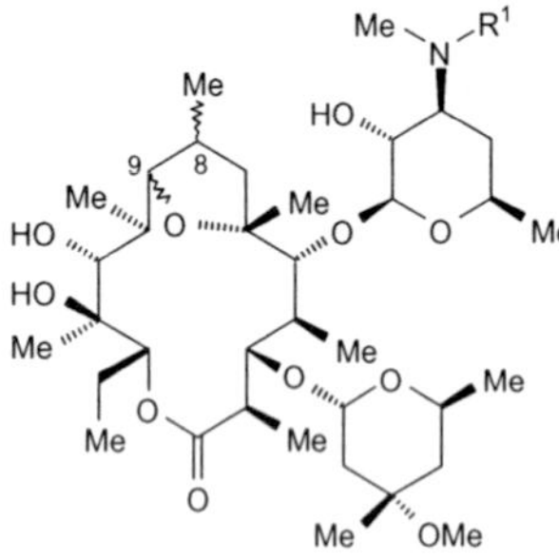

(23)

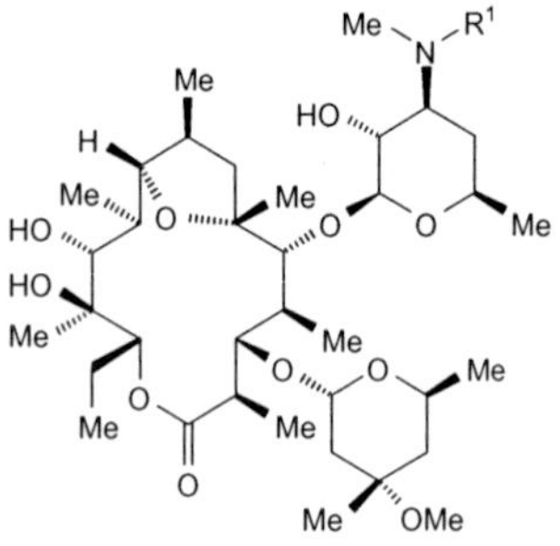

(24) R^1 = Et
(25) R^1 = ^{i}Pr
(26) R^1 = -CH$_2$CH$_2$OH

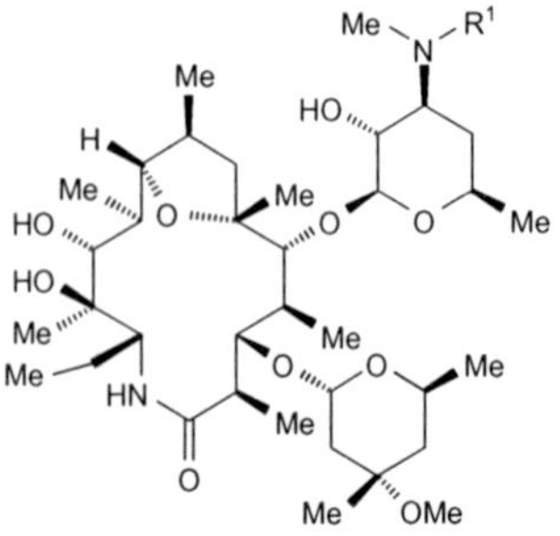

(27) R^1 = Me
(28) R^1 = ^{i}Pr

In the early studies on macrolide antibiotics, Itoh *et al* showed that certain 16-membered macrolides did not show prokinetic activity and proposed that the 14-membered ring structure was required [26]. As described earlier, the activity of oleandomycin was some 20-fold less than that of erythromycin-A, and other 14-membered antibiotics such as clarithromycin and roxithromycin also showed much reduced levels of smooth muscle contraction [61]. The structures of these latter compounds and oleandomycin are such that the cyclic enol ether derivatives cannot be formed as happens with erythromycin-A. Therefore it appears that the enol ether moiety is important for enhanced GI prokinetic activity. Indeed, in their early studies [31] Peeters and co-workers showed that the cyclic enol ether derivative of pseudoerythromycin-A (29), a 12-membered macrolide, showed greater binding affinity to the motilin receptor in rabbit gastric antrum tissue homogenate (Table 2.1) and greater contractile activity of rabbit duodenum smooth muscle tissue than erythromycin-A (Table 2.4). *In vivo* prokinetic activity of compound (29) in the dog has been disclosed by Eli Lilly (compound code-named LY267108) [62], but it also possesses prokinetic activity in mice, suggesting that it is non-selective as motilin receptors are not thought to be present in this species (see Introduction).

(29) R^1 = Me
(30) R^1 = ^{i}Pr

Researchers at Solvay have disclosed data for the 3'-*N*-isopropylmethyl analogue of (29), compound (30), code-named KC11458 [63]. This

Table 2.4   SUMMARY OF RABBIT DUODENUM SMOOTH MUSCLE CONTRACTION ASSAY DATA FOR MOTILIDES (29)–(46)

Cpd	Cpd result	Motilin result	Erythromycin-A result	References
29	pEC_{50} 6.09	pEC_{50} ~8.7	pEC_{50} 5.16	[31]
30	pEC_{50} 7.9	pEC_{50} 8.1	pEC_{50} 6.0	[63]
44	EC_{50} 180 nM			[69]
46	EC_{50} 58 nM		EC_{50} 1.2 μM	[70]

compound is more potent than erythromycin-A in binding to rabbit antral tissue (pIC_{50} 8.3) (Table 2.1) and in causing contractions of rabbit duodenal tissue (Table 2.4). Furthermore, this motilide has also been tested at the cloned human motilin receptor expressed in a CHO cell line using an aequorin assay, where it gave a pEC_{50} of 7.6 (Table 2.5). Incidentally, this appears to be the first instance of published functional activity data for a motilide at the cloned human receptor; all previously published SAR data have been generated using predominantly native antral and duodenal tissues since the receptor for motilin was not identified until 1999. Intravenous administration of (30) to conscious dogs accelerated gastric emptying of both the liquid and solid phases of a radiolabelled meal. Compound (30) was also progressed to development and like many other motilides showed increased gastric emptying in healthy human volunteers [64]. The compound was also evaluated in diabetic gastroparesis and FD patients but no effects were observed [64]. The study was very limited in that only one dose was tested but no further data or development has been reported.

Kosan Biosciences have also been very active in the area of motilide research and although none of their work has been published in the scientific literature, several patent applications have appeared and key compounds are presented below. 6,9-Cyclic enol ethers, 6,9-epoxides and furans were covered in a case from 2001 [65]. In particular, the 11-hydroxy and 11-oxo analogues (31) and (32) are specifically claimed, where the key difference is the 13-*n*-propyl group. This is introduced via feeding the appropriate diketide intermediate to the producing organism, resulting in the synthesis of 13-*n*-propyl-6-deoxyerythronolide B, which is then processed through further biosynthetic and synthetic methods to the target compounds. The

Table 2.5 SUMMARY OF HUMAN RECOMBINANT MOTILIN RECEPTOR AGONIST FUNCTIONAL ASSAY DATA FOR MOTILIDES (30)–(44)

Cpd	Cell line/assay type	Cpd result	Motilin result	Erythromycin-A result	References
30	CHO cell line, aequorin assay	pEC_{50} 7.6	pEC_{50} 9.0	pEC_{50} 5.5	[63]
36 37	HEK293 cell line, aequorin assay	EC_{50} 1.4 μM EC_{50} 0.16 μM		EC_{50} 1.1 μM	[66]
38 39	HEK293 cell line, aequorin assay	EC_{50} 0.7 μM EC_{50} 2.1 μM		EC_{50} 2.0 μM	[67]
40 41 42 43	HEK293 cell line, aequorin assay	EC_{50} 42 nM EC_{50} 13 nM EC_{50} 45 nM EC_{50} 230 nM			[68]
44	HEK293 cell line, aequorin assay	EC_{50} 1.2 μM			[69]

corresponding saturated epoxide analogues (33) and unsaturated furan analogues (34) are also specifically claimed as are the ring contracted 12-membered macrolides (35). No data are presented for compounds in this case.

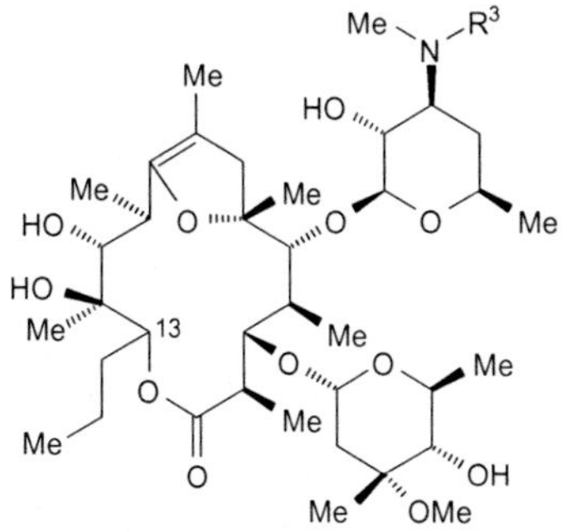

(31) R^3 = Et, ^{i}Pr

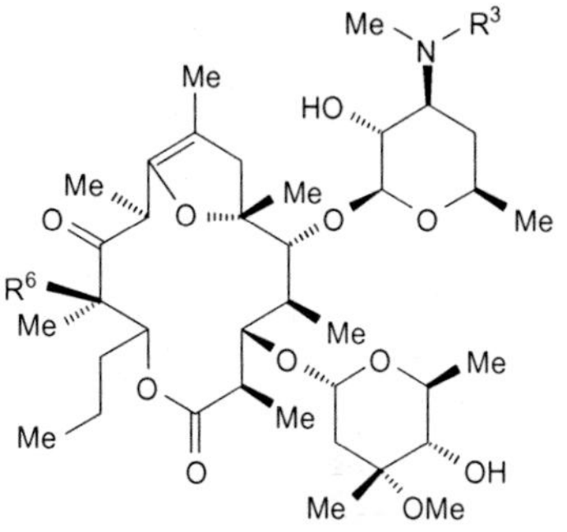

(32) R^3 = Et, ^{i}Pr;
R^6 = OH, OMe

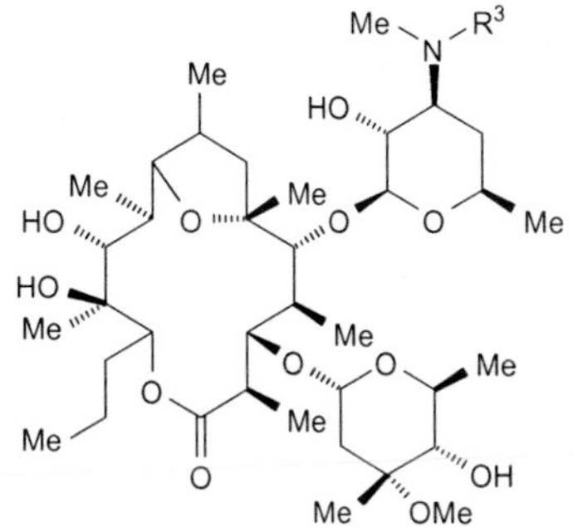

(33) R^3 = ^{i}Pr

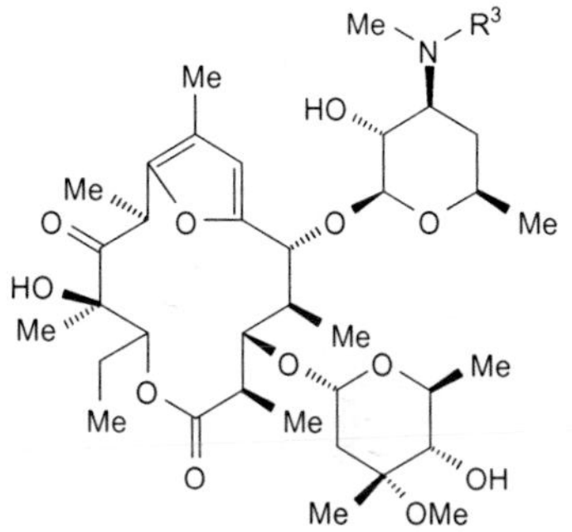

(34) R^3 = Et, ^{i}Pr

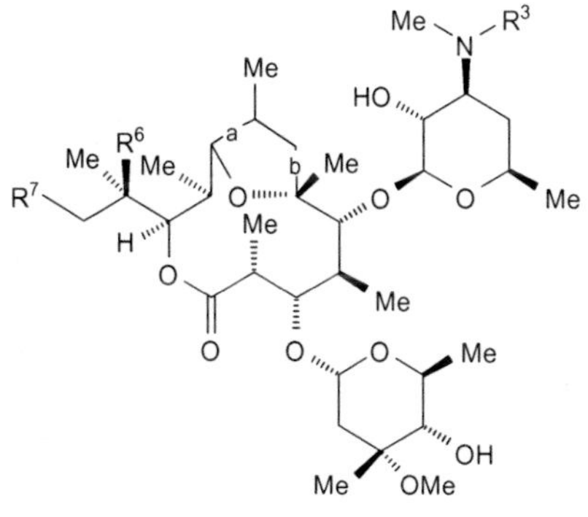

(35) a = b = single; a = b = double
a = double, b = single
R^3, R^6 as above, R^7 = Me, Et

9-Hydroxy-erythromycin-A and erythromycin-B analogues are claimed in a case from 2004 [66]. Data for key compounds are presented including agonism of the cloned human motilin receptors expressed in HEK293 cells using an aequorin assay (Table 2.5). The compounds possessing the best overall profiles (potency, antibacterial activity and desensitisation profile) are the 9-hydroxy-13-n-propyl erythromycin-A analogue (36) (EC_{50} 1.4 µM) and 9-hydroxy-11-deoxy-erythromycin-B analogue (37) (EC_{50} 0.16 µM), both of which also contain the ^{i}PrMeN-substituent in the desosamine sugar at C-5. Further 11-deoxy-erthromy-cin-B analogues are disclosed in a 2005 [67] case where (38) and (39) are specifically claimed with EC_{50} values in the above-mentioned assay of 0.7 and 2.1 µM respectively (Table 2.5). 11-Deoxy-6,9-epoxide derivatives of erythromycin-B have also been claimed [68]; compounds (40)–(42), which do not possess the 4″-OH in the cladinose sugar, show the highest potency at the human motilin receptor, whereas the 4″-OH containing (43) is 5–17-fold less potent but shows a better *in vitro* desensitisation profile in a cell-based assay, where repeat dosing resulted in a smaller decease in agonist activity compared to compounds (40)–(42) (see section on tachyphylaxis and desensitisation for a discussion of this and other methods).

9-Desoxo analogues of erythromycin-A are claimed in another case from 2005 [69]. In addition to potency data for (44) (EC_{50} 1.2 µM) at the motilin receptor, desensitisation and hERG channel inhibition data are also presented for compounds (44) and (45), with the latter showing the better profile in these latter assays although its potency is not disclosed.

The most recent case [70] discloses a series of derivatives of 9-hydroxy-erythromycin-A, and motilin receptor agonist potency data are given for a range of compounds; the most potent possess EC_{50} values of less than 50 nM combined with no antibacterial activity. Kosan have progressed one of their motilides to Phase I clinical trials in partnership with Pfizer (compound code-named KOS-2187 or PF-04548043) but its structure has not appeared in the public domain. However, a recent Pfizer patent application [71] on polymorphic forms of a particular example from this recent Kosan patent application [70], compound (46), may indicate that this is the development compound. It has recently been reported that the development compound has a significant effect on gastric half-emptying time and gastric lag time in healthy volunteers when dosed b.i.d. for nine days with no tachyphylaxis observed [72]. However, it has also been disclosed that development has now been discontinued.

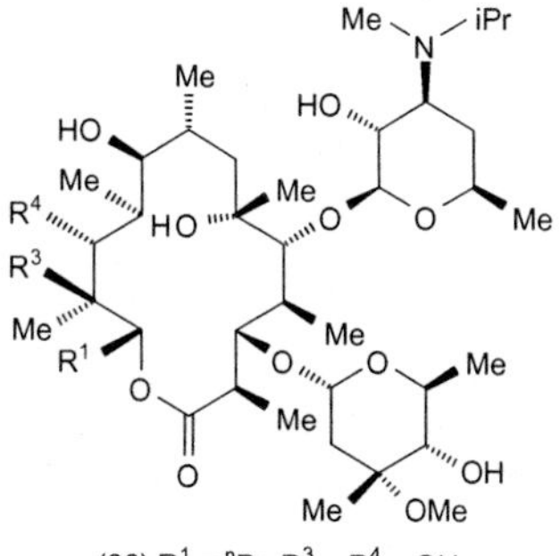

(36) $R^1 = {}^nPr$, $R^3 = R^4 = OH$
(37) $R^1 = Et$, $R^3 = R^4 = H$

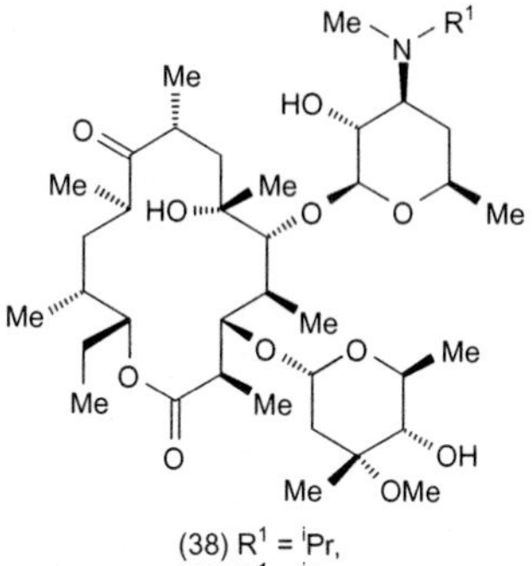

(38) $R^1 = {}^iPr$,
(39) $R^1 = {}^iBu$

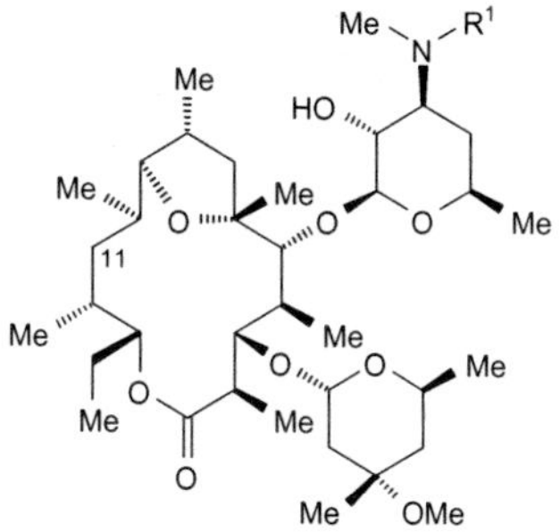

(40) $R^1 = Me$,
(41) $R^1 = {}^iPr$,
(42) $R^1 = -CH_2CH_2F$

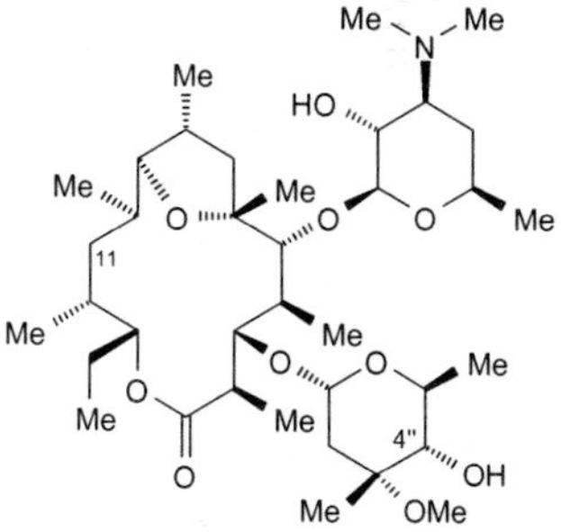

(43)

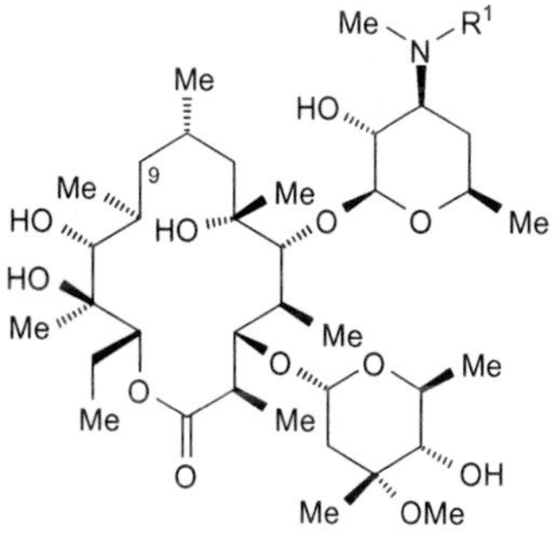

(44) $R^1 = Et$
(45) $R^1 = -CH_2CH_2OH$

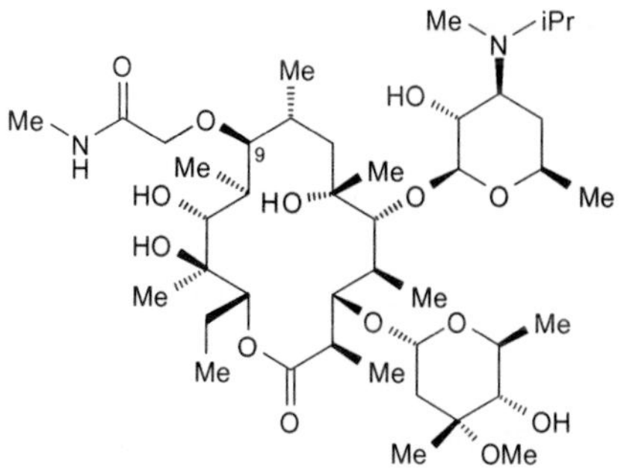

(46)

PEPTIDE AGONISTS

Early SAR studies on the 22-amino acid motilin peptide (1) were undertaken through the preparation of various fragments of the N-terminal, C-terminal and central regions of (Leu^{13})motilin, in which the readily oxidisable 13-methionine residue was substituted by leucine, resulting in a more stable but equipotent analogue [73]. The results showed that the pharmacophoric region for motilin receptor binding in rabbit gastric antrum tissue and contractile activity of rabbit duodenal tissue lies principally in the N-terminal region and that the C-terminal region contributes little with respect to these activities. The truncated version of this peptide consisting of residues 1–14 (47) gave a binding pIC_{50} value of 8.36 (Table 2.6). Alanine scanning together with D-amino acid substitution determined that the phenylalanine, isoleucine and tyrosine residues at positions one, four and seven respectively were of considerable importance for binding and functional activity [74]. A later study on intact porcine motilin, i.e. with the 13-methionine residue still present, confirmed these results [75]. More recently, similar results were obtained on analysing the functional activities of (Leu^{13})motilin fragments ranging in size from residues 1–5 to residues 1–19, and by using alanine scanning on the 1-14 fragment, at the cloned human motilin receptor utilising an aequorin assay [76]. It should be noted, however, that none of the above studies took into account the possibility that in contrast to the motilides and small molecule motilin receptor agonists, motilin peptide structures may occupy an additional binding site, arguably to facilitate desensitisation of the response to activation of the receptor (see Section, Advances in Motilin Receptor Neuropharmacology, for references and discussion). Accordingly, in terms of generating clinically useful gastric prokinetic agents, such studies start with a great handicap.

H$_2$N-Phe·Val-Pro-Ile·Phe-Thr–Tyr–Gly–Glu–Leu·Gln
|
HO$_2$C – Gln–Gly–Lys–Asn·Arg-Glu–Lys-Glu–Gln–Met–Arg

(1)

H$_2$N–Phe·Val – Pro – Ile · Phe-Thr–Tyr–Gly–Glu–Leu·Gln–Arg– *Leu* ·Gln–CO$_2$H

(47)

$\overset{+}{Me_3N}$ - Phe·Val–Pro–Ile·Phe·Thr–Tyr–Gly–Glu–Leu·Gln– *D-Arg – Leu – Lys* –CONH$_2$

X$^-$ (48)

Natural motilin (1) is unsuitable for use as a therapeutic agent due to its instability under physiological conditions; its elimination half-life has been

Table 2.6 SUMMARY OF PEPTIDE AGONIST BINDING AFFINITY DATA TO
MOTILIN RECEPTORS IN RABBIT TISSUE

Cpd	Tissue type	Cpd result	Motilin result	References
47	Gastric antrum	pIC_{50} 8.36	pIC_{50} 9.18	[73, 76]
48	Gastric antrum	pIC_{50} 8.93	pIC_{50} 9.18	[79]
49	Small intestine	IC_{50} 0.37 µM	–	[81]
50		IC_{50} 0.23 µM		

measured at between 4 and 10 min following i.v. infusion [77, 78]. Efforts towards more stable analogues have been limited but have resulted in the discovery of atilmotin (48), an analogue of residues 1–14 of the natural peptide [79, 80]. This compound has been shown to be a potent agonist in the rabbit duodenal contractility assay (pEC_{50} 7.6) (Table 2.7) and to bind to the motilin receptor with high affinity (pIC_{50} 8.9) (Table 2.6). In comparison (Leu^{13})motilin had a pEC_{50} value in the contractility assay of 8.1 and a pIC_{50} value in the binding assay of 9.2.

Incubation in 2% hog kidney homogenate was chosen as a method to assess the relative stability of the peptides, and it was shown that atilmotin (48) showed far superior stability under these conditions with a half-life of 162 min; half lives for (Leu^{13})motilin and (Leu^{13})motilin [1–14] (47) were only 30 and 7.5 min, respectively [79]. However, this improved stability *in vitro* did not translate *in vivo* as atilmotin has been shown to possess a half-life of less than 10 min in humans after intravenous infusion [80]. Nonetheless, it has demonstrated efficacy in the clinic when dosed intravenously to healthy volunteers, showing increased gastric emptying [80].

Haramura *et al.* have reported the design and synthesis of novel tetrapeptides based on the N-terminus of the motilin peptide incorporating various amino acids at position 3 [81]. The tryptophan analogue (49) showed reasonably high binding affinity to the motilin receptors in rabbit small intestinal homogenate with a pIC_{50} value of 0.37 µM (Table 2.6) but it was inactive in the rabbit duodenal contractility assay (Table 2.7). Substitution at the 2′ position of the tryptophan residue with (S)-alkyl groups gave compounds such as (50) with similar levels of binding affinity (IC_{50} 0.23 µM) and a moderate level of contractile activity (EC_{50} 4.6 µM), indicating an additional interaction with the receptor resulting in signal transduction.

$H_2N-Phe-Val- Trp - Ile - CONH_2$ $H_2N - Phe \cdot Val - Trp \ (2'-S^{\eta}Pr) - Ile - CONH_2$

(49)   (50)

Table 2.7 SUMMARY OF RABBIT DUODENUM SMOOTH MUSCLE
CONTRACTION ASSAY DATA FOR PEPTIDE AGONISTS (47)–(50)

Cpd	Cpd result	Motilin result	Erythromycin-A result	References
47	pEC_{50} 7.55	pEC_{50} 8.13	pEC_{50} 5.16	[76]
48	pEC_{50} 7.60			
49	EC_{50} > 100 μM	–	–	[81]
50	EC_{50} 4.6 μM			

A variant of (Leu^{13})motilin with an additional homoserine residue at the C-terminus has also been extensively evaluated *in vivo* in healthy volunteers but its pharmacokinetic profile, with an elimination half-life of 4.6–5.6 min following intravenous infusion, offered no improvement over motilin [82, 83].

SMALL MOLECULE AGONISTS

The advent of small molecule motilin agonist research followed the identification of the human motilin receptor in 1999, since molecular biology techniques allowed overexpression of the receptor in appropriate cell lines and thus development of assays suitable for high-throughput screening.

Solvay has applied for a patent on a series of 1-amidomethylcarbonyl piperidine derivatives as exemplified by (51), claiming they possess agonist activity at the human motilin receptor expressed in a CHO cell line, using an aequorin assay [84]. No data for the examples in this case are presented and no development has been reported.

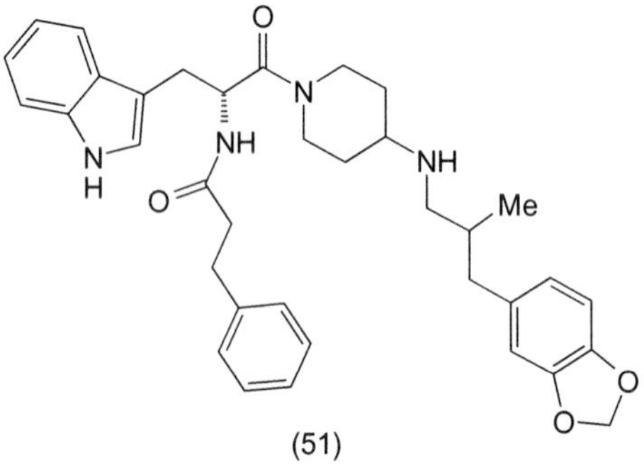

(51)

Researchers at Bristol Myers Squibb discovered a series of dihydrotria-zolopyridazine-1,3-dione-based amino acid derivatives [85]. Compound (52) was identified from high-throughput screening with an EC_{50} of 560 nM (Table 2.8) versus the human motilin receptor (FLIPR assay using the receptor expressed in a HeLa-MR9 cell line). This was a single enantiomer but the stereochemistry at C-8′ was not defined. From the SAR data presented, the presence of the terminal amide was crucial for activity since the corresponding ester (53) was inactive. The epimeric amino acid side-chain analogue was also inactive and the phenylalanine analogue (54) was 7.5-fold less potent. Reduction of the core pyridazine ring and addition of a further phenethylglycinamide unit to the side chain of the C-8′ epimeric mixture containing (52) led to an approximately 1,900-fold increase in potency (EC_{50} 0.35 nM c.f. 660 nM). The most potent C-8′ epimer from this dipeptide mixture, (55) (stereochemistry not defined) was extremely active, possessing an EC_{50} of 0.047 nM in the FLIPR assay, and was therefore analysed for its effects in a cell-based tachyphylaxis assay. The compound showed a more favourable profile than ABT-229 (21) although it was not as good as motilin or erythromycin-A. This series has also been the subject of a patent application [86] and three specific compounds are claimed; the saturated analogue (56) of high-throughput screening hit (52), the isoleucine

Table 2.8 SUMMARY OF HUMAN RECOMBINANT MOTILIN RECEPTOR AGONIST FUNCTIONAL ASSAYS FOR SMALL MOLECULE AGONISTS (52)–(65)

Cpd	Cell line/assay type	Cpd result	Motilin result	Erythromycin-A result	References
52	HeLA-MR9 cell line,	EC_{50} 560 nM	EC_{50} 0.15 nM	EC_{50} 400 nM	[85]
53	FLIPR assay	EC_{50} >10 µM			
54		EC_{40} 4.2 µM			
55		EC_{50} 0.047 nM			
58	CHO cell line, FLIPR assay	pEC_{50} 8.0	pEC_{50} 10.4	pEC_{50} 7.3	[87]
59	HEK293 cell line, FLIPR assay	pEC_{50} 7.7	pEC_{50} 9.3	pEC_{50} 6.2	[88]
59	CHO cell line,	pEC_{50} 8.4	pEC_{50} 10.4	pEC_{50} 7.3	[89]
60	FLIPR assay	pEC_{50} 7.9			
61	CHO cell line, FLIPR assay	pEC_{50} 10.3	pEC_{50} 10.4	pEC_{50} 7.3	[91]
65	CHO cell line, FLIPR assay	pEC_{50} 8.2	pEC_{50} 10.4	pEC_{50} 7.3	[95]

analogue (57) and the less potent phenylalanine analogue (54), but no development of these compounds has been reported. Although no data are presented for the parent acid or corresponding primary amide, the SAR for the first amino amide side chain indicates that the peptidic side chain is crucial for motilin receptor agonist activity so these molecules could be considered as small molecule/peptide hybrids.

(52) x = double bond
(56) x = single bond

(53) R^1 = -CH$_2$CH$_2$Ph, R^2 = OMe
(54) R^1 = -CH$_2$Ph, R^2 = NH$_2$
(57) R^1 = -CH$_2$^{i}Pr, R^2 = NH$_2$

(55)

Researchers at GlaxoSmithKline have embarked on an extensive programme to discover small molecule motilin receptor agonists. High-throughput screening and early high-throughput chemistry efforts led to the identification of lead urea (58) containing a key cis-dimethylpiperazine warhead [87]. This compound showed good potency at the cloned human motilin receptor expressed in a CHO cell line, with a pEC_{50} of 8.0 in a FLIPR assay (Table 2.8), and efficacy in a model of prokinetic-like activity in isolated rabbit gastric antrum. This is an alternative model to those used previously by other groups (where compounds were tested for an ability to cause contraction of rabbit isolated duodenal muscle via direct stimulation of motilin receptors on the muscle itself). For reasons described earlier (see earlier Section, Advances in Motilin Receptor Neuropharmacology), this assay was deemed to be unsuitable because (a) the receptors were located on a cell type (muscle) that did not have clear therapeutic and hence, translatable value and (b) activation of the 'muscle receptors' was subject to rapid desensitisation, a phenomenon not clearly observed when the receptors present on the nerves within the stomach are activated. Since motilin receptor agonists are known to increase gastric motility by stimulating cholinergic activity [17], an assay that tries to mimic this type of activity was thought to have greater translatable value. In addition, the chances of successful translation were further enhanced by using rabbit gastric antrum (the therapeutic organ) rather than duodenum.

In the new assay, motilin receptor activation by the compounds principally facilitates cholinergically mediated contractions of the stomach evoked by electrical field stimulation (EFS). Interestingly, the level of efficacy exhibited by (58) in this assay was partial in comparison with erythromycin-A and motilin: $182 \pm 56\%$ potentiation of the EFS-mediated contractions at $10 \,\mu M$ compared with $490 \pm 117\%$ at $3 \,\mu M$ for erythromycin-A and $506 \pm 112\%$ at $0.3 \,\mu M$ for motilin (these differences in intrinsic activity immediately highlight the advantage of using an assay with translatable value, as differences in intrinsic activity have not previously been reported for other motilin receptor agonists in other assays). A fairly high concentration of $10 \,\mu M$ was required to elicit the maximum effect but nevertheless this compound represented a new and exciting small molecule lead for further optimisation. The *in vitro* ADME profile of (58) was not ideal, with high microsomal clearance and a very poor cytochrome P450 inhibition profile. This was attributed to the high molecular weight and lipophilicity of the molecule and a programme of work targeting reduction of these properties was embarked upon. This led to the discovery of (59), a biarylcarboxamide possessing the key dimethylpiperazine warhead but with molecular weight 501 and *c*LogP 4.1 [88]. The potency of (59) at the recombinant human motilin receptor in the same assay (FLIPR, CHO cell line) was similar to that of the urea lead (pEC$_{50}$ 8.4) (Table 2.8) but *in vitro* microsomal clearance and CYP inhibition profiles were much improved. The level of contractile activity in rabbit gastric antrum tissue was higher than for (58) and more similar to that of erythromycin-A with $343 \pm 120\%$ potentiation of the EFS-mediated contractions at $3 \,\mu M$. Compound (59) was also selective and in *in vivo* pharmacokinetic studies it showed moderate clearance and some oral bioavailability in the rat (F_{po} 13%). However on further assessment (59) was found to cause time-dependent inhibition (TDI) of cytochrome P450 3A4. Since this is one of the key metabolising P450 enzymes, the level of inhibition observed was not ideal for further progression of (59) due to the increased potential for drug–drug interactions in the clinic.

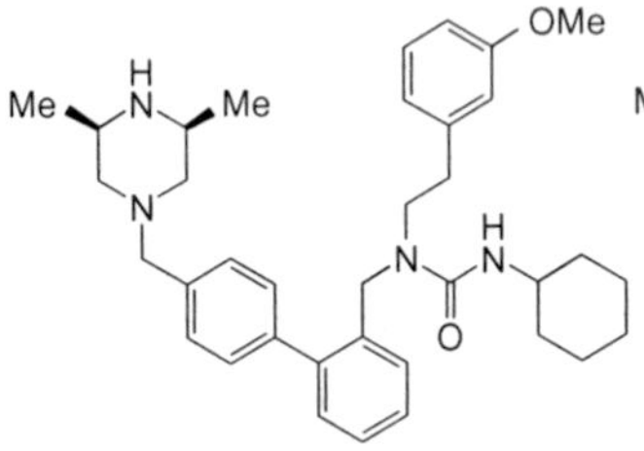

(58)

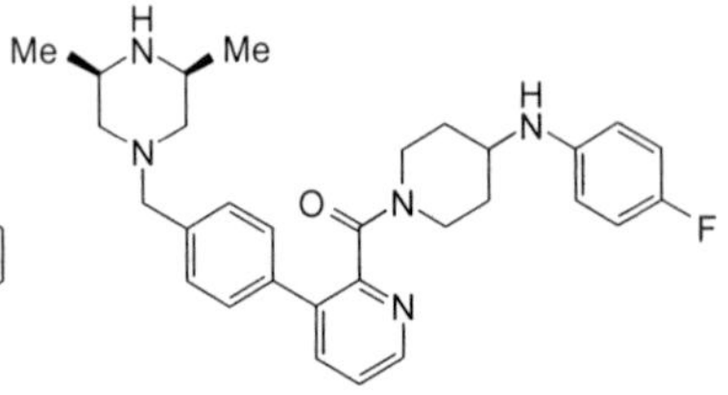

(59)

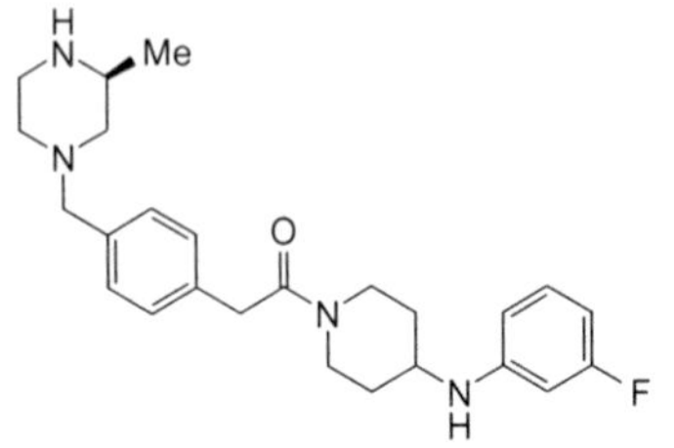

(60) GSK962040

To address this finding, a further reduction of lipophilicity and molecular weight was targeted and this strategy resulted in the identification of (60), code-named GSK962040, in which the *cis*-2,6-dimethylpiperazine warhead was replaced with (2*S*)-methylpiperazine and the biaryl carboxamide core was replaced with a phenyl acetamide core [20, 89]. Compound (60) was potent at the cloned human receptor with a pEC_{50} of 7.9 in the FLIPR assay (Table 2.8), possessed a much improved ADME profile, including no TDI at CYP 3A4, and good oral bioavailability in both rat and dog (F_{po} 48% and 51% respectively). Efficacy in the rabbit gastric antrum contractility assay was slightly lower than for (59) with a maximum potentiation of $248 \pm 47\%$ of the EFS-mediated contractions at $3 \mu M$. Compound (60) also showed a long duration of action in this assay indicating a potential for a lower desensitisation liability. The contractile activity of (60) in human gastric tissue was also examined [20], although due to issues with accessing tissue with a viable nervous system, the effects of (60), together with erythromycin-A and motilin, were tested on the direct contraction of the muscle; (60) evoked contractile activity similar to erythromycin-A at a concentration of $10 \mu M$ and this was comparable to that evoked by (Nle[13]) motilin at 100 nM. Interestingly, although the rank order of potency was the same as that obtained previously in the rabbit stomach assay, the effective concentration of (60) was somewhat higher than that required to elicit a strong cholinergically mediated response in the rabbit gastric antrum. Arguably, this difference may indicate the possibility of separating the desirable prokinetic effects from unwanted effects mediated by motilin receptors located on gastric smooth muscle (such as increased fundic muscle tone leading to reduced gastric accommodation with high doses). In an initial *in vivo* experiment, (60) was shown to increase GI transit in the conscious rabbit, resulting in increased faecal output over a 2 h period. On the basis of these data, (60) was progressed to preclinical development and is currently in Phase I clinical trials which will include an assessment of gastroprokinetic activity in healthy volunteers.

A further series of bi-heteroaryl piperidines closely related to (59) has also been disclosed [90, 91]. The most active of these compounds are highly potent in the FLIPR assay (recombinant human receptor expressed in a CHO cell line) with bipyridyl analogue (61) possessing a pEC_{50} value of 10.3 and this also translated into excellent potency in the rabbit gastric antrum contractility assay.

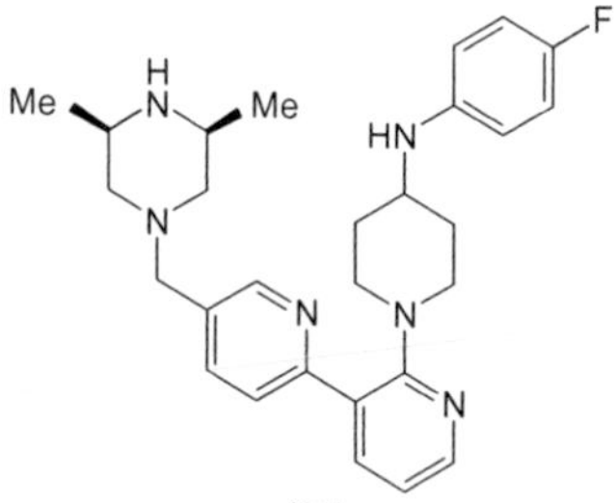

(61)

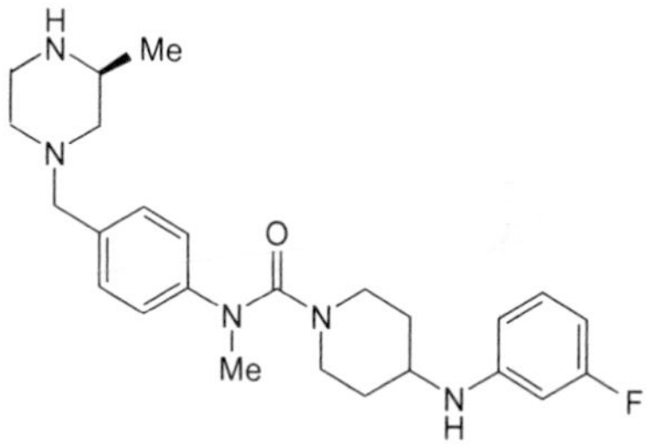

(62)

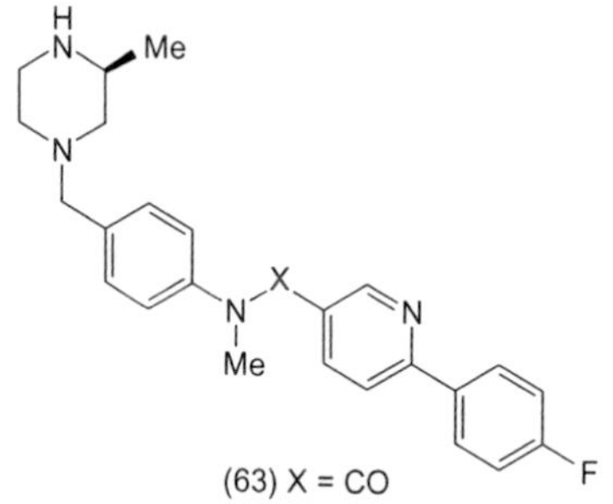

(63) X = CO
(64) X = SO$_2$

Three further GlaxoSmithKline patent applications claiming benzylpi-
perazine compounds have also appeared. The first [92] contains urea
derivatives as exemplified by (62); the second [93], nicotinamide derivatives
as exemplified by (63) and the third [94], sulfonamide derivatives
exemplified by (64).

A second high-throughput screen of the GSK compound collection
revealed an additional novel series of motilin receptor agonists based on a
benzazepine core [95]. A range of amides, sulfonamide and sulfones were
prepared with and without head-groups on the benzazepine nitrogen. The
compound possessing the best overall profile in this series was amide (65)
with a pEC_{50} of 8.2 at the recombinant human motilin receptor (Table 2.8),
together with good selectivity and *in vitro* clearance, but its strong inhibition
of CYP 2D6 precluded further progression.

(65)

THE QUESTION OF DESENSITISATION AND TACHYPHYLAXIS

As mentioned above, one of the significant issues that has been faced by all
those working on motilin receptor agonists is the problem of tachyphylaxis
or desensitisation. As an endogenous peptide acting to promote the
migrating, neuronally mediated and vigorous, but short-lasting, contractile
activity which defines the Phase III component of a MMC, motilin seems
to self-regulate by promoting rapid loss of its own activity at least partly
via receptor desensitisation. Such activity has been demonstrated in
recombinant receptor and in isolated tissue assays that measure directly
evoked muscle contractions or changes in neuronally mediated responses.
Given the characteristics of the Phase III MMC activity, this property of
motilin seems reasonable (see Section, Advances in Motilin Receptor
Neuropharmacology).

The abilities of non-peptide motilin receptor agonists (such as erythromycin) to activate the recombinant motilin receptor and cause muscle
contraction *in vitro* are also each subject to rapid desensitisation. However,

this property is not true for the ability of low concentrations of erythromycin-A (or other non-peptide compounds that have been tested, such as GSK962040 (60)) to facilitate cholinergically mediated activity in rabbit isolated gastric antrum, where the rate of loss of activity during the continual presence of the compound is considerably slower. This maintained response to erythromycin-A is consistent with the cholinergically mediated mechanism by which erythromycin-A increases gastric emptying, and with a maintained clinical benefit observed with repeated administration of low doses of erythromycin-A (see Section, Advances in Motilin Receptor Neuropharmacology, for references and speculations on possible mechanisms). These kinetic variations illustrate the need for medicinal chemistry programmes to be supported by models that have proven translational value and underline the important need to choose a clinical dose with great care.

Several groups have established *in vitro* models to examine the desensitisation profiles of a variety of motilin receptor agonists; such activity was especially prevalent following the failure of ABT-229 (21) in the clinic, as desensitisation of the effects of this molecule were argued to play a significant role in this failure. These methods involve the use of human recombinant receptors [96] and the rabbit duodenal muscle preparation [97], where both motilin and non-peptide motilin receptor agonists evoke muscle contractions that fade rapidly during the continued presence of the agonist. In both models, desensitisation experiments then typically consist of repeat dosing (with washout periods) until the functional response is lost. However, as noted above, whilst these methods appear a valid experiment, the translational value of the data can be questioned.

When using recombinant G-protein coupled receptors (GPCRs), a common feature of all agonists following initial dosing is subsequent desensitisation of the receptor through phosphorylation and internalisation, and certainly it has been shown that ABT-229 has a far greater potential to desensitise this form of the motilin receptor than erythromycin-A and other motilides such as EM574 (7) and EM523 (5) [96]. However, the results for mitemcinal (13) are somewhat confusing, with Carreras *et al.* reporting that the compound causes tachyphylaxis in the rabbit duodenal muscle preparation to a slightly greater extent than ABT-229 [97]. In these experiments the compounds were dosed four times, half-hourly, at their EC_{90} concentrations with washout periods in between. On the second dose, the muscle contractions evoked by mitemcinal were less than 20% of the initial contraction and by the fourth dose the contraction was only 9% of that of the initial response (c.f. 18% for ABT-229). Conversely, in a recent review on mitemcinal [45], it is reported that the desensitising effect of mitemcinal is much less than that of ABT-229 in an *in vitro* assay using CHO cells expressing the human motilin receptor. Depoortere and co-workers have demonstrated that ABT-229 induces internalisation and trafficking of motilin receptors expressed in a CHO cell line in a different manner and to a greater extent than motilin and erythromycin-A, and that the recovery period was significantly longer for ABT-229, with a $T_{1/2}$ value of 26 h compared with 3 h for motilin and 1 h for erythromycin-A [98]. Unfortunately, no data from testing mitemcinal in this assay are available for comparison to those generated by Carreras *et al.* in their rabbit duodenal muscle contractility assay [97]. In this study, the time taken for contractile activity to return when an EC_{90} concentration is dosed to rabbit duodenal strips was measured. ABT-229 showed only 20% recovery and mitemcinal showed 25% recovery after 1 h, whereas erythromycin-A and motilin showed >90% recovery. After 2 h, ABT-229 had recovered approximately 90% of its activity whereas mitemcinal was still giving only 31% of its initial activity. These results show that the desensitisation profile of motilin receptor agonists can vary according to the type of assay employed, even in these non-neuronal preparations where desensitisation is normally rapid. However, since mitemcinal shows effective symptom relief in a subset of diabetic gastroparesis patients in a Phase II clinical trial [45], it is not clear whether any of these *in vitro* studies are relevant in predicting tachyphylaxis *in vivo*. This doubt is reinforced by the clinical studies that report a maintained clinical efficacy with low doses of erythromycin-A (see Section, Advances in Motilin Receptor Neuropharmacology, for references). With a new small molecule motilin receptor agonist, GSK962040, now in the clinic, it will be of great interest to see how a non-motilide compound behaves in this respect. This compound was not assessed by repeat dosing in the rabbit duodenum muscle assay (where loss of activity is likely to occur) but instead, was assessed in terms of its ability to evoke a maintained facilitation of cholinergically mediated contractions in rabbit isolated gastric antrum, a property shared by erythromycin-A and thought to have greater relevance to the clinical situation [20].

In summary, the experience with motilin receptor agonists indicates a critical need to assess GPCR desensitisation liability by using models that have therapeutic relevance (in this case, looking at models where the receptor is functionally coupled to a cholinergic nerve), rather than easier-to-use models that use different tissues or even recombinant receptors where the relevance of the adopted receptor-coupling pathway is unknown. This conclusion is currently being tested by the clinical evaluation of non-peptide compounds such as mitemcinal and GSK962040, using dosing regimens that take into account the pharmacokinetic and pharmacodynamic properties of the molecules, to avoid tiring the stomach by continual stimulation of motility.

THE NEED FOR NEW GASTRIC PROKINETIC DRUGS AND COMPARISONS WITH OTHER APPROACHES

There is a high unmet clinical need for drugs that promote motility in the upper regions of the gut [58, 99]. Conditions where such drugs are needed include:

- patients needing improved clearance of acid and refluxed material from the oesophagus
  ⇕ reflux esophagitis is usually well served by drugs that inhibit gastric acid secretion, but this does not reduce reflux itself, leading to difficulties in a significant group of patients
- patients requiring increased gastric emptying
  ⇕ when requiring rapid intubation or endoscopy, or removal of gastric contents prior to endoscopy or surgery
  ⇕ when enteral feeding is delivered during intensive care, a situation where the physiological delay in gastric emptying limits the ability to deliver adequate nutrition
  ⇕ in diabetics where improved glucose control can be achieved by better regulation of gastric emptying
- patients where increased gastric emptying may improve symptoms
  ⇕ this includes certain patients with type I (especially those with evidence of autonomic neuropathy) and type II diabetes, where gastric emptying may be delayed
  ⇕ a subgroup of patients with FD, where gastric emptying may be delayed

The clinical need for upper GI prokinetic drugs is driven not just by the range of patients who might benefit from such drugs but also by the suboptimal efficacy and safety profile of those drugs that are currently available. These include: the peripherally restricted dopamine D_2 receptor antagonist domperidone (a weak, short-lived prokinetic agent that also increases blood prolactin concentrations [99, 100]); metoclopramide (which increases gastric emptying by acting as a $5\text{-}HT_4$ receptor agonist, but which also antagonizes the D_2 receptor, leading to increased blood prolactin concentrations and, after crossing the blood–brain barrier, induces akathisia or other extra-pyramidal movement disorders [101, 102]); and erythromycin-A (an antibiotic and motilin receptor agonist that may also cause QT-prolongation and cytochrome P450 inhibition, which is used off-label to facilitate gastric emptying but commonly, at the high doses needed for antibiotic activity, leads to nausea, causes tolerance with repeat dosing and generates concerns over the exacerbation of antibiotic drug resistance [9, 20, 55, 103]).

It has been reported that the gastric prokinetic efficacy of erythromycin-A may be greater than that achieved by domperidone and by 5-HT$_4$ receptor agonists such as metoclopramide and cisapride [104, 105], although this has not been a consistent observation [106, 107]. Caution must, nevertheless, be exercised when making clinical comparisons, as the optimal dose of erythromycin-A as a gastric prokinetic agent has not been rigorously defined (see reference [16] for a discussion and for a clear demonstration of greater prokinetic-like efficacy over the 5-HT$_4$ receptor agonist tegaserod, in rabbit isolated stomach).

Head-to-head comparisons between the gastric prokinetic activity of motilin and ghrelin receptor agonists have not yet been performed. As a gastric prokinetic, motilin is generally thought to operate predominantly via the enteric nervous system, perhaps with a smaller contribution via activation of the vagal nerve (see above); the opposite seems to be true for ghrelin, which depends more on vagal, rather than enteric nerve stimulation for its prokinetic activity and its ability to stimulate appetite [108–110]. Ghrelin receptor agonists may also markedly increase the release of various endocrine substances, possibly leading to reductions in insulin sensitivity, increased appetite, and reduced emesis, and evoking a small change in cardiovascular functions [111–114]. These additional actions are likely to lead to different clinical indications for motilin and ghrelin receptor agonists, perhaps with the latter more focused on conditions where there is a need to influence appetite and glycaemic control [115].

MOTILIN RECEPTOR ANTAGONISTS

PEPTIDE ANTAGONISTS

The earliest antagonists were peptides derived from motilin. Replacement of the proline residue with a phenylalanine in position three of both (Leu^{13}) motilin and the 1–12 fragment of motilin, gave compounds (66) (code-named OHM11526) [116] and (67) (code-named ANQ11125) [117] respectively. These compounds showed high binding affinity to motilin receptors in rabbit gastric antrum homogenate with the full-length peptide (66) giving a pK_D value of 9.26, while the fragment was approximately 10-fold less potent with pK_D 8.24 (Table 2.9). Both compounds also inhibited

the contractile response of rabbit duodenal muscle strips elicited by motilin
and the motilide compound (5) (EM523) (Table 2.10). Interestingly, the full-
length peptide (66) showed agonist activity against the avian motilin
receptor present in a small intestinal preparation [116] and at high
concentrations it also showed agonistic properties in rabbit duodenum

Table 2.9 SUMMARY OF BINDING AFFINITY DATA TO RABBIT
AND HUMAN MOTILIN RECEPTORS FOR ANTAGONISTS (66)–(74)

Cpd	Binding assay tissue type	Cpd result	Motilin result	References
66	Rabbit gastric antrum	pK_D 9.26	pK_D 9.11	[116]
67		pK_D 8.24		
67	Rabbit gastric antrum	pK_D 8.16	–	[117]
68	Rabbit small intestine	pK_i 7.99	pK_i 9.25	[122, 123]
68	Rabbit duodenum	IC_{50} 12 nM	–	[125, 126]
69		IC_{50} 1.9 nM		
70		IC_{50} 4.3 nM		
71	Rabbit duodenum	IC_{50} 0.27 nM		[127]
72	Rabbit colon	pK_i 8.58	pK_D 9.03	[129]
	Human recombinant, HEK293 cell line	pK_i 8.39		
73	Rabbit duodenum	IC_{50} 1.4 nM		[128]
74		IC_{50} 0.52 nM		

Table 2.10 SUMMARY OF FUNCTIONAL ASSAY DATA
FOR MOTILIN RECEPTOR ANTAGONISTS (66)–(74)

Cpd	Functional assay type	Result	References
66	Rabbit duodenum contraction versus motilin/EM523 (5)	pA_2 7.79/8.10	[116]
67	Rabbit duodenum contraction versus motilin/EM523 (5)	pA_2 7.03/7.55	[117]
68	Rabbit duodenum contraction versus motilin	pA_2 7.37	[122, 123]
68	Rabbit duodenum contraction versus motilin	pA_2 7.4	[125, 126]
69		pA_2 8.4	
70		pA_2 8.6	
71	Rabbit duodenum contraction versus motilin	pA_2 9.8	[127]
72	Rabbit duodenum contraction versus motilin	pA_2 9.17	[129]
73	Rabbit duodenum contraction versus motilin	pA_2 9.9	[128]
74		pA_2 8.2	

[118]. Its properties in dog and human are not clear; one study indicated potential agonistic effects on dog and human jejunal circular muscle cells [119] and in another study [120] the compound was shown to abolish motilin and erythromycin-A-induced tonic contractions of human colon circular muscle strips, although there was evidence for weak agonist activity in the myocyte cells. Weak agonist activity *in vivo* in the dog has also been described [121].

H₂N –Phe ·Val – *Phe* -Ile ·Phe ·Thr -Tyr -Gly -Glu –Leu ·Gln

HO₂C –Gln -Gly –Lys -Asn ·Arg -Glu –Lys ·Glu -Gln –*Leu* ·Arg

(66)

H₂N —Phe ·Val –*Phe* -Ile ·Phe ·Thr -Tyr -Gly -Glu –Leu ·Gln –Arg -CO₂H

(67)

GM109 (68) is a cyclic peptide derivative prepared by researchers at Chugai and has been shown to bind strongly to rabbit small intestinal tissue homogenate (Table 2.9) and to competitively inhibit motilin-induced contractions in rabbit duodenal smooth muscle (pA_2 7.37) (Table 2.10) [122, 123]. The compound is claimed to be selective but its activity in mice [124] (where the motilin receptor is not thought to exist, see Section, Advances in Motilin Receptor Neuropharmacology) questions this assertion. Furthermore, (68) has no oral bioavailability [125] and this has led to the development of further acyclic and cyclic peptidic antagonists by the Chugai group, for example, compounds (69)–(74) [126–128]. The profiles of tripeptides (69) and (70) have recently been published (Tables 2.9 and 2.10) and the compounds show good levels of oral bioavailability and decreased gastric and colonic motility in dogs [125]. A close analogue, MA-2029 (72), has demonstrated *in vivo* activity in the rabbit on oral dosing; where it reduced motilin-induced colonic contractions [129]. Furthermore, (72) reduced abdominal muscle contractions induced by administration of motilin in combination with colorectal distension, indicating a reduction in the visceral pain experienced by these animals that had been exacerbated by motilin [129]. Orally dosed (72) has also been shown to inhibit motilin induced increases in fundic tone, GI transit and diarrhoea in conscious dogs without effects on basal GI tone or motility [130]. These experiments demonstrate the pharmacodynamic characteristics of the molecule but do not yet demonstrate a pathophysiological role for motilin.

(68)

(69) R^1 = R^2 = R^3 = H, R^4 = Me
(70) R^1 = R^2 = H, R^3 = Me, R^4 = CONH$_2$
(71) R^1 = H, R^2 = F, R^3 = H, R^4 = 2-oxadiazole
(72) R^1 = Me, R^2 = F, R^3 = H, R^4 = CONHEt

(73) R = Me, n = 1
(74) R = H, n = 0

SMALL MOLECULE ANTAGONISTS

The first non-peptide motilin receptor antagonists disclosed were cyclo-pentenyl derivatives [131] identified by searching a compound collection using a pharmacophore model generated from the 3D-structure of the motilin peptide [132]. These compounds showed good binding affinity (Table 2.11), together with reasonable activity in abolishing motilin-induced contractions of rabbit duodenal smooth muscle (Table 2.12). The most potent compound, (75) (as a racemate) had a K_i value of 40 nM in the rabbit colon binding assay and an IC_{50} value of 287 nM in the rabbit duodenal functional assay. This compound was separated into its two components and the (R)-enantiomer (75a), was shown to possess the higher levels of activity with a K_i of 18 nM in the rabbit colon binding assay, and a pA_2 of 6.96 in the rabbit duodenal muscle functional assay.

Other cases have appeared claiming cyclohexene [133] and cyclobutene [134] analogues but, from the limited data presented, it appears that the cyclopentene compounds possess the highest potency. Compound (75a), code-named RWJ68023, was also shown to strongly inhibit the binding of motilin to both endogenous (gastric antrum) and recombinant human motilin receptors (expressed in a HEK293 cell line) with K_i values of 31 and 113 nM respectively [132].

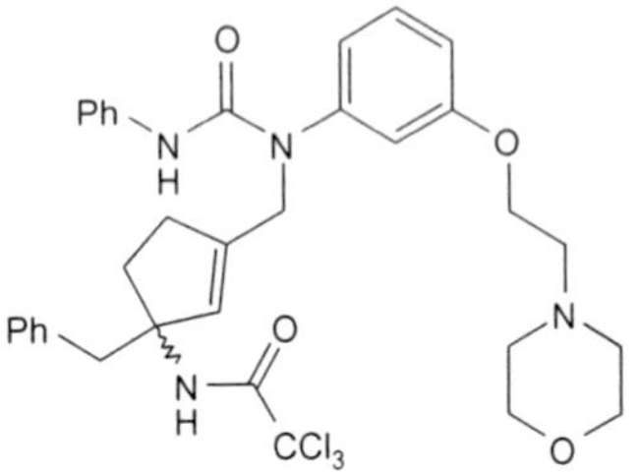

(75)

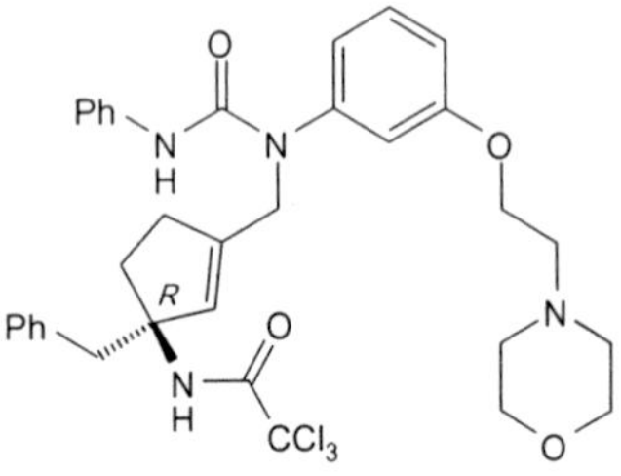

RWJ68023 (75a)

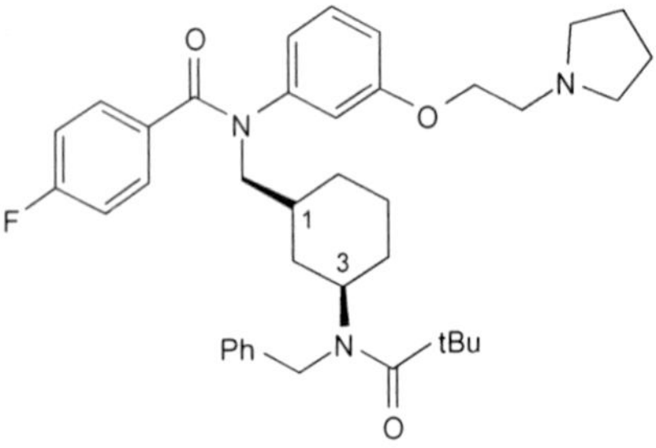

(76) racemate
(76a) (1S,3R) enantiomer

Since (75a) showed a favourable pharmacokinetic profile on i.v. dosing, it was studied in healthy human volunteers for its effects on basal and motilin-stimulated gastric volume. However, despite plasma levels being sufficient to block motilin receptors, (75a) did not affect basal proximal gastric muscle tone, and at high doses only a partial antagonistic effect was observed when the tonic condition of the stomach was modulated by motilin. The observation that (75a) did not affect proximal gastric volume should, therefore, be treated with caution [135].

Table 2.11 SUMMARY OF BINDING AFFINITY DATA TO RABBIT AND HUMAN MOTILIN RECEPTORS FOR ANTAGONISTS (75)–(81)

Cpd	Binding assay type	Cpd result	Motilin result	References
75	Rabbit colon	K_i 40 nM	K_i 0.9 nM	[132]
75a		K_i 18 nM		
75	Human gastric antrum	K_i 60 nM	K_i 1.0 nM	[132]
75a		K_i 31 nM		
75	Human recombinant, HEK293 cell line	K_i 102 nM	K_i 3.5 nM	[132]
75a		K_i 113 nM		
76	Rabbit colon	IC_{50} 29 nM		[136]
76a		IC_{50} 10 nM		
77	Human recombinant, CHO cell line	K_i 8 nM	K_i 0.36 nM	[140]
78	Human recombinant, CHO cell line	K_i 4.6 nM		[141]
79		K_i 6 nM		
80		K_i 6 nM		
81	Human recombinant, CHO cell line	$K_i \leq$ 10 nM		[138]

Table 2.12 SUMMARY OF FUNCTIONAL ASSAY DATA FOR MOTILIN RECEPTOR ANTAGONISTS (65)–(81)

Cpd	Functional assay type	Result	References
75	Rabbit duodenum contraction	K_i 18 nM	[132]
75a	versus motilin	K_i 5.2 nM, pA_2 6.96	
75	Human recombinant, HEK293 cell line,	K_i 53 nM	[132]
75a	FLIPR assay, versus motilin	K_i 20 nM	
76	Rabbit duodenum contraction versus motilin	IC_{50} 24 nM	[136]
76a		IC_{50} 8 nM	
76	Human recombinant, HEK293 cell line,	IC_{50} 0.99 μM	[136]
76a	FLIPR assay versus motilin	IC_{50} 0.39 μM	
77	Rabbit duodenum contraction versus motilin	pA_2 7.0	[140]
	Human recombinant, CHO cell line, aequorin assay versus motilin	IC_{50} 23 nM	
78	Human recombinant, CHO cell line,	IC_{50} 2 nM	[141]
79	aequorin assay versus motilin	IC_{50} 3 nM	
80		IC_{50} 28 nM	
81	Human recombinant, CHO cell line, aequorin assay versus motilin	$IC_{50} \leq$ 1 nM	[138]

A series of cis-1,3-disubstituted cyclohexane compounds has also been prepared [136]; racemic (76) strongly inhibited binding of radiolabelled motilin to the receptor in a rabbit colon tissue homogenate (Table 2.11) and also inhibited functional activity at the rabbit duodenal smooth muscle receptor and the cloned human receptor (Table 2.12). On separation of the enantiomers, it was found that the (1S), (3R)- isomer (76a) was the most potent with a binding IC_{50} value of 10 nM.

Tranzyme Pharma has worked extensively on motilin receptor antagonists based on peptide-derived macrocyclic templates. A number of patent applications [137–139] have appeared and SAR data around some of these series have been published [140] with compounds such as (77) showing good antagonist activity *in vitro* using the rabbit duodenum smooth muscle contraction assay (Table 2.12). Further optimization, including replacing the norvaline side chain with a basic or heterocyclic substituent [141], resulted in compounds such as (78)–(80) with improved binding affinity at the human receptor (Table 2.11) and increased antagonist activity in a cell-based human motilin receptor functional assay (Table 2.12).

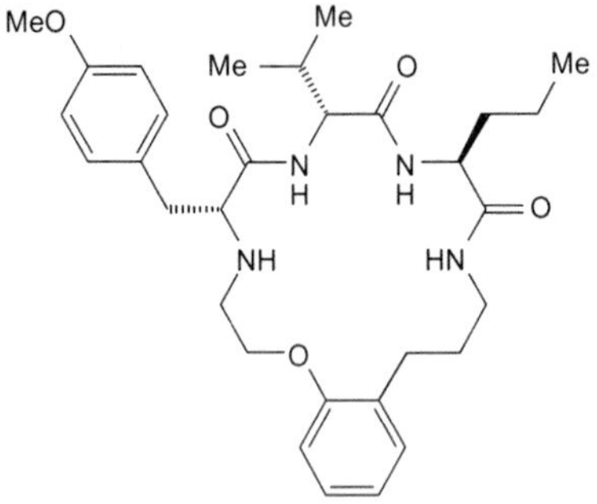

(77)

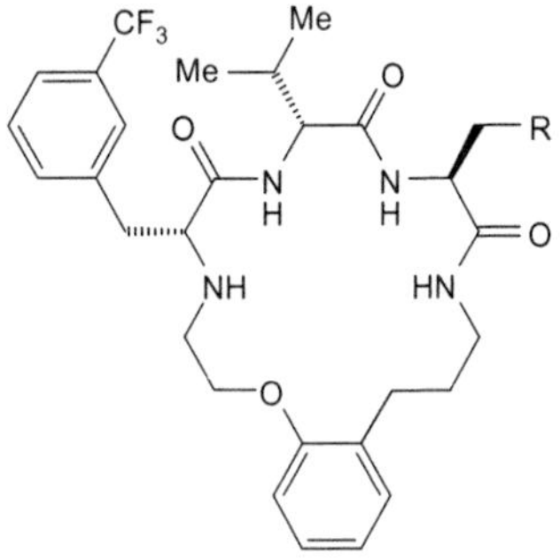

(78) R = 1-pyrrolidinyl
(79) R = NHMe
(80) R = NH(C=NH)NH$_2$

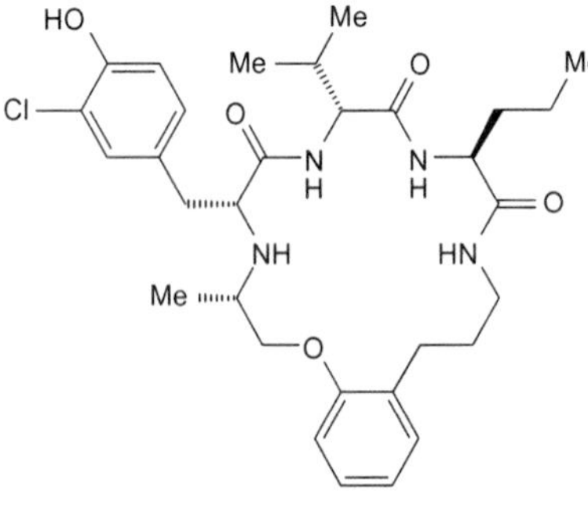

(81)

One compound, TZP-201, has been reported to be in Phase I clinical trials for conditions associated with GI hypermotility such as chemotherapy-induced diarrhoea (Tranzyme website). A recent publication [142] has disclosed an *in vitro* study in which TZP-201 was shown to antagonize motilin-induced contractile activity of the colon of the musk shrew, *Suncus murinus,* and these effects were found to be mediated through cholinergic enteric neurons. However, neither motilin receptor activation nor antagonism changed the electrogenic transport of water and electrolytes across the mucosa. Nevertheless, this species represents a new alternative for evaluation of motilin receptor agonists and antagonists, as it represents a small laboratory animal that actually possesses a functional motilin receptor and, in contrast to rodents (but similar to humans) has the capacity to vomit and hence, provides a more relevant gastric physiology.

The structure of TZP-201 is not yet in the public domain but data for the compound have been disclosed [143] and the above paper [142] references a recent patent application [138] in which the *in vitro* and *in vivo* effects of one particular example (compound 552) (81), are extensively described. These include an *in vitro* study examining the effects of (81) on the (Leu^{13})motilin-induced contractile activity of musk shrew colon, and the data presented are very similar to those presented in the above paper, indicating that compound (81) may be TZP-201. Similarly, in the patent application, the data for an *in vivo* study in the dog examining the efficacy of (81) in treating chemotherapy (irinotecan) induced diarrhoea are similar to those presented elsewhere for TZP-201 [143]. The results look very promising; i.v. administration of (81) at 2.5 and 7.5 mg/kg/day gave a superior level of efficacy compared to loperamide and octreotide. A further study with twice-daily dosing has also been presented and published as an abstract [144]. If confirmed, these data represent an important breakthrough in establishing a potential therapeutic use for motilin receptor antagonists.

THE CLINICAL NEED FOR MOTILIN RECEPTOR ANTAGONISTS

At the present time this is not clear, although several interesting threads of data exist to suggest that clinical utility for motilin receptor antagonists should be explored. Increased blood concentrations of motilin have been reported in patients with irritable bowel syndrome (IBS) and FD, suggesting that the use of a selective motilin receptor antagonist could be useful. In the IBS studies, the high concentrations of motilin occurred in patients defined as suffering from diarrhoea- or constipation-predominant forms of IBS [145]. Also in IBS patients, psychological stress may increase the release of motilin as well as the colonic motility [146]. However, induction of a gastrocolic reflex by infusion of intra-duodenal lipids, increased blood plasma concentrations of motilin only in the constipation-predominant IBS patients; there was a tendency for the motilin concentrations to be reduced in the diarrhoea-predominant group [147]. Nevertheless, it is claimed that the motilin receptor antagonist TZP-201 may reduce anticancer chemotherapy-induced diarrhoea in dogs [143, 144]. If confirmed, these exciting data are consistent with an ability of erythromycin-A and other motilin receptor agonists to increase defecation in conscious rabbits and dogs [20, 148–150], and if translated to humans, this finding will represent a significant medical advance. However, until more studies are conducted it must remain a possibility that such activity may be species-dependent, as in contrast to dogs and rabbits, erythromycin-A has not been found to consistently stimulate human colonic function (see reference [20] for discussion and references).

In patients with FD, a low blood plasma concentration of motilin has been reported [151] and during a stress interview, the release of motilin was found to be negatively correlated with 'indigestion symptoms' [152].

Finally, if motilin receptor antagonists are shown to be of therapeutic value, a theoretical possibility that such antagonism may also cause GI-related adverse events must be considered. For example, low blood plasma concentrations of motilin have been associated with gastroesophageal reflux disease [153]. Such data point to a need to look for potential changes in GI function when developing such compounds.

CONCLUSIONS

Identification of the pro-motility effects of erythromycin-A focused the main work around modulators of the motilin receptor on the design and synthesis of agonists. Initial efforts were focused predominantly on erythromycin analogues, the so-called 'motilides', and a few compounds have been progressed to the clinic. Whilst the earliest compounds such as EM523 (5) and EM574 (7) showed positive effects on gastric motility in healthy volunteers, development was halted, presumably due to issues with acid instability and oral bioavailability. ABT-229 (21) also showed a promising affect in human volunteers but its inability to effect gastric emptying at a later meal was perhaps, with hindsight, an indication that the clinical studies in patients would not be straightforward. Given the clinical effectiveness of erythromycin-A, the failure of ABT-229 to relieve symptoms in patients with diabetic gastroparesis and FD came as a severe blow to the area of motilin agonist research especially as some symptoms appeared to be worsened. There has been much debate around the reasons for this failure and tachyphylaxis has been viewed as a major problem with the compound. Perhaps this, combined with its long elimination half-life (20 h), is the cause of the inability of ABT-229 to affect gastric emptying of a second meal in the healthy volunteer studies. There has also been debate regarding its selectivity of action; the compound is reported to be effective in rats [154], a species where a functional motilin receptor is not thought to exist [9]. Could a lack of selectivity contribute to symptom worsening? The study design has also been called into question since the compound was dosed b.i.d. in spite of its 20 h plasma half-life [56].

Conversely, mitemcinal (13) has shown long-term therapeutic benefit to some patients with diabetic gastroparesis and clinical trials with this compound are continuing. The positive results with mitemcinal have highlighted the need for great care when designing therapeutically relevant assays of tachyphylaxis. Thus, in one of these assays (rabbit duodenal smooth muscle contraction), mitemcinal possessed a somewhat worse profile with regard to both the level of desensitisation and recovery period. In the other assay (recombinant human receptor), it has been stated that mitemcinal possesses a better profile than ABT-229 but the full data have not yet been published. In the clinic, no tachyphylaxis was observed over the three-month treatment period for the subset of patients in which mitemcinal improved symptoms. It is likely that the shorter elimination half-life of mitemcinal (10.5 h compared with 20 h for ABT-229) has played a significant role in the clinical outcome. A new class of motilin receptor agonist, GSK962040 (60), is currently in Phase I trials and is being evaluated for its safety, pharmacokinetic and pharmacodynamic profiles in human volunteers. As such, it is the first small molecule to progress to this stage of development and represents a new chapter in the discovery and development of motilin receptor agonists.

The clinical indication for motilin receptor antagonists is not yet clear but compounds are being identified to examine this area. The most notable compound is TZP-201, reported to reduce chemotherapy-induced diarrhoea in dogs [143, 144]. If confirmed and translated to humans, these studies would represent an important breakthrough.

REFERENCES

[1] Sanger, G.J. and Lee, K. (2008) Nat. Rev. Drug Discov. 7, 241–2540.
[2] Baggio, L.L. and Drucker, D. (2006) Gastroenterology 132, 2131–2157.
[3] Feighner, S.D., Tan, C.P., McKee, K.K., Palyha, O.C., Hreniuk, D.L., Pong, S.-S., Austin, C.P., Figueroa, D., MacNeil, D., Cascieri, M.A., Nargund, R., Bakshi, R.,

Abramovitz, A.A., Stocco, R., Kargman, S., O'Neill, G., Van Der Ploeg, L.H.T., Evans, J., Patchett, A.A., Smith, R.G. and Howard, A.D. (1999) *Science (Washington D.C.)* **284**, 2184–2188.

[4] Peeters, T.L., Matthijs, G., Depoortere, I., Cachet, T., Hoogmartens, J. and Vantrappen, G. (1989) *Am. J. Physiol.* **257**, G469–G474.

[5] Sarna, S.K. (1985) *Gastroenterology* **89**, 894–913.

[6] Nieuwenhuijs, V.B., Verheem, A., van Duijvenbode-Beumer, H., Visser, M.R., Verhoef, J., Gooszen, H.G. and Akkermans, L.M. (1998) *Ann. Surg.* **228**, 188–193.

[7] Husebye, E. (1999) *Neurogastroenterol. Motil.* **11**, 141–161.

[8] Peeters, T.L. (1993) *Gastroenterology* **105**, 1886–1899.

[9] Sanger, G.J. (2008) *Drug Discov. Today* **13**, 234–239.

[10] Rohm, K.D., Bolt, J. and Piper, S.N. (2009) *Curr. Opin. Clin. Nutr. Metab. Care* **12**, 161–167.

[11] DiBaise, J.K. and Quigley, E.M. (1999) *J. Clin. Gastroenterol.* **28**, 131–134.

[12] Richards, R.D., Davenport, K. and McCallum, R.W. (1993) *Am. J. Gastroenterol.* **88**, 203–207.

[13] Ueno, N., Inui, A., Asakawa, A., Takao, F., Tani, S., Komatsu, Y., Itoh, Z. and Kasuga, M. (2000) *Diabetologia* **43**, 411–415.

[14] Dass, N.B., Hill, J., Muir, A., Testa, T., Wise, A. and Sanger, G.J. (2003) *Br. J. Pharmacol.* **140**, 948–954.

[15] Takeshita, E., Matsuura, B., Dong, M., Miller, L.J., Matsui, H. and Onji, M. (2006) *J. Gastroenterol.* **41**, 223–230.

[16] Jarvie, E.M., North Laidler, V., Corcoran, S., Bassil, A. and Sanger, G.J. (2007) *Br. J. Pharmacol.* **150**, 455–462.

[17] Coulie, B., Tack, J., Peeters, T. and Janssens, J. (1998) *Gut* **43**, 395–400.

[18] Boivin, M.A., Carey, M.C. and Levy, H. (2003) *Pharmacotherapy* **23**, 5–8.

[19] Cuomo, R., Vandaele, P., Coulie, B., Peeters, T., Depoortere, I., Janssens, J. and Tack, J. (2006) *Am. J. Gastroenterol.* **101**, 804–811.

[20] Sanger, G.J., Westaway, S.M., Barnes, A.A., MacPherson, D.T., Muir, A.I., Jarvie, E. M., Bolton, V.N., Cellek, S., Naeslund, E., Hellstroem, P.M., Borman, R.A., Unsworth, W.P., Matthews, K.L. and Lee, K. (2009) *Neurogastroenterol. Motil.* **21**(6), 657–664.

[21] Matsuura, B., Dong, M. and Miller, L.J. (2002) *J. Biol. Chem.* **277**, 9834–9839.

[22] Mathis, C. and Malbert, C.H. (1998) *Am. J. Gastroenterol.* **274**, G80–G86.

[23] Witteman, B.J.M., Edwards-Teunissen, K., Hopman, W.P.M. and Jansen, J.B.M.J. (1994) *Clin. Neuroendocrinol.* **60**, 452–456.

[24] Masclee, A.A.M., Gielkens, H.G., Ledeboer, M.L., van der Kleij, F.G.H., Jebbink, M. C.W. and Lamers, C.B.H.W. (1995) *Reg. Peptides* **58**, 157–161.

[25] Suzuki, H., Kuwano, H., Mochiki, E., Haga, N., Shimura, T., Nomoto, K., Tanaka, T., Mizumoto, A. and Itoh, Z. (2003) *Dig. Dis. Sci.* **48**, 2263–2270.

[26] Itoh, Z., Suzuki, T., Nakaya, M., Inoue, M., Arai, H. and Wakabayashi, K. (1985) *Am. J. Physiol.* **248**(3 (Pt. 1)), G320–G325.

[27] Omura, S., Tsuzuki, K., Sunazuka, T., Marui, S., Toyoda, H., Inatomi, N. and Itoh, Z. (1987) *J. Med. Chem.* **30**, 1941–1943.

[28] Kondo, Y., Torii, K., Omura, S. and Itoh, Z. (1988) *Biochem. Biophys. Res. Commun.* **150**(2), 877–882.

[29] Tsuzuki, K., Sunazuka, T., Marui, S., Toyoda, H., Omura, S., Inatomi, N. and Itoh, Z. (1989) *Chem. Pharm. Bull.* **37**, 2687–2700.

[30] Sunazuka, T., Tsuzuki, K., Marui, S., Toyoda, H., Omura, S., Inatomi, N. and Itoh, Z. (1989) *Chem. Pharm. Bull.* **37**, 2701–2709.
[31] Depoortere, I., Peeters, T.L., Matthijs, G., Cachet, T., Hoogmartens, J. and Vantrappen, G. (1989) *J. Gastrointest. Motil.* **1**, 150–159.
[32] Depoortere, I., Peeters, T.L. and Vantrappen, G. (1990) *Peptides* **11**, 515–519.
[33] Satoh, T., Inatomi, N., Satoh, H., Marui, S., Itoh, Z. and Omura, S. (1990) *J. Pharmacol. Exp. Ther.* **254**, 940–944.
[34] Inatomi, N., Satoh, H., Maki, Y., Hashimoto, N., Itoh, Z. and Omura, S. (1989) *J. Pharmacol. Exp. Ther.* **251**, 707–712.
[35] Ohtawa, M., Mizumoto, A., Hayashi, N., Yanagida, K., Itoh, Z. and Omura, S. (1993) *Gastroenterology* **104**, 1320–1327.
[36] Shiba, Y., Mizumoto, A., Inatomi, N., Haga, N., Yamamoto, O. and Itoh, Z. (1995) *Gastroenterology* **109**, 1513–1521.
[37] Kawamura, O., Sekiguchi, T., Itoh, Z. and Omura, S. (1993) *Dig. Dis. Sci.* **38**, 1026–1031.
[38] Okano, H., Inui, A., Ueno, N., Morimoto, S., Ohmoto, A., Miyamoto, M., Aoyama, N., Nakajima, Y. and Baba, S. (1996) *Peptides* **17**, 895–900.
[39] Sato, F., Sekiguchi, M., Marui, S., Inatomi, N., Shino, A., Itoh, Z. and Omura, S. (1997) *Eur. J. Pharmacol.* **322**, 63–71.
[40] Inatomi, N., Sato, F., Marui, S., Itoh, Z. and Omura, S. (1996) *Jpn. J. Pharmacol.* **71**, 29–38.
[41] Satoh, M., Sakai, T., Sano, I., Fujikura, K., Koyama, H., Ohshima, K., Itoh, Z. and Omura, S. (1994) *J. Pharmacol. Exp. Ther.* **271**, 574–579.
[42] Tanaka, T., Mizumoto, A., Mochiki, E., Suzuki, H., Itoh, Z. and Omura, S. (1998) *J. Pharmacol. Exp. Ther.* **287**, 712–719.
[43] Choi, M.-G., Camilleri, M., Burton, D.D., Johnson, S. and Edmonds, A. (1998) *J. Pharmacol. Exp. Ther.* **285**, 37–40.
[44] Koga, H., Sato, T., Tsuzuki, K., Onoda, H., Kuboniwa, H. and Takanashi, H. (1994) *Bioorg. Med. Chem. Lett.* **4**, 1347–1352.
[45] Takanashi, H. and Cynshi, O. (2009) *Reg. Peptides* **155**, 18–23 and references therein.
[46] Mccallum, R.W. and Cynshi, O.US Investigative Team. (2007) *Aliment. Pharmacol. Ther.* **26**, 107–116.
[47] Koga, H., Sato, T., Tsuzuki, K. and Takanashi, H. (1994) *Bioorg. Med. Chem. Lett* **4**, 1649–1654.
[48] Koga, H., Tsuzuki, K., Sato, T., Yogo, K. and Takanashi, H. (1995) *Bioorg. Med. Chem. Lett.* **5**, 835–838.
[49] Koga, H. and Sato, T. (1996). *Jpn. Pat. Appl.* JP 08231580; *Chem. Abstr.*, **125**, 329280.
[50] Lartey, P.A., Nellans, H.N., Faghih, R., Petersen, A., Edwards, C.M., Freiberg, L., Quigley, S., Marsh, K., Klein, L.L. and Plattner, J.J. (1995) *J. Med. Chem.* **38**, 1793–1798.
[51] Verhagen, M.A.M.T., Samsom, M., Maes, B., Geypens, B.J., Ghoos, Y.F. and Smout, A.J.P.M. (1997) *Aliment. Pharmacol. Ther.* **11**, 1077–1086.
[52] Talley, N.J., Verlinden, M., Geenen, D.J., Hogan, R.B., Riff, D., McCallum, R.W. and Mack, R.J. (2001) *Gut* **49**, 395–401.
[53] Talley, N.J., Verlinden, M., Snape, W., Beker, J.A., Ducrotte, P., Dettmer, A., Brinkhoff, H., Eaker, E., Ohning, G., Miner, P.B., Mathias, J.R., Fumagalli, I., Staessen, D. and Mack, R.J. (2000) *Aliment. Pharmacol. Ther.* **14**, 1653–1661.

[54] Chen, C.L., Orr, W.C., Verlinden, M.H., Dettmer, A., Brinkhoff, H., Riff, D., Schwartz, S., Soloway, R.D., Krause, R., Lanza, F. and Mack, R.J. (2002) *Aliment. Pharmacol. Ther.* **16**, 749–757.

[55] Peeters, T.L. (2006) *Neurogastroenterol. Motil.* **18**, 1–5.

[56] Tack, J. and Peeters, T. (2001) *Gut* **49**, 317–318.

[57] Poitras, P. and Peeters, T.L. (2008) *Curr. Opin. Endocrinol. Diabetes Obes.* **15**, 54–57.

[58] Sanger, G.J. and Alpers, D.H. (2008) *Neurogastroenterol. Motil.* **20**, 177–184.

[59] Faghih, R., Lartey, P.A., Nellans, H.N., Seif, L., Burnell-Curty, C., Klein, L.L., Thomas, P., Petersen, A., Borre, A., Pagano, T., Kim, K.H., Heindel, M., Bennani, Y.L. and Plattner, J.J. (1998) *J. Med. Chem.* **41**, 3402–3408.

[60] Faghih, R., Nellans, H.N., Lartey, P.A., Petersen, A., Marsh, K., Bennani, Y.L. and Plattner, J.J. (1998) *Bioorg. Med. Chem. Lett.* **8**, 805–810.

[61] Nakayoshi, T., Izumi, M. and Tatsuta, K. (1992) *Drugs Exp. Clin. Res.* **18**, 103–109.

[62] Kirst, H.A., Greenwood, B. and Gidda, J.S. (1992) *Drugs Fut.* **17**, 18–20.

[63] Sann, H. (2005) *In* "Functional Disorders of the Gastrointestinal Tract". Krause, G., Malagelada, J.R. and Preuschoff, U. (eds)IOS Press, Amsterdam.

[64] Krause, G. (2005) *In* "Functional Disorders of the Gastrointestinal Tract". Krause, G., Malagelada, J.R. and Preuschoff, U. (eds)IOS Press, Amsterdam.

[65] Ashley, G., Burlingame, M., Carreras, C. and Santi, D. (2001) PCT Int. Appl. WO 2001 060833; *Chem. Abstr.* **135**, 180924.

[66] Santi, D., Metcalf, B., Carreras, C., Liu, Y., McDaniel, R. and Rodriguez, J. (2004) PCT Int. Appl. WO 2004 019879; *Chem. Abstr.* **140**, 217956.

[67] Carreras, C. and Liu, Y. (2005) PCT Int. Appl. WO 2005 018576; *Chem. Abstr.* **142**, 240672.

[68] Carreras, C., Liu, Y. and Myles, D.C. (2005) PCT Int. Appl. WO 2005 018577; *Chem. Abstr.* **142**, 261736.

[69] Liu, Y., Carreras, C. and Myles, D.C. (2005) PCT Int. Appl. WO 2005 060693; *Chem. Abstr.* **143**, 91046.

[70] Liu, Y., Carreras, C., Myles, D.C., Li, Y., Shaw, S.J., Fu, H., Chen, Y., Zheng, H., Li, Y. and Burlingame, M.A. (2006) PCT Int. Appl. WO 2006 127252; *Chem. Abstr.* **146**, 28008.

[71] Licari, P., Galazzo, J., Buchanan, G., Erberlin, A.R. and Eddleston, M. (2008) PCT Int. Appl. WO 2008 068593; *Chem. Abstr.* **149**, 38659.

[72] Gale, J.D., Colman, P.J., Kantaridis, C., Claes, C. and Tutian, R. (2009) Digestive Diseases Week, Chicago, USA, Poster 235; *Gastroenterology* **136**(5 Suppl. 1), pp. A-45.

[73] Macielag, M.J., Peeters, T.L., Konteatis, Z.D., Florance, J.R., Depoortere, I., Lessor, R.A., Bare, L.A., Cheng, Y.S. and Galdes, A. (1992) *Peptides* **13**, 565–569.

[74] Peeters, T.L., Macielag, M.J., Depoortere, I., Konteatis, Z.D., Florance, J.R., Lessor, R.A. and Galdes, A. (1992) *Peptides* **13**, 1103–1107.

[75] Haramura, M., Tsuzuki, K., Okamachi, A., Yogo, K., Ikuta, M., Kozono, T., Takanashi, H. and Murayama, E. (1999) *Chem. Pharm. Bull.* **47**, 1555–1559.

[76] Thielemans, L., Depoortere, I., Van den Broeck, J. and Peeters, T.L. (2002) *Biochem. Biophys. Res. Commun.* **293**, 1223–1227.

[77] Mitznegg, P., Bloom, S.R., Domschke, W., Domschke, S., Wuensch, E. and Demling, L. (1977) *Gastroenterology* **72**, 413–416.

[78] Kamerling, I.M.C., van Haarst, A.D., Burggraaf, J., de Kam, M., Biemond, I., Jones, R., Cohen, A.F. and Masclee, A.A.M. (2002) *Dig. Dis. Sci.* **47**, 1732–1736.

[79] Macielag, M.J., Dharanipragada, R., Florance, J.R., Marvin, M.S. and Galdes, A. (1995) Can. Pat. Appl. CA2127330, US Pat. Appl US 5422341; *Chem. Abstr.* **139**, 53309.

[80] Park, M.-I., Ferber, I., Camilleri, M., Allenby, K., Trillo, R., Burton, D. and Zinsmeister, A.R. (2006) *Neurogastroenterol. Motil.* **18**, 28–36.

[81] Haramura, M., Tsuzuki, K., Okamachi, A., Yogo, K., Ikuta, M., Kozono, T., Takanashi, H. and Murayama, E. (2002) *Bioorg. Med. Chem.* **10**, 1805–1811.

[82] Tsukamoto, K., Tagi, Y., Nakazawa, T. and Takeda, M. (2001) *Pharmacology* **63**, 95–102.

[83] Furuta, Y., Nakayama, Y., Nakashima, M. and Suzuki, Y. (2003) *Clin. Pharmacokin.* **42**, 575–584.

[84] Jasserand, D., Antel, J., Preuschoff, U., Brueckner, R., Sann, H., Wurl, M. and Eickelmann, P. (2002) PCT Int. Appl. WO 2002 092592; *Chem. Abstr.* **137**, 384755.

[85] Li, J.J., Chao, H.-G., Wang, H., Tino, J.A., Lawrence, R.M., Ewing, W.R., Ma, Z., Yan, M., Slusarchyk, D., Seethala, R., Sun, H., Li, D., Burford, N.T., Stoffel, R.H., Salyan, M.E., Li, C.Y., Witkus, M., Zhao, N., Rich, A. and Gordon, D.A. (2004) *J. Med. Chem.* **47**, 1704–1708.

[86] Li, J.J., Chao, H.J., Tino, J.A. and Ewing, W.R. (2005) US Pat. Appl. Publ. US 2005065156; *Chem. Abstr.* **142**, 336372.

[87] Heightman, T.D., Conway, E., Corbett, D.F., Macdonald, G.J., Stemp, G., Westaway, S.M., Celestini, P., Gagliardi, S., Riccaboni, M., Ronzoni, S., Vaidya, K., Butler, S., McKay, F., Muir, A., Powney, B., Winborn, K., Wise, A., Jarvie, E.M. and Sanger, G.J. (2008) *Bioorg. Med. Chem. Lett.* **18**, 6423–6428.

[88] Westaway, S.M., Brown, S.L., Conway, E., Heightman, T.D., Johnson, C.N., Lapsley, K., Macdonald, G.J., MacPherson, D.T., Mitchell, D.J., Myatt, J.W., Seal, J.T., Stanway, S.J., Stemp, G., Thompson, M., Celestini, P., Colombo, A., Consonni, A., Gagliardi, S., Riccaboni, M., Ronzoni, S., Briggs, M.A., Matthews, K.L., Stevens, A.J., Bolton, V.J., Boyfield, I., Jarvie, E.M., Stratton, S.C. and Sanger, G.J. (2008) *Bioorg. Med. Chem. Lett.* **18**, 6429–6436.

[89] Westaway, S.M., Brown, S.L., Fell, S.C.M., Johnson, C.N., MacPherson, D.T., Mitchell, D.J., Myatt, J.W., Stanway, S.J., Seal, J.T., Stemp, G., Thompson, M., Lawless, K., McKay, F., Muir, A.I., Barford, J.M., Cluff, C., Mahmood, S.R., Matthews, K.L., Mohamed, S., Smith, B., Stevens, A.J., Bolton, V.J., Jarvie, E.M. and Sanger, G.J. (2009) *J. Med. Chem.* **52**, 1180–1189.

[90] MacDonald, G.J., Mitchell, D.J., Thompson, M. and Westaway, S.M. (2007) PCT Int. Appl., WO 2007 065669; *Chem. Abstr.* **147**, 72799.

[91] Brown, S., Bolton, V., Boyfield, I., Jarvie, E., Lawless, K., MacKay, F., Matthews, K., Mitchell, D., Muir, A., Myatt, J., Sanger, G., Seal, J., Stemp, G., Stevens, A., Stratton, S., Thompson, M. and Westaway, S. (2008) Poster at XXth International Symposium on Medicinal Chemistry, Vienna, Austria.

[92] Seal, J.T., Stemp, G., Thompson, M. and Westaway, S.M. (2007) PCT Int. Appl. WO 2007 144400; *Chem. Abstr.* **148**, 79067.

[93] Mitchell, D.J., Seal, J.T., Thompson, M., Westaway, S.M. and Brown, S.L. (2008) PCT Int. Appl. WO 2008 000729; *Chem. Abstr.* **148**, 121732.

[94] Mitchell, D.J., Seal, J.T., Stemp, G., Thompson, M. and Westaway, S.M. (2009) PCT Int. Appl. WO 2009 068552; *Chem. Abstr.* **151**, 8540.

[95] Bailey, J.M., Scott, J.S., Basilla, J.B., Bolton, V.J., Boyfield, I., Evans, D.G., Fleury, E., Heightman, T.D., Jarvie, E.M., Lawless, K., Matthews, K.L., McKay, F., Mok, H.,

Muir, A., Orlek, B.S., Sanger, G.J., Stemp, G., Stevens, A.J., Thompson, M., Ward, J., Vaidya, K. and Westaway, S.M. (2009) *Bioorg. Med. Chem. Lett.* doi:10.1016/j. bmcl.2009.09.027.

[96]   Thielemans, L., Depoortere, I., Perret, J., Robberecht, P., Liu, Y., Thijs, T., Carreras, C., Burgeon, E. and Peeters, T.L. (2005) *J. Pharmacol. Exp. Ther.* **313**, 1397–1405.

[97]   Carreras, C.W., Claypool, M., Santi, D.V., Schuukes, J., Peeters, T.L. and Johnson, R.G. (2004) Digestive Diseases Week, New Orleans, USA, Poster M997; *Gastroenterology* **126** (4 Suppl. 2), pp A-276.

[98]   Mitselos, A., Van den Berghe, P., Peeters, T.L. and Depoortere, I. (2008) *Biochem. Pharmacol.* **75**, 1115–1128.

[99]   Karamanolis, G. and Tack, J. (2006) *Dig. Dis.* **24**, 297–307.

[100] Prakash, A. and Wagstaff, A.J. (1998) *Drugs* **56**, 429–445.

[101] Cellek, S., John, A.K., Thangiah, R., Dass, N.B., Bassil, A.K., Jarvie, E.M., Lalude, O., Vivekanandan, S. and Sanger, G.J. (2006) *Neurogastrol. Motil.* **18**, 853–861.

[102] Pasricha, P.J., Pehlivanov, N., Sugumar, A. and Jankovic, J. (2006) *Nat. Clin. Practice Gastroenterol. Hepatol.* **3**, 138–148.

[103] Hawkyard, C.V. and Koerner, R.J. (2007) *J. Antimicrob. Chemother.* **59**, 347–358.

[104] Annese, V., Lombardi, G., Frusciante, V., Germani, U., Andriulli, A. and Bassotti, G. (1997) *Aliment. Pharmacol. Ther.* **11**, 599–603.

[105] Sturm, A., Holtmann, G., Goebell, H. and Gerken, G. (1999) *Digestion* **60**, 422–427.

[106] Erbas, T., Varoglu, E., Erbas, B., Tastekin, G. and Akalin, S. (1993) *Diabetes Care* **16**, 1511–1514.

[107] Zatman, F., Hall, J.E. and Harmer, M. (2001) *Br. J. Anaesth.* **86**, 869–871.

[108] Dass, N.B., Munonyara, M., Bassil, A.K., Hervieu, G.J., Osbourne, S., Corcoran, S., Morgan, M. and Sanger, G.J. (2003) *Neuroscience* **120**, 443–453.

[109] Edholm, T., Levin, F., Hellstrom, P.M. and Schmidt, P.T. (2004) *Reg. Peptides* **121**, 25–30.

[110] le Roux, C.W., Neary, N.M., Halsey, T.J., Small, C.J., Martinez-Isla, A.M., Ghatei, M. A., Theodorou, N.A. and Bloom, S.R. (2005) *J. Clin. Endocrinol. Metabol.* **90**, 4521–4524.

[111] Rudd, J.A., Ngan, M.P., Wai, M.K., King, A.G., Witherington, J., Andrews, P.L.R. and Sanger, G.J. (2006) *Neurosci. Lett.* **392**, 79–83.

[112] Shimizu, Y., Chang, E.C., Shafton, A.D., Sanger, G.J., Witherington, J. and Furness, J. B. (2006) *J. Physiol.* **576**, 329–338.

[113] Leite-Moreira, A.F. and Soares, J.-B. (2007) *Drug Disc. Today* **12**, 276–288.

[114] Vestergaard, E.T., Hansen, T.K., Gormsen, L.C., Jakobsen, P., Moller, N., Christiansen, J.S. and Jorgensen, J.O.L. (2007) *Am. J. Physiol.* **292**, E1829–E1836.

[115] Camilleri, M., Papathanasopoulos, A. and Odunsi, S.T. (2009) *Nat. Rev. Gastroenterol. Hepatol.* **6**, 343–352.

[116] Depoortere, I., Macielag, M.J., Galdes, A. and Peeters, T.L. (1995) *Eur. J. Pharmacol.* **286**, 241–247.

[117] Peeters, T.L., Depoortere, I., Macielag, M.J., Dharanipragada, R., Marvin, M.S., Florance, J.R. and Galdes, A. (1994) *Biochem. Biophys. Res. Commun.* **198**, 411–416.

[118] Clark, M.J., Wright, T., Bertrand, P.P., Bornstein, J.C., Jenkinson, K.M., Verlinden, M. and Furness, J.B. (1999) *Clin. Exp. Pharmacol. Physiol.* **26**, 242–245.

[119] Farrugia, G., Macielag, M.J., Peeters, T.L., Sarr, M.G., Galdes, A. and Szurszewski, J. H. (1997) *Am. J. Physiol.* **273**(2 (Pt. 1)), G404–G412.

[120] Van Assche, G., Depoortere, I., Thijs, T., Missiaen, L., Penninckx, F., Takanashi, H., Geboes, K., Janssens, J. and Peeters, T.L. (2001) *Neurogastroenterol. Motil.* **13**, 27–35.

[121]   Matsuo, H., Peeters, T.L., Janssens, J. and Depoortere, I. (1994) *Neurogastroenterol. Motil.* **6**, 148 (Abstract).

[122]   Takanashi, H., Yogo, K., Ozaki, K., Ikuta, M., Akima, M., Koga, H. and Nabata, H. (1995) *J. Pharmacol. Exp. Ther.* **273**, 624–628.

[123]   Haramura, M., Okamachi, A., Tsuzuki, K., Yogo, K., Ikuta, M., Kozono, T., Takanashi, H. and Murayama, E. (2002) *J. Med. Chem.* **45**, 670–675.

[124]   Momose, K., Inui, A., Asakawa, A., Ueno, N., Nakajima, M. and Kasuga, M. (1998) *Peptides* **19**(10), 1739–1742.

[125]   Taka, N., Matsuoka, H., Sato, T., Yoshino, H., Imaoka, I., Sato, H., Kotake, K., Kumagai, Y., Kamei, K., Ozaki, K., Higashida, A. and Kuroki, T. (2009) *Bioorg. Med. Chem. Lett.* **19**, 3426–3429.

[126]   Kotake, K., Kozono, T., Sato, T. and Takanashi, H. (1999) PCT Int. Appl. WO 99 09053; *Chem. Abstr.* **130**, 196958.

[127]   Matsuoka, H., Sato, T., Takahashi, T., Kim, D.I., Jung, K.Y. and Park, C.H. (2000) PCT Int. Appl. WO 2000 044770; *Chem. Abstr.* **133**, 150920.

[128]   Matsuoka, H. and Sato, T. (2002) PCT Int. Appl. WO 2002 016404; *Chem. Abstr.* **136**, 217049.

[129]   Sudo, H., Yoshida, S., Ozaki, K., Muramatsu, H., Onoma, M., Yogo, K., Kamei, K., Cynshi, O., Kuromaru, O., Peeters, T.L. and Takanashi, H. (2008) *Eur. J. Pharmacol.* **581**, 296–305.

[130]   Ozaki, K., Onoma, M., Muramatsu, H., Sudo, H., Yoshida, S., Shiokawa, R., Yogo, K., Kamei, K., Cynshi, O., Kuromaru, O., Peeters, T.L. and Takanashi, H. (2009) *Eur. J. Pharmacol.* **615**, 185–192.

[131]   Chen, R.H., Xiang, M., Moore J.B. Jr. and Beavers, M.P. (1999) PCT Int. Appl. WO 99 21846; *Chem. Abstr.* **130**, 296443.

[132]   Beavers, M.P., Gunnet, J.W., Hageman, W., Miller, W., Moore, J.B., Zhou, L., Chen, R.H.K., Xiang, A., Urbanski, M., Combs, D.W., Mayo, K.H. and Demarest, K.T. (2001) *Drug Des. Discov.* **17**, 243–251.

[133]   Chen, R.H. and Xiang, M.A. (2001) PCT Int. Appl. WO 2001 068621; *Chem. Abstr.* **135**, 241947.

[134]   Chen, R.H. and Xiang, M.A. (2001) PCT Int. Appl. WO 2001 068622; *Chem. Abstr.* **135**, 241935.

[135]   Kamerling, I.M.C., van Haarst, A.D., Burggraaf, J., Schoemaker, R.C., de Kam, M.L., Heinzerling, H., Cohen, A.F. and Masclee, A.A.M. (2004) *Br. J. Clin. Pharmacol.* **57**(4), 393–401.

[136]   Johnson, S.G., Gunnet, J.W., Moore, J.B., Miller, W., Wines, P., Rivero, R.A., Combs, D. and Demarest, K.T. (2006) *Bioorg. Med. Chem. Lett.* **16**, 3362–3366.

[137]   Fraser, G., Marsault, E., Peterson, M., Hoveyda, H., Beaubien, S. and Benakli, K. (2004) PCT Int. Appl. WO 2004 111077; *Chem. Abstr.* **142**, 74840.

[138]   Marsault, E., Fraser, G., Benakli, K., St-Louis, C., Rouillard, A. and Thomas, H. (2008) PCT Int. Appl. WO 2008 033328; *Chem. Abstr.* **148**, 379966.

[139]   Fraser, G.L., Marsault, E., Peterson, M., Hoveyda, H., Beaubien, S., Benakli, K. and Deziel, R. (2009) US Pat. Appl. US 7521420; *Chem. Abstr.* **150**, 448321.

[140]   Marsault, E., Hoveyda, H.R., Peterson, M.L., Saint-Louis, C., Landry, A., Vezina, M., Ouellet, L., Wang, Z., Ramaseshan, M., Beaubien, S., Benakli, K., Beauchemin, S., Deziel, R., Peeters, T. and Fraser, G.L. (2006) *J. Med. Chem.* **49**(24), 7190–7197.

[141]   Marsault, E., Benakli, K., Beaubien, S., Saint-Louis, C., Deziel, R. and Fraser, G. (2007) *Bioorg. Med. Chem. Lett.* **17**, 4187–4190.

[142] Venkova, K., Thomas, H., Fraser, G.L. and Greenwood-Van Meerveld, B. (2009) *J. Pharm. Pharmacol.* **61**, 367–373.

[143] AACR-NCI-EORTC Conference on molecular targets and cancer therapeutics, October 2007 (slidepack available at http://www.tranzyme.com/publications.html).

[144] Thomas, H., Chen, C. and Marsault, E. (2007) Oral presentation at the 21st International Symposium on Neurogastroentrology and Motility, Jeju Island, Korea; (2007) *Neurogastroenterol. Motil.* **19**(Suppl. 3), 35 (abstract).

[145] Simrén, M., Bjornsson, E.S. and Abrahamsson, H. (2005) *Neurogastroenterol. Motil.* **17**, 51–57.

[146] Fukudo, S. and Suzuki, J. (1987) *Tohoku J. Exp. Med.* **151**, 373–385 PubMed 3617051.

[147] Simrén, M., Abrahamsson, H. and Bjornsson, E.S. (2001) *Gut* **48**, 20–27.

[148] Costa, A., De Ponti, F., Gibelli, G., Crema, F. and d'Angelo, L. (1997) *Pharmacology* **54**, 64–75.

[149] Sudo, H., Ozaki, K., Muramatsu, H., Kamei, K., Yogo, K., Cynshi, O., Koga, H., Itoh, Z., Omura, S. and Takanashi, H. (2007) *Neurogastroenterol. Motil.* **19**, 318–326.

[150] Hirabayashi, T., Morikawa, Y., Matsufuji, H., Hoshino, K., Hagane, K. and Ozaki, K. (2009) *Neurogastroenterol. Motil.* doi:10.1111/j.1365-2982.2009.01441.x.

[151] Kusano, M., Sekiguchi, T., Kawamura, O., Kikuchi, K., Miyazaki, M., Tsunoda, T., Horikoshi, T. and Mori, M. (1997) *Am. J. Gastroenterol.* **92**, 481–484.

[152] Jonsson, B.H. and Hellstrom, P.M. (2000) *Integr. Physiol. Behav. Sci.* **35**, 256–265.

[153] Gadenstätter, M., Prommegger, R., Klinger, A., Schwelberger, H., Weiss, H.G., Glaser, K. and Wetscher, G.J. (2001) *Am. J. Surg.* **180**, 483–487.

[154] Nieuwenhuijs, V.B., Van Duijvenbode-Beumer, H., Verheem, A., Visser, M.R., Verhoef, J., Gooszen, H.G. and Akkermans, L.M.A. (1999) *Eur. J. Clin. Invest.* **29**, 33–40.

3 Progress in the Design and Development of Phosphoinositide 3-Kinase (PI3K) Inhibitors for the Treatment of Chronic Diseases

STEPHEN SHUTTLEWORTH[1], FRANCK SILVA[1], CYRILLE TOMASSI[1], ALEXANDER CECIL[1], THOMAS HILL[1], HELEN ROGERS[1] and PAUL TOWNSEND[2]

[1]*Karus Therapeutics Ltd., 2 Venture Road, Southampton Science Park, Southampton, SO16 7NP, UK*

[2]*School of Medicine, Southampton General Hospital, University of Southampton, SO16 6YD, UK*

Progress in Medicinal Chemistry – Vol. 48   81
Edited by G. Lawton and D.R. Witty
DOI: 10.1016/S0079-6468(09)04803-6

INTRODUCTION

The phosphoinositide 3-kinases (PI3Ks) constitute a family of lipid kinases involved in the regulation of a network of signal transduction pathways that control a range of cellular processes [1–4]. Over the last 15 years, considerable progress has been made in the development of small molecules that target PI3K signalling for the treatment of a spectrum of chronic diseases, including cardiovascular and immune-inflammatory disorders [5] and cancer. A number of ATP-competitive, small molecule PI3K inhibitors with distinct sub-type selectivity profiles have recently entered clinical trials. Progress in the design, development and clinical evaluation of these compounds is outlined in this chapter.

THE PHOSPHOINOSITIDE 3-KINASE (PI3K) FAMILY

PI3Ks are classified into four distinct sub-families – commonly referred to as classes I, II, III and IV – based upon their substrate specificities, primary sequences, modes of regulation and domain structures (Figure 3.1) [6].

The principal structural features of the classes I, II and III PI3Ks are shown in Figure 3.2. These isoforms each possess a kinase domain that is located towards the C-terminus, and they share a helical region that acts as a scaffold around which the enzyme is folded. The classes I, II and III enzymes also possess a C2 domain that, in other proteins, has been shown to bind phospholipids. The N-terminus, where interactions with adaptor or regulatory sub-units and other proteins are believed to occur, is highly variable amongst the isoforms.

Class I PI3Ks exist as heterodimers, and are categorised into two sub-families, designated IA and IB. Class IA PI3Ks possess a p110α, p110β, or p110δ catalytic sub-unit complexed with one of the three regulatory sub-units, p85α, p85β or p85δ. A single class IB PI3K exists, comprising a p110γ catalytic and a p101 regulatory sub-unit. Class IA PI3Ks are activated by receptor tyrosine kinases, antigen receptors and cytokine receptors. Evidence has emerged that p110β can also be activated by G-protein coupled receptors (GPCRs) [7, 8], as in the case for the sole class IB PI3K, p110γ [9]. The class I PI3Ks primarily phosphorylate phosphatidylinositol-4,5-bisphosphate, PIP2, to generate the lipid second messenger phosphatidylinositol-3,4,5-trisphosphate, PIP3, which then activates the down-stream target AKT, regulating proliferation, growth, survival and glucose homoeostasis. The class IA PI3Ks p110α and p110β are ubiquitously expressed, whilst p110δ and p110γ are mostly expressed in the haematopoetic system, though p110γ has been detected in the endothelium, brain and heart [9].

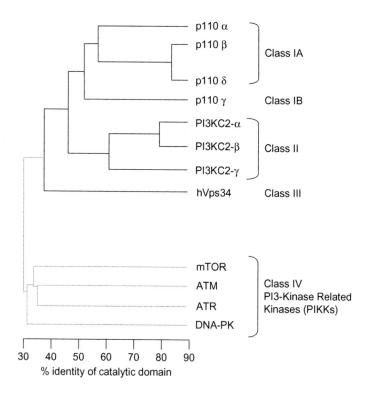

Fig. 3.1   The PI3K family.

The class II PI3Ks comprise three isoforms, namely C2-α, -β and -γ. These enzymes are characterised by a carboxyl-terminal phospholipid-binding domain, are predominantly membrane-bound and are activated by a diverse range of receptor types, notably receptor tyrosine kinases and integrin receptors. It has been reported that class II PI3Ks bind clathrin-coated pits, suggesting that they have a role in membrane trafficking and receptor internalisation [10, 11]. Both C2-α and C2-β are ubiquitously expressed, whilst C2-γ is mainly expressed in the liver.

There is one class III PI3K in mammals, Vps34, which represents the only mammalian PI3K that is also conserved in yeast. Vps34 phosphorylates phosphatidylinositol to generate phosphatidylinositol-3-phosphate. Vps34 has been shown to play an essential role in trafficking of proteins from the Golgi apparatus in yeast [12]. More recently, Vps34 has been linked to autophagy [13] and the activation of the mammalian target of rapamycin (mTOR), discussed below, by amino acids [14, 15].

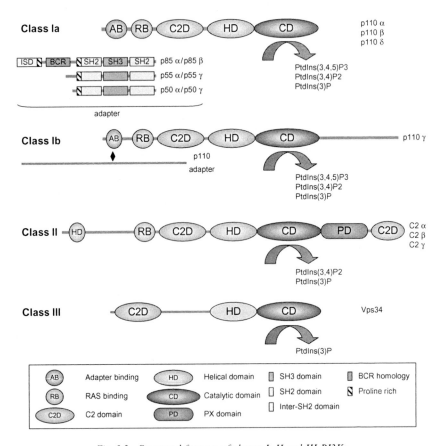

Fig. 3.2   Structural features of classes I, II and III PI3Ks.

Finally, there is also a fourth family of serine/threonine kinases known as the phosphatidylinositol-3-kinase-related kinases (PIKKs), which contain a catalytic core similar to classes I–III PI3Ks [16]. The PIKK family includes enzymes involved in signal transduction and DNA damage response, such as mTOR, DNA-dependent protein kinase (DNA-PK), ataxia telangiectasia mutated (ATM), ataxia telangiectasia and Rad3 related (ATR), and human suppressor of morphogenesis in genitalia-1 (hSMG-1). Of the PIKKs, mTOR is the most extensively studied [17–19]. mTOR has been the subject of many literature review articles, and a detailed historical over-view of mTOR inhibitors is beyond the scope of this chapter. However, a summary of recent progress made in the development of small molecule dual class I/mTOR inhibitors and of heterocyclic ATP-competitive mTOR inhibitors is outlined below.

THE PI3K SIGNALLING PATHWAY: DOWN-STREAM TARGETS AND FUNCTIONS

Phosphorylation of AKT results in its translocation to the nucleus, from which it promotes cell proliferation through regulation of the cell cycle (Figure 3.3). An important down-stream target of AKT is the mTOR complex 1 (mTORC1), which is a critical regulator of translation initiation, ribosome biogenesis and cell growth control. AKT indirectly activates mTORC1 through inhibition of tuberous sclerosis complex 2 (TSC2). TSC2, in complex with TSC1, acts as a GTPase activating protein for the Ras-related small G-protein Rheb which, in its GTP-bound active form, activates

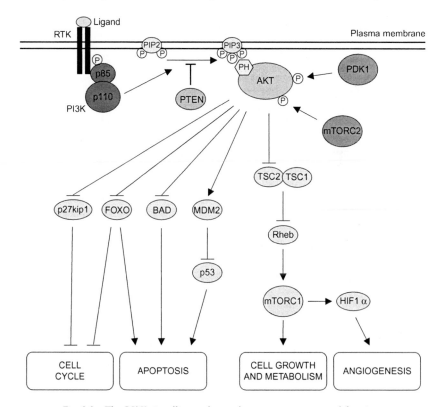

Fig. 3.3   The PI3K signalling pathway, down-stream targets and functions.

mTORC1. AKT also has an important role in promoting cell survival through negative regulation of pro-apoptotic proteins and processes. AKT directly phosphorylates and inhibits the pro-apoptotic Bcl-2-associated death (BAD) promoter via its Bcl-2 homology domain 3 (BH3). It also indirectly inhibits other BH3 proteins by regulation of transcription factors such as forkhead box O (FOXO) proteins and p53. AKT phosphorylates FOXO proteins leading to their export from the nucleus, thus blocking transcription of their target genes. AKT regulates p53 through phosphorylation of Mdm2, which promotes p53 degradation. Other target proteins include the cell cycle inhibitor p27^{kip1} and HIF1α. AKT induces increased expression of the latter, in part through mTORC1-mediated translation. This leads to expression and secretion of angiogenic factors such as VEGF. Activation of AKT is reversed by the mixed function phosphatase, PTEN.

TARGETING PI3Ks: THERAPEUTIC POTENTIAL OF SMALL MOLECULE INHIBITORS

To date, the majority of activity in the design and development of small molecule inhibitors of PI3Ks has focused on class I modulators, notably for the treatment of cancer. It has been established that the class I PI3K/AKT pathway is dysregulated in many human cancers [1–4], whilst PTEN, the negative regulator of PI3K, is one of the most commonly mutated proteins in human malignancy. Additionally, the gene encoding the p110α sub-unit, PIK3CA, is amplified and over-expressed in several tumour types, and has also been found to be frequently mutated in a diverse range of human cancers [20–25]. These PIK3CA mutations have been shown to reduce cellular dependence on growth factors, to attenuate apoptosis, and to facilitate tumour invasiveness. More recently, the distinct functions of p110α and p110β in cancer have been elucidated: it has been demonstrated that p110α is essential for the growth of tumours driven by PIK3CA mutations, oncogenic receptor tyrosine kinases and Ras, whilst p110β is the major isoform involved in mediating PTEN-deficient tumourigenesis [26–28]. It has also been established that p110δ is a therapeutic target for haematological malignancies [29–32], notably acute myeloid leukaemia (AML). The p110δ isoform is expressed at elevated levels in blast cells from AML, and contributes to the survival and proliferation of these cells. It has been reported that p110δ is up-regulated in melanoma and breast cancer [33], and is over-expressed in neuroblastoma [34]. Consequently, inhibition of class IA PI3Ks represents a potentially important strategy for the development of novel cancer therapeutics, and, moving forward, will have a significant impact in the discovery and development of personalised medicines in the oncology setting.

It has been established that p110β is an important target for anti-thrombotic therapy [35–37], and there is also emerging evidence that modulation of this isoform may have therapeutic potential in the treatment of immune and inflammatory disorders [38]. Pharmacological and genetic analyses of p110β function have revealed that p110β catalytic activity is required to sustain long-term insulin signalling, and may have a role in diabetes [39].

p110δ has been shown to play a key role in the recruitment and activation of immune and inflammatory cells [40, 41], and strong evidence has emerged to suggest that this isoform is an important target for therapeutic intervention in allergy and mast cell-related pathologies [42]. Additionally, the p110δ isoform has a critical function in neutrophil degranulation and migration [43, 44], in macrophage migration [45], in B-cell and IL4-receptor signalling [46, 47], and in NK-cell development and cytokine secretion [48]. From a therapeutic perspective, p110δ is also being pursued for the treatment of airway hyper-responsiveness [49, 50] and rheumatoid arthritis [51].

Inhibition of p110γ, a critical down-stream signalling protein that relays chemokine receptor signals, has emerged as a particularly attractive approach to facilitate perturbation of the chemokine network, together with other cellular processes that are implicated in the onset and progression of immune and inflammatory disorders. This isoform has been shown to be an essential amplifier of mast cell function [52], to regulate the recruitment and survival of eosinophils [53], to govern dendritic cell migration [54], to regulate T-cell chemotaxis [55], and, with p110δ, to control reactive oxygen species production from neutrophils [56, 57]. Targeting p110γ has attracted attention in the development of therapeutics to treat rheumatoid arthritis, systemic lupus erythematosus, lung injury, asthma, heart ischaemia, pancreatitis and hypertension [9].

The classes II and III isoforms have received comparatively limited attention from a small molecule perspective, and an understanding of the therapeutic potential of modulating their activity is, in contrast with class I and IV PI3Ks, in its infancy. However, there is emerging evidence that regulation of their activity could present opportunities for the development of new therapeutics to treat chronic diseases. Of the class II enzymes, it is known that C2-α is ubiquitously expressed and preferentially activated by insulin, and that, *in vitro*, through the use of sense- and anti-sense-oligonucleotides based on the sequence of the C2 domain of the C2-α gene, apoptotic cell death could be conferred, indicating that this isoform is a crucial survival factor [58]. This observation was supported by a further study involving the use of C2-α-specific RNAi reagents in which it was also discovered that down-regulation of C2-α in a range of cancer cell lines reduced their proliferation and viability, and that knockdown of this isoform sensitised cells to taxol (paclitaxel) [59]. C2-α has also been shown to play a critical role during neuroexocytosis [60]. Both C2-α and C2-β have been discovered to be overexpressed in human small cell lung carcinoma lines [61], and C2-β has been implicated in LPA-induced migration of ovarian and cervical cancer cells [62]. The recent elucidation of the crystal structure of the PX domain of C2-α may provide some structural insights to enable future design and development of synthetic modulators [63], potentially for the treatment of cancer [11]. It has been postulated that Vps34 plays a role in the activation of mTOR down-stream of amino acids, resulting in mTOR-dependent phosphorylation of p70S6K and 4E-BP1 [64–66] and its regulation has been strongly linked to autophagic responses as a consequence of its association with the autophagic gene Beclin 1 [67].

Substantial progress has been made in the development of agents that target the PIKK sub-family, most notably mTOR [16–19, 68]. This protein was discovered in the 1990s, when the mechanism of action of rapamycin (Sirolimus/Rapamune), a macrolide-based natural product with immuno-suppressant activity, was elucidated [69]. mTOR complexes with raptor (regulatory-associated protein of mTOR) and rictor (rapamycin-insensitive companion of mTOR) to form mTOR complex-1, mTORC1 and mTORC2. As described above, mTORC1 is down-stream to AKT, and is susceptible to inhibition by rapamycin and derivatives thereof; mTORC2, however, is an upstream regulator of AKT, and the activity is up-regulated in certain circumstances as a compensatory response to mTORC1 inhibition. Rapamycins bind with high affinity to the immunophilin FK506-binding protein-12 (FKBP12), forming a complex that selectively inhibits mTORC1 down-stream signalling to elements involved in growth control, and have been evaluated as agents for the treatment of solid tumours [70, 71]. Recent progress has, however, been made in targeting the ATP-binding site of mTOR with small molecule inhibitors, for the treatment of cancer, as it has with the other PIKK sub-family members, DNA-PK and ATM, whose inhibition confers chemo- and radio-sensitisation [72].

p110α, PAN-CLASS I, AND DUAL CLASS I/IV INHIBITORS FOR THE TREATMENT OF CANCER

One of the most extensively studied small molecule ATP-competitive PI3K inhibitors is the chromenone LY294002, (1), derived from the flavinoid quercetin, which was first disclosed as a PI3K inhibitor by researchers at Eli Lilly in 1994 [73]. Through a conventional SAR exercise, it was established that the exocyclic phenyl group located at the 8-position of (1) and the morpholine group located at the 2-position were critical structural elements for PI3K activity. Removal of the 8-phenyl, and replacement of the 2-morpholine with a thiomorpholine, resulted in a 4-fold and 100-fold decrease in activity, respectively. It was subsequently confirmed through X-ray crystallographic analysis that the morpholine oxygen of (1) forms a key hydrogen bond with the Val-882 residue located in the hinge region of p110γ [74]. Furthermore, the bound crystal structure revealed that the 8-phenyl occupies a region of the protein adopted by the ribose of ATP, and that the carbonyl forms a putative hydrogen bond with Lys-833. The discovery of morpholine as a 'privileged' hinge-region binding motif, conferring potent and selective PI3K inhibition, was a crucial development in the field, and subsequently had a significant impact on the design of small molecule inhibitors of this enzyme family, a fact reflected in the large number of arylmorpholine-based PI3K inhibitors that have been described since the publication of the liganded crystal structure.

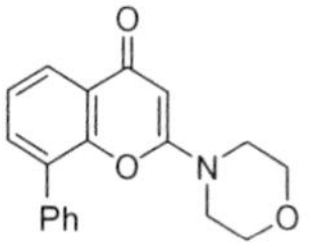

(1)
LY294002

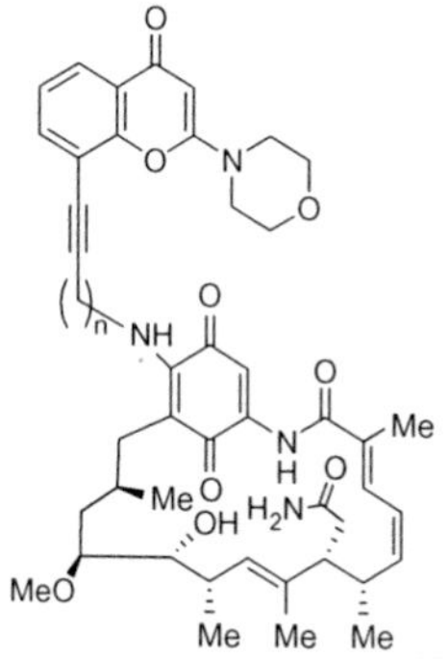

(2)   n = 1 or 3

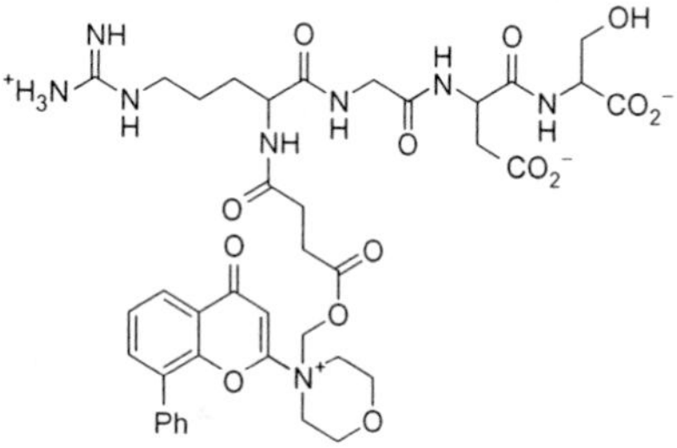

(3)
SF1126

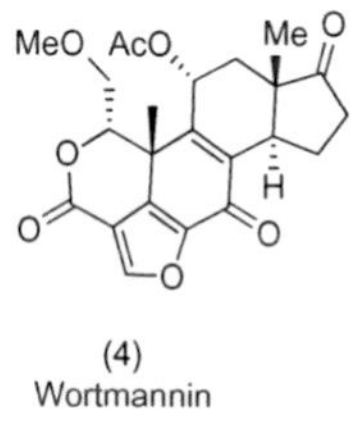

(4)
Wortmannin

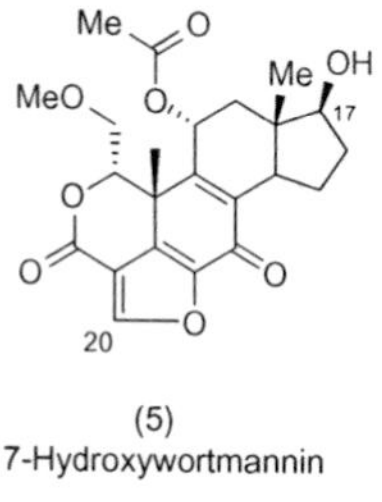

(5)
17-Hydroxywortmannin

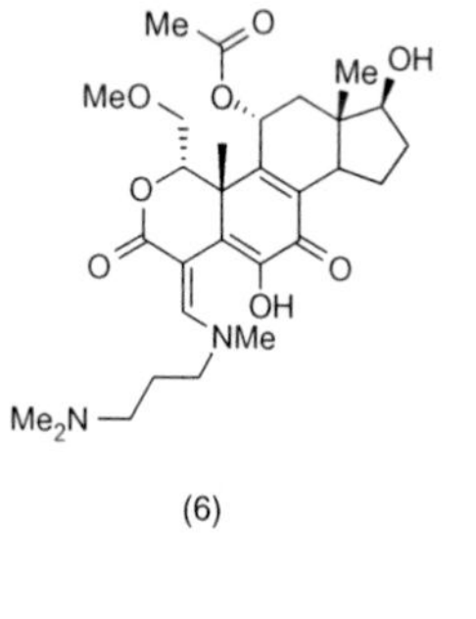

(6)

LY294002 displayed activity against all of the class I PI3K isoforms, with IC_{50}s of 0.55, 11, 1.6 and 12 μM against p110α, β, δ and γ, respectively [75]. Additional selectivity studies have confirmed that the compound had a number of off-target activities [76], including against casein kinase 2 (CK2)

[77], phosphodiesterase-2 (PDE-2) [78], and the proto-oncogene PIM-1. Interestingly, in PIM-1 X-ray co-crystallisation studies conducted by researchers at Vertex, LY294002 was seen to adopt a radically different hinge-region binding mode to that seen for p110γ with hydrogen bonds involving the hinge- and the 6- and 7-position protons of the chromenone core being observed [79].

A number of synthetic methodologies to facilitate the preparation of analogues of (1) have been developed over recent years with the aim of augmenting potency, selectivity and physiochemical properties [80]. In one example, heterodimers of (1) and geldanamycin, exemplified by compound (2), were seen to exhibit activity for PI3K and HSP-90 [81]. The most significant development in this area, however, has been the design of a series of prodrugs of LY294002 by Semafore Pharmaceuticals, in which a series of derivatives was designed to increase solubility and enhance tumour and tumour-vasculature delivery of the compound [82, 83]. One key example is SF1126 (3), in which the morpholine nitrogen of LY294002 is quaternised with an oxobutanoic acid-linked αvβ3/αvβ1 integrin-targeted peptide. (3) is water soluble, and in animal models displayed a favourable pharmacokinetic profile and was well-tolerated. In pre-clinical xenograft models, (3) was seen to confer enhanced U87MG and PC3 tumour growth inhibition, understood to be a consequence of the pharmacokinetic accumulation of (3) in tumour tissue, and the concomitant pharmacodynamic knockdown of phosphory-lated AKT. SF1126 was subsequently advanced into clinical development and Phase I trials were initiated in patients with solid tumours in 2007.

Together with LY294002, wortmannin (4) is another benchmark PI3K inhibitor that has been the subject of intensive research over the last 15 years [84]. Wortmannin was seen to display pan-class I activity, with IC_{50}s of 4, 1, 4 and 9 nM against p110α, β, δ and γ, respectively [75]. Unlike LY294002, however, this compound is an irreversible inhibitor, forming a covalent bond with a key Lys-833 residue located in the catalytic site [74, 85]. Owing to toxicity, poor chemical stability and pharmacokinetic properties, wortman-nin has, historically, not been considered as a viable drug candidate. However, over recent years, there has been considerable effort geared towards the development of formulated derivatives and synthetically modified analogues of wortmannin for clinical development. In one example, analogues of 17-hydroxywortmannin (5), (prepared by nucleophilic ring opening of the furan ring at position C-20 with primary or secondary amines), were developed by researchers at Wyeth with the aim of improving chemical stability and solubility, whilst maintaining the high PI3K potency of 17-hydroxywortmannin ($IC_{50} = 2.7$ nM) [86–88]. It was discovered that insertion of secondary amines at the C-20 position gave rise to derivatives with optimum potency, with the most potent analogues possessing acyclic diamine residues, as exemplified by (6) ($IC_{50} = 6.4$ nM). Importantly, compound (6) exhibited good aqueous solubility, and its plasma stability was significantly improved compared with 17-hydroxywortmannin. In mouse xenograft models, compound (6) conferred potent tumour growth inhibition and, crucially, exhibited a 2-fold greater therapeutic index when compared with 17-hydroxywortmannin, together with improved pharmaco-kinetic properties. In a separate report, Wyeth also disclosed the discovery of PWT-458, (7), a pegylated derivative of 17-hydroxywortmannin, which also displayed an improved tolerability profile [89].

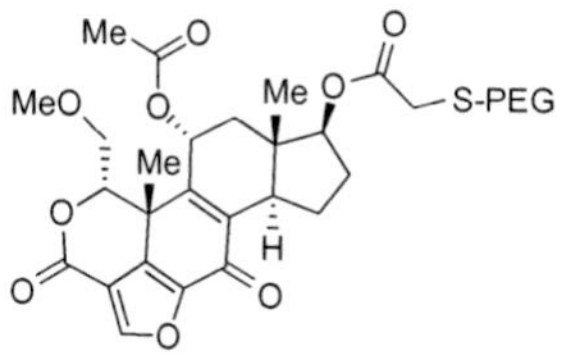

(7)
PWT-458

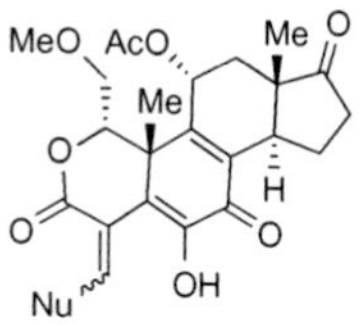

(8)

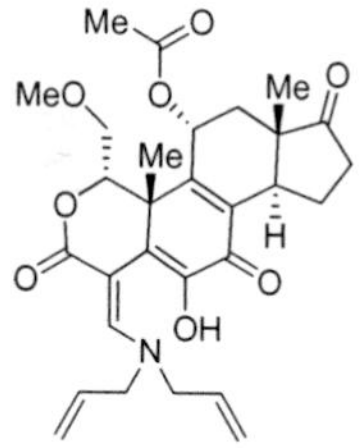

(9)
PX-866

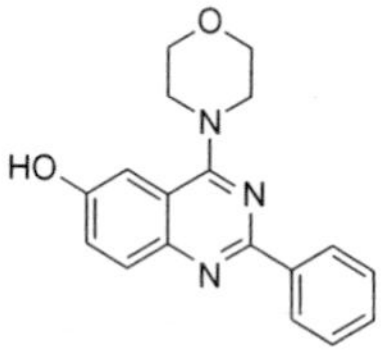

(10)

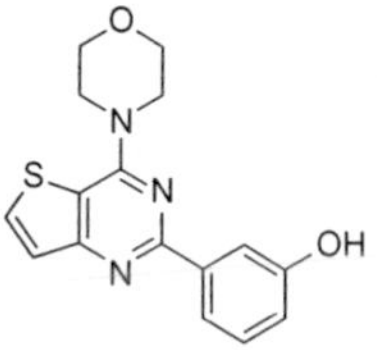

(11)

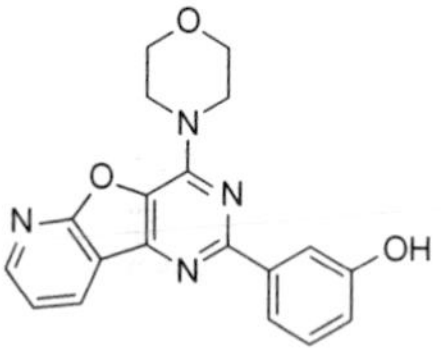

(12)
PI-103

Researchers at ProlX Pharmaceuticals have designed a series of wortmannin derivatives, obtained through the nucleophilic ring opening at the C-20 position. These compounds, of the generic structure (8), were seen to inhibit class I PI3Ks at concentrations equal to or lower than that for wortmannin; derivatives with biochemical IC_{50} values as low as 100 pM, and improved tumour cell proliferation inhibition activities over wortman-nin, were developed [90–92]. One such compound, PX-866 (9), which displayed selectivity for p110α, δ and γ (IC_{50}s of 6, 3 and 9 nM, respectively) over p110β ($IC_{50} > 300$ μM) [75], conferred potent single agent *in vivo* efficacy in ovarian (OVCAR-3) and lung (A-549) tumour xenograft models. In addition, (9) was seen to display synergy with cisplatin in these animal models. A Phase I clinical trial of PX-866 (9) in patients with advanced metastatic cancer was subsequently initiated in 2008 under the auspices of Oncothyreon. Related conjugates, with improved physicochemical, pharmacokinetic and tumour delivery properties compared with wortmannin, have since been disclosed by other groups [93, 94].

Both LY294002 and wortmannin have been used extensively as biochemical tool compounds in a plethora of *in vitro* and *in vivo* mechanistic studies, principally in the oncology arena [1–4]. However, over recent years, attention has been focused increasingly on the development of a diverse range of small molecule heterocyclic inhibitors with distinct isoform selectivity profiles for the treatment of cancer and other chronic diseases. Amongst the first groups to initiate activities in this area was Yamanouchi (now Astellas), who, in collaboration with the Ludwig Institute of Cancer Research and Cancer Research, UK, developed a series of fused pyrimidines as class I PI3K inhibitors [95–99]. Compound (10) was identified from high throughput screening with a confirmed IC_{50} of 1.3 μM against p110α. Lead optimisation resulted in the discovery of the thienopyrimidine (11), which displayed high biochemical potency (IC_{50} p110α = 2.5 nM) and promising *in vitro* anti-proliferative activity in the A375 melanoma cell line (IC_{50} = 580 nM) [96]. In a separate report, the discovery of PI-103 (12) was described by Astellas. This compound exhibited very potent biochemical activity against the class IA isoforms (IC_{50}s for p110α = 3.6 and p110β = 3 nM), though was weaker against p110γ (IC_{50} = 250 nM), and was inactive against a series of protein kinases including KDR, PKA, PKCα and cyclin E/CDK_2 when screened at 100 μM [97]. PI-103 also displayed potent anti-proliferative activity in several tumour cells, including the MCF7 ADR-res breast line (IC_{50} = 130 nM), and conferred *in vivo* efficacy in a HeLa human cervical cancer xenograft model, showing good tolerability and not affecting body weight. In 2006, scientists at University of California, San Francisco, discovered that PI-103 also potently inhibited mTOR (IC_{50}s for mTORC1 = 20 nM, mTORC2 = 80 nM) [98], and PI-103 has since been the subject of a number of reports citing its potential as a dual class IA/mTOR inhibitor for the treatment of cancer, notably in the context of leukemia [99], PTEN-mutant gliomas [100, 101] and Kaposi's sarcoma [102].

Piramed have disclosed the design and development of a series of fused pyrimidine-based class I-selective inhibitors [103, 104]. Functionalised thienopyrimidine derivatives with high p110α activity were disclosed, notably the orally bioavailable inhibitor, GDC-0941 (13). This compound displayed potent biochemical activity against all class I isoforms (IC_{50}s for p110α, β, δ, γ = 3, 33, 3 and 75 nM respectively) and equipotently inhibited the wild-type and the activating oncogenic mutant forms of p110α. GDC-0941 displayed potent *in vitro* tumour growth inhibition in a range of

PIK3CA- and PTEN-null-expressing tumour cell lines, conferred a tumour growth inhibition of 86% when dosed daily at 75 mg/kg and showed potent pharmacodynamic biomarker modulation in the U87MG PTEN-null glioma mouse xenograft model. Additionally, GDC-0941 exhibited promising pharmacokinetic properties, with an oral bioavailability of 71%, clearance below half of liver blood flow in dogs, and encouraging *in vivo* tolerability. In 2008, GDC-0941 was taken into Phase I clinical trials in partnership with Genentech for the treatment of patients with advanced solid tumours. Further derivatives of this compound have also been disclosed by Piramed and Genentech [105].

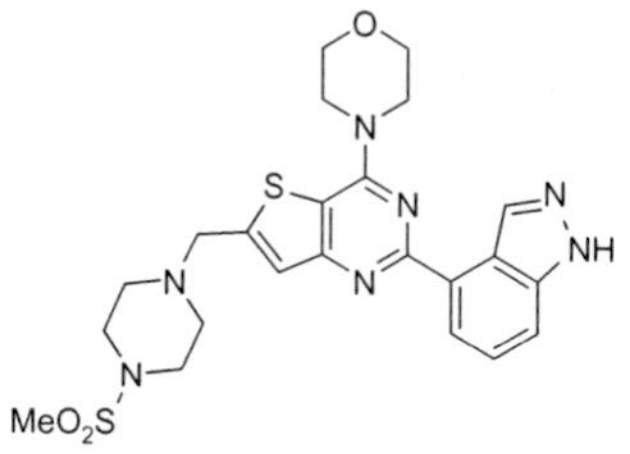

(13)
GDC-0941

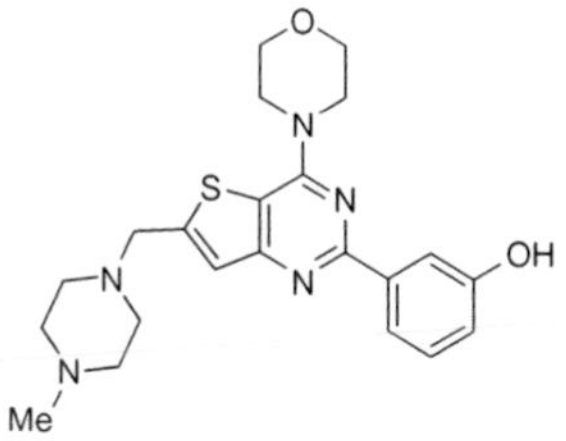

(14)

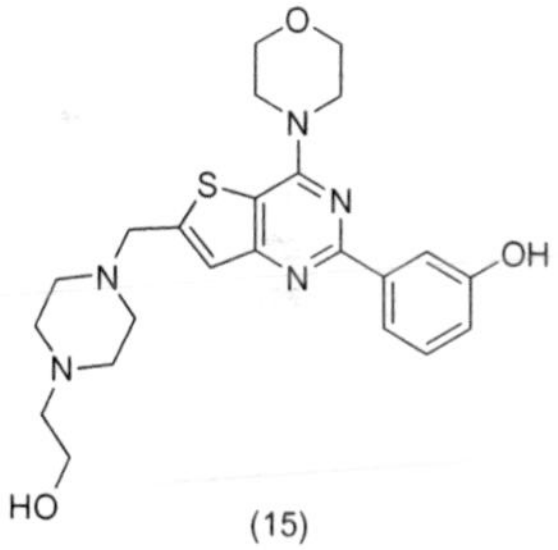

(15)

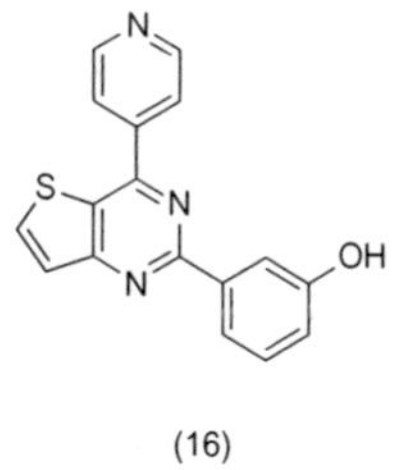

(16)

Piramed have also reported the discovery and development of a series of class IA-selective inhibitors related to GDC-0941, exemplified by the 3-(thieno[3,2-*d*]pyrimidin-2-yl)phenols (14) and (15) [106], the biochemical

IC_{50}s of which were 10 and 7 nM respectively against p110α, and 330 nM and 670 nM respectively against p110γ. In a separate report, in partnership with the Institute of Cancer Research, it was revealed that (14) and (15) displayed potent *in vitro* anti-proliferative activity in a range of tumour cell lines; these compounds also conferred potent pharmacodynamic biomarker modulatory effects, inhibited forkhead transcription factor translocation, and potently induced apoptosis. In the U87MG glioblastoma xenograft model, both (14) and (15) elicited potent tumour growth inhibition and a dose-dependent inhibition of p-AKT in tumours was observed. Intriguingly, compound (14) was shown to be more potent against mTOR ($IC_{50} = 60$ nM) than either (15) (230 nM) or GDC-0941 (580 nM) [107].

An additional series of fused pyrimidine-based inhibitors has been disclosed by Piramed, exemplified by compound (16), in which the morpholine of (11) was replaced by an unsaturated heterocyclic 'surrogate', capable of forming similar hydrogen-bonding interactions with the hinge region [108]. These compounds were reported to be class IA-selective inhibitors.

Novartis have developed a series of novel imidazo[4,5-c]quinoline-based modulators of the PI3K/AKT pathway [109, 110]. Initially, compound (17) was identified as a dual inhibitor of class I PI3Ks (IC_{50}s = 56, 446, 35 and 117 nM for p110α, p110β, p110δ and p110γ respectively) and of PDK1 ($IC_{50} = 245$ nM), and elicited a dose-dependent decrease of p-AKT in PTEN-null tumour cells. Optimisation of (17) led to the discovery of the *N*-methyl imidazolinone, (18), which, unlike (17), was shown to be inactive against PDK1 ($IC_{50} > 25,000$ nM), although it retained activity for class I PI3Ks (IC_{50}s = 72 nM, 2.3 μM, 201 nM and 382 nM for p110α, β, δ and γ respectively). This loss of PDK1 affinity was rationalised to stem from the electrostatic mismatch caused by the close proximity of the carbonyl of Leu-88 in the p-loop and the carbonyl located in the 2-position of (18). Compound (18) displayed potent *in vitro* pharmacodynamic modulatory effects, as confirmed through the down-regulation of p-AKT ($IC_{50} = 33$ nM), inhibition of p70S6K phosphorylation ($IC_{50} = 11$ nM), and concomitantly displayed potent *in vitro* tumour cell proliferation inhibition ($IC_{50} = 13$ nM). In a PC3M prostate tumour xenograft model, (18) conferred a tumour growth inhibition of 91% when administered twice daily at 60 mg/kg. Further optimisation of (18) led to the discovery of (19), NVP-BEZ235 [111–113], a potent dual class I/mTOR inhibitor. The antiproliferative activity of (19) was reported to be superior to that of the allosteric selective mTOR complex inhibitor everolimus in a panel of 21 varied cancer cell lines and, similar to GDC-0941, elicited anti-proliferative activity in both wild-type- and mutant p110α-expressing tumours. NVP-BEZ235 is currently in Phase I clinical trials for the treatment of patients with advanced solid tumours.

(17)   (18)   (19)
                                               NVP-BEZ235

(20)   (21)   (22)
ZSTK474

By using a COMPARE analysis of chemosensitivity measurements from 39 human cancer cell lines, ZSTK474 (20) was discovered by researchers at Zenyaku Kogyo as a potent pan-class I PI3K inhibitor (IC$_{50}$s for p110α = 16 nM, p110β = 44 nM, p110δ = 5 nM, and p110γ = 49 nM) [114–120]. ZSTK474 displayed potent anti-tumour efficacy and anti-angiogenic activity *in vivo*, as determined in the RXF-631L renal cancer xenograft model [116]. Unlike PI-103 and NVP-BEZ235, ZSTK474 was seen to be selective for class I PI3K isoforms over class IV, notably mTOR and DNA-PK [121, 122]; through proteomic analysis, however, (20) was shown to inhibit phosphorylation of 4E-BP1 [122].

Astellas have disclosed a series of p110α-selective inhibitors based on an imidazo[1,2-*a*] pyridine scaffold, which displayed > 100-fold specificity over p110β and p110γ [123–125]. Principal examples include (21), and a more chemically stable derivative, (22) (p110α = 3 nM, p110β = 170 nM, p110γ = 230 nM) which showed promising *in vivo* efficacy in a mouse HeLa human cervical cancer xenograft model. In a separate disclosure from researchers at the University of Auckland, the SAR of a series of related compounds possessing alternative heterocyclic 'head' groups to imidazo [1,2-*a*] pyridine, though retaining the sulfonolohydrazide motif present in (22), were prepared and evaluated for activity against the class IA PI3K isoforms [126]. Of particular note is the isoquinoline derivative (23) that, despite being significantly less potent than (22), displayed selectivity for p110α (IC$_{50}$ = 0.8 µM) over p110β (IC$_{50}$ > 10 µM), and inhibited the growth of NZB5 (medulloblastoma) and NZOV9 (ovarian) tumour cell lines with IC$_{50}$s of 890 and 500 nM respectively. In the same report, docking studies using the p110γ crystal structure [74] revealed that (22) was deeply inserted in the ATP-binding pocket, and formed a network of hydrogen bonds with the catalytic site, including a key interaction between *N*-1 and Val-882 in the hinge, and additional interactions between the nitro group and Asp-884, and between the arylsulfonyl moiety and Ala-885. By contrast, docking studies confirmed that isoquinoline derivatives such as (23) were less deeply inserted within the ATP-binding pocket, and the hydrogen bonds with Ala-884 and Ala-885 formed by (22) were lost for (23) and derivatives thereof, thus contributing to their lower intrinsic activity.

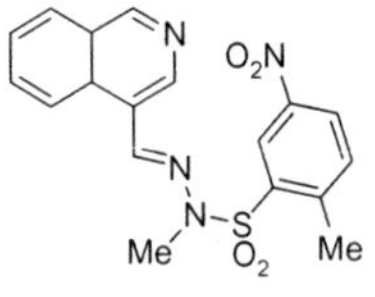

(23)

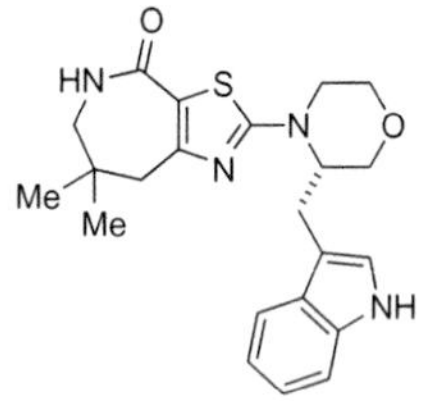

(24)

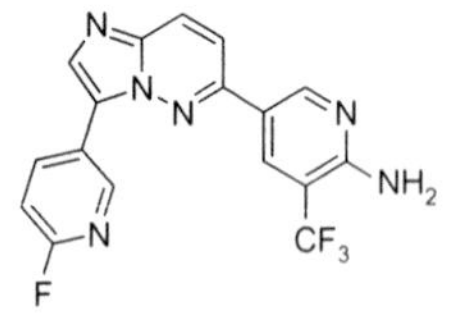

(25)

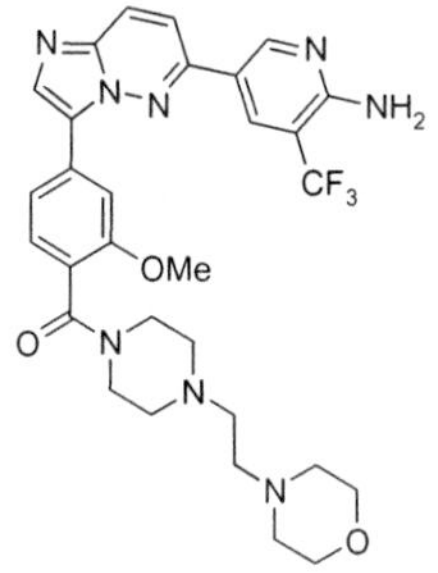

(26)

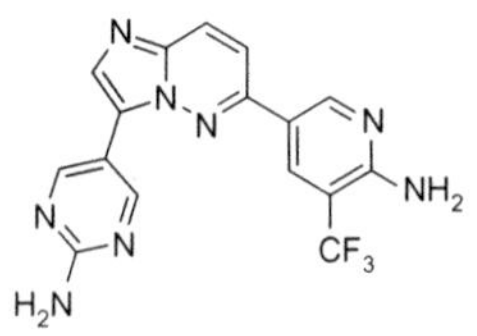

(27)

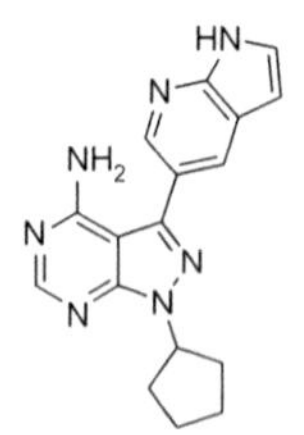

(28)

UCB have developed a series of class I PI3K inhibitors based on 5, 5-dimethyl-5,6-dihydrobenzo[d]thiazol-7(4H)-one, exemplified by (24) [127–131]. Compound (24) displayed selectivity for p110α, p110δ and p110γ (IC_{50}s = 59, 18, 31 nM respectively) over p110β (1 μM), and elicited *in vitro* anti-proliferative activity in PTEN-null and p110α mutant-expressing cell lines, and down-regulated p-AKT *in vitro* (IC_{50} for MCF7 = 862 nM) and *in vivo*. In a U87MG glioma xenograft model, compound (25) displayed 75% tumour growth inhibition when dosed orally at 200 mg/kg bi-daily [131].

Novartis have disclosed a series of imidazopyridazine- and pyrrolopyrimidine-based class I PI3K inhibitors with distinct sub-type selectivities. Compound (25) displayed high potency against p110α (IC_{50} = 1.4 nM), though other examples of structurally related compounds with both dual p110α/δ activity and displaying class IA/IB selectivity were reported, exemplified by (26) (IC_{50}s for p110α = 8 nM, p110β = 178 nM, p110δ = 6 nM, and p110γ = 405 nM) and (27) (IC_{50}s for p110α = 15 nM, p110β = 4 nM, p110δ = 9 nM, and p110γ = 737 nM) [132].

Researchers at the University of California, San Francisco, have disclosed a novel series of 1H-pyrazolo[3,4-d]pyrimidin-4-amine derivatives with dual p110α/class IV activity, exemplified by (28). Compound (28) was reported to have IC_{50} values against p110α, DNA-PK and mTOR of 52, 60 and 10 nM, respectively, whilst its activity against p110β and p110γ was significantly lower, with IC_{50}s of 1.4 and 1.1 μM respectively. Despite its specificity within the class I sub-family, (28) also displayed potent inhibition of several protein kinases, including Abl, Hck, Sec PDGFR and VEGFR [133, 134]. Further derivatives from this structural class displayed distinct PI3K superfamily selectivities that were governed by the nature of substitution patterns present on the pyrazolo[3,4-d]pyrimidine core, and which are described in further detail below.

Wyeth have also reported the development of a series of PI3K pathway inhibitors based on the pyrazolopyrimidine template. Dual p110α/mTOR inhibitors displaying potent tumour cell proliferation inhibition were reported, exemplified by (29) (IC_{50}s for p110α = 11 nM, mTOR = 17.5 nM, LNCaP = 450 nM, and MDA468 = 1.7 μM). In addition, p110α-selective inhibitors from the same chemical series were disclosed, including (30) (IC_{50}s for p110α = 18 nM and mTOR = 3.75 μM), in which the 6-position phenol group present in (29) had been replaced with an aminopyrimidine. Interestingly this compound conferred potent tumour cell proliferation inhibition *in vitro* (IC_{50}s for LNCaP = 80 nM and MDA468 = 400 nM). By contrast, compound (31), an mTOR-selective inhibitor, and three orders of magnitude less potent against p110α (IC_{50}s for mTOR = 1.6 nM and p110α = 1.65 μM), exhibited comparatively weaker *in vitro* anti-tumour activity in the same cell lines (IC_{50}s for LNCaP = 2.5 μM and MDA468 = 10.5 μM) [135].

(29)　　　　　　　　　　(30)　　　　　　　　　　(31)

(32)　　　　　　　　　　(33)　　　　　　　　　　(34)

Intellikine have developed a new class of quinazolines with distinct PI3K isoform specificity profiles. Specific activities were not revealed, though (32) was reported as having an IC_{50} of < 100 nM against p110α, δ and γ, displaying selectivity over p110β and mTOR, for which IC_{50}s above 1 μM were disclosed. Compound (33) displayed more pan-class I selectivity over mTOR, whilst (34) was reported to show dual p110α/mTOR inhibitory activity [136].

Amgen have disclosed a novel class of class I PI3K inhibitors based on a benzothiazole core. Examples include the p110α-selective inhibitor (35) (IC_{50} p110α = 29 nM, p110β = 5.2 μM) which inhibited p-AKT in the HCT116 (colorectal) tumour cell line *in vitro* with an IC_{50} of 2.5 μM, and the dual p110α/β inhibitor (36) (IC_{50}s for p110α = 3 nM and p110β = 3.8 nM), which displayed a significantly more potent pharmacodynamic marker response in HCT116 cells (IC_{50} p-AKT = 3.8 nM) [137].

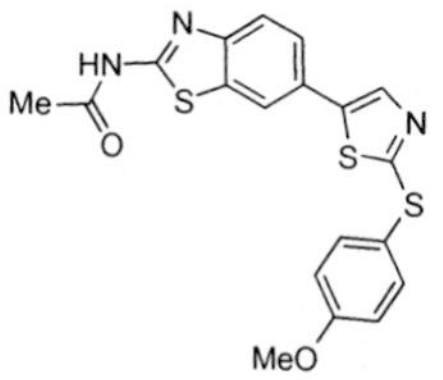

(35)

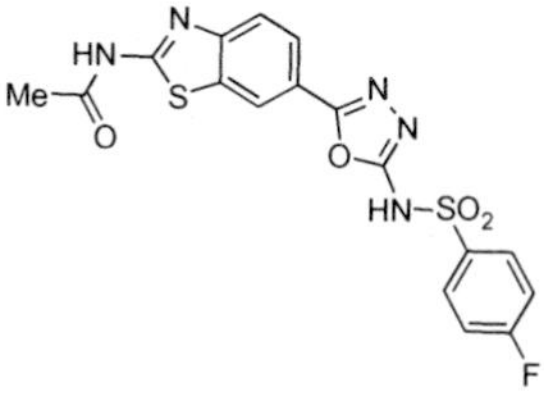

(36)

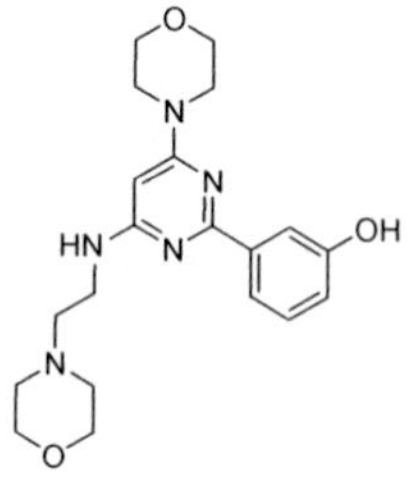

(37)

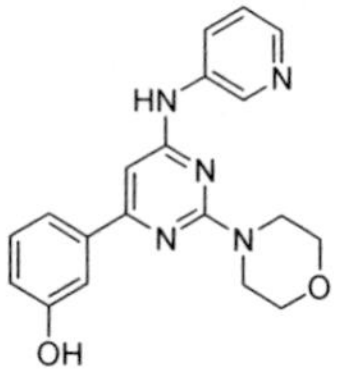

(38)

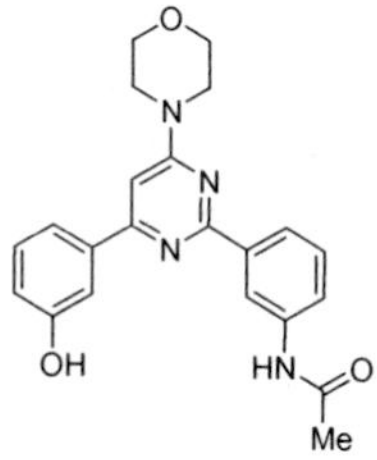

(39)

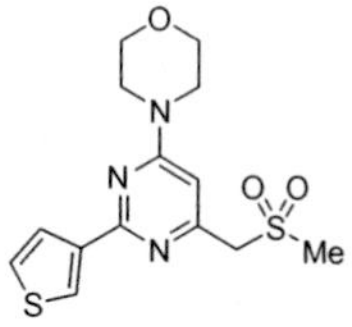

(40)

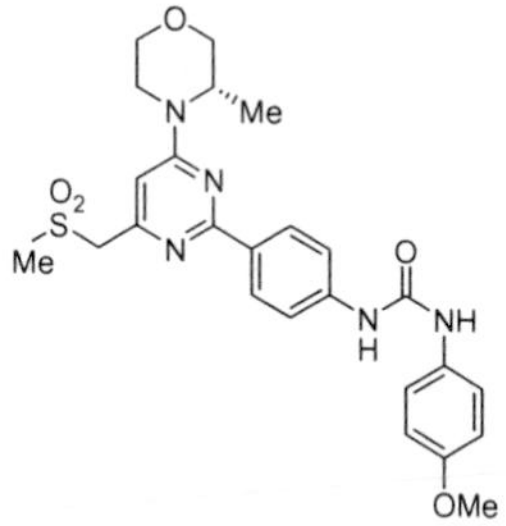

(41)

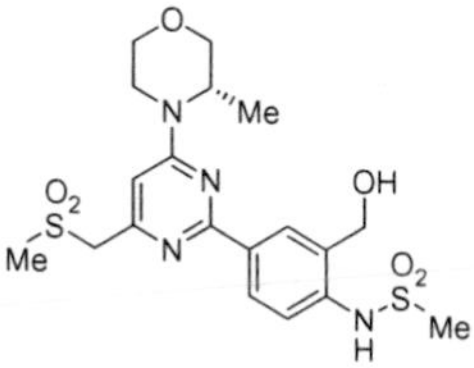

(42)

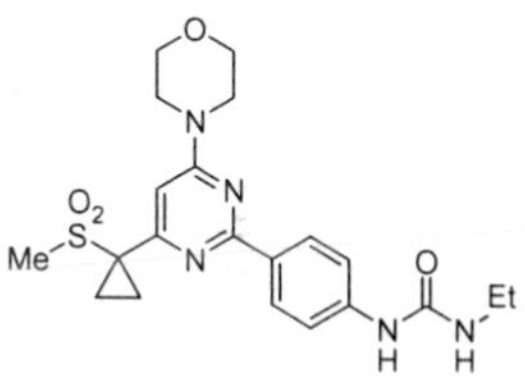

(43)

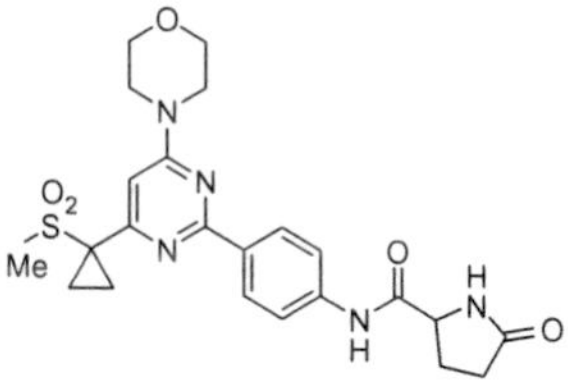

(44)

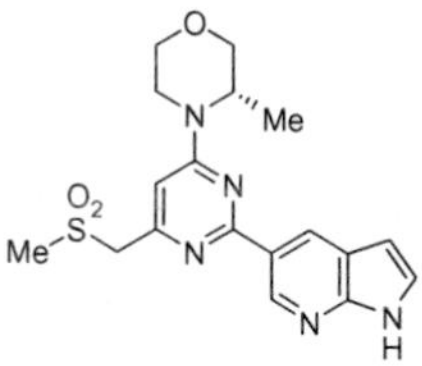

(45)

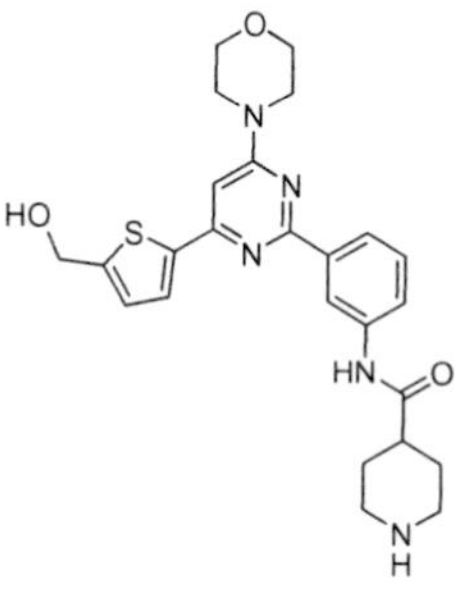

(46)

Piramed, Chiron and AstraZeneca have, individually, reported the development of class I and dual class I/mTOR-targeted pyrimidylmorpholine derivatives. Piramed have, in partnership with the Institute of Cancer Research, reported class IA over IB selectivity for a series of pyrimidine derivatives, exemplified by (37) [138]. Chiron have reported the development of a series of structurally related 4-(pyrimidin-2-yl)morpholine derivatives, exemplified by (38), which target PI3K, for the treatment of cancer. A combination of solid and solution phase synthetic approaches was used by Chiron to prepare these target compounds; however, no details of sub-type selectivity or activities were disclosed [139]. Finally, AstraZeneca have developed a number of pyrimidine-based inhibitors, exemplified by (39), for the treatment of cancer. No data for these compounds were disclosed, though it was revealed that specific derivatives displayed greater selectivity for class IA isoforms over p110γ [140]. A similar class of morpholinopyrimidines, namely the 4-(6-(alkylsulfonylmethyl)pyrimidin-4-yl)morpholine derivatives, exemplified by (40), were disclosed by AstraZeneca. These compounds were reported to exhibit potent mTOR and class I PI3K activity [141]. In a further report, AstraZeneca summarised the development of related compounds with potent mTOR and class I PI3K activity, exemplified by (41)–(45) [142–147]. Compounds (43), (44) and (45) were reported to have IC_{50}s against mTOR of 2.7, 36.9 and 37.7 nM, respectively. A series of tri-substituted pyrimidine-based PI3K inhibitors, exemplified by (46), has also been disclosed by AstraZeneca [148, 149].

Exelixis have developed a series of quinoxaline-based p110α inhibitors, such as (47), for the treatment of cancer [150]. In a separate disclosure, Exelixis reported the discovery of pyrido[2,3-d]pyrimidin-7(8H)-one-based p110α inhibitors for the treatment of cancer, for example (48) [151]. Data were not provided for these compounds, though Exelixis have reported the development of an orally bioavailable small molecule ATP-competitive inhibitor, XL-147, which is currently in clinical trials for the treatment of cancer [4]. XL-147 displays class I selectivity (IC_{50}s for p110α, β, δ, γ = 39, 383, 36 nM and 23 nM respectively), and was reported to be inactive against mTOR [4]. A further compound, XL-765, which inhibits both p110α and mTOR, is also currently in clinical trials for anti-tumour efficacy [4].

(47)   (48)   (49)

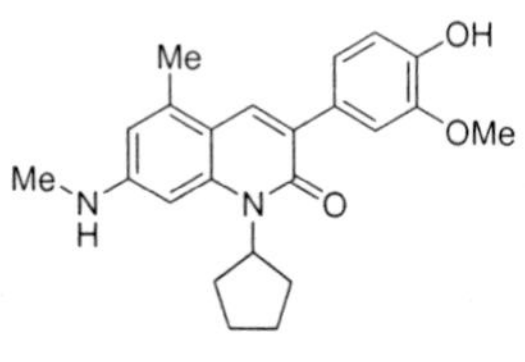

(50)

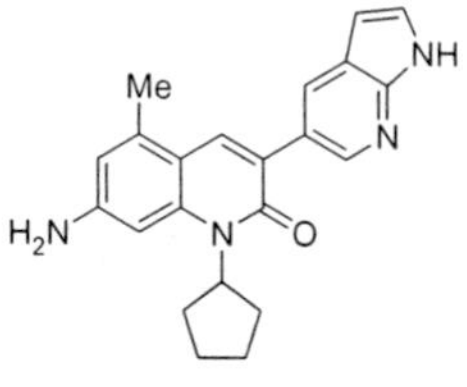

(51)

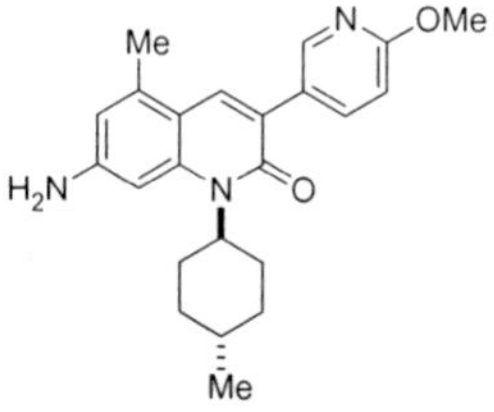

(52)

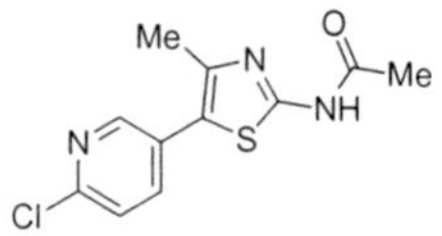

(53)

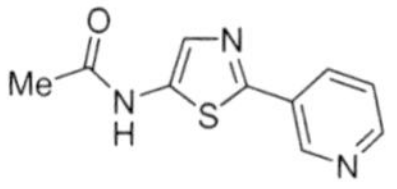

(54)

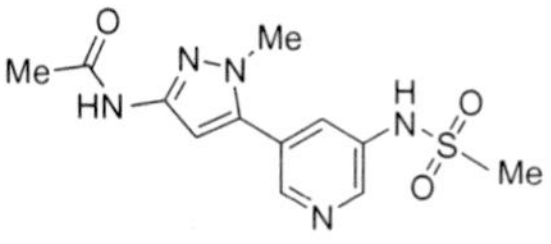

(55)

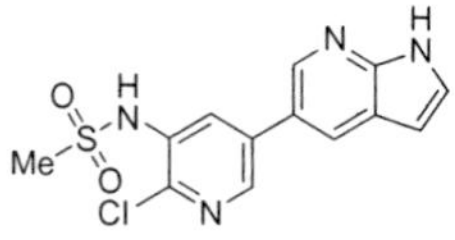

(56)

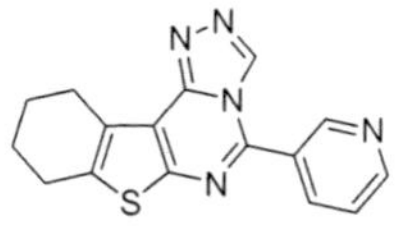

(57)

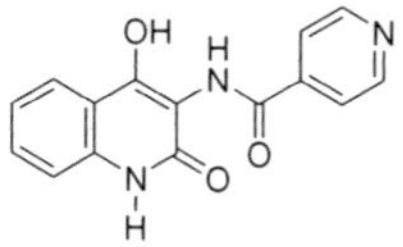

(58)

Piramal have disclosed a series of p110α inhibitors, exemplified by (49). Compound (49) is reported to have an IC_{50} of 1 μM against p110α in an enzyme assay, and of 5 μM in an mTOR cellular assay [152].

A series of p110α inhibitors based on the pyrido[2,3-d]pyrimidine-7-one core has been developed by Pfizer [153]. Potent biochemical and cellular activity was disclosed for a number of analogues, notably (50) and (51) (IC_{50}s against p110α = 1.5 nM and 500 pM respectively). Compound (52) (IC_{50} p110α = 24 nM) showed 89% *in vivo* tumour growth inhibition at a daily dose of 1 mpk in a PC3 prostate mouse xenograft model; concomitant dose proportional inhibition of p-AKT in tumours in this model was confirmed (0.25 – 3 mpk) [153, 154].

AstraZeneca have disclosed the discovery of a class of 5-heteroaryl-thiazoles, for example (53), with dual class I PI3K/mTOR inhibition, as potential therapeutics for the treatment of cancer [155]. AstraZeneca reported the discovery of an additional series of thiazole derivatives such as (54), which also elicit dual class I PI3K/mTOR activity [156]. Two series of class IA-active PI3K inhibitors based on functionalised pyrazoles, including (55) [157], and indoles, such as (56) [158], have also been reported by AstraZeneca, though data were not provided for these compounds.

Iconix reported the discovery of a series of thieno[3,2-e][1,2,4]triazolo[4,3-c] pyrimidine- and 4-hydroxyquinolin-2(1H)-one-based p110α inhibitors, including

(57) and (58), respectively [159, 160]. Data were not provided for these compounds.

Echelon developed class IA PI3K inhibitors based on a pyrazolo[1,5-*a*] pyrimidine core [161]. Specific data were not disclosed, though (59) was reported to have a biochemical activity of < 10 μM and a tumour cell antiproliferative IC$_{50}$ of 1 μM.

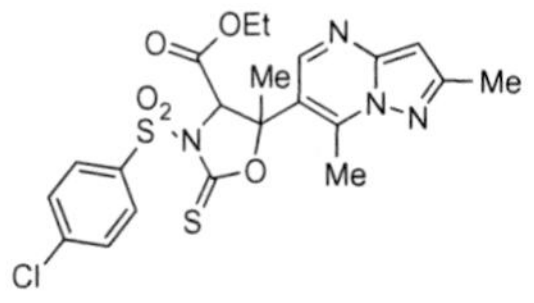

(59)

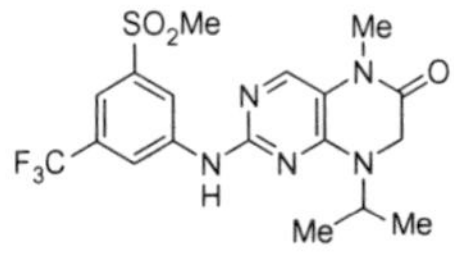

(60)

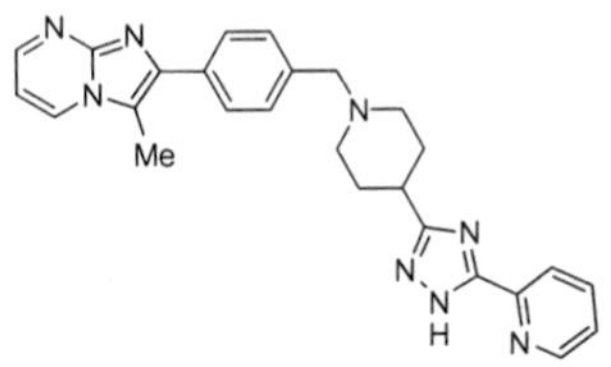

(61)

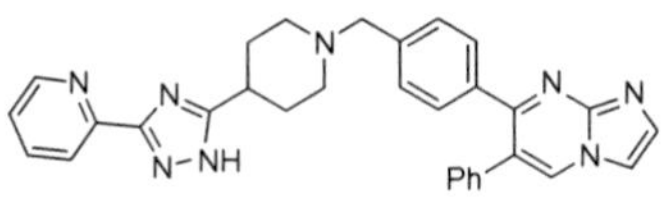

(62)

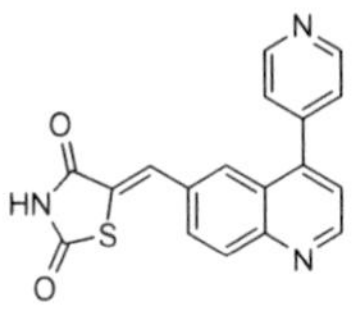

(63)

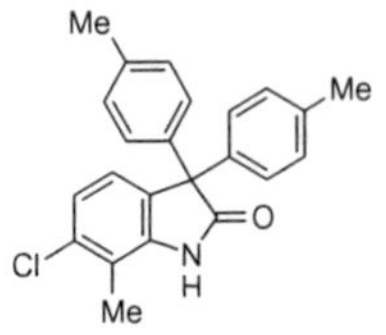

(64)

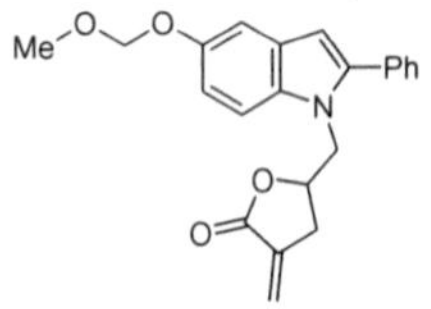

(65)
S9

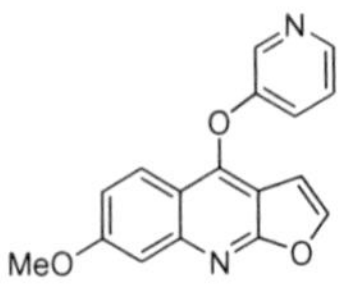

(66)

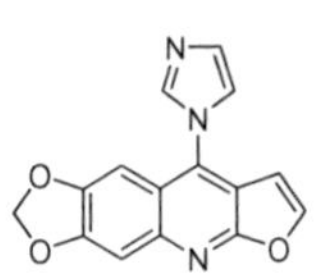

(67)

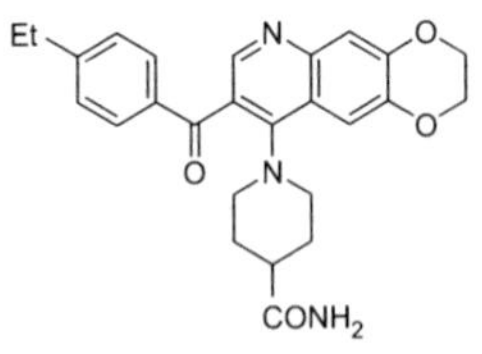

(68)
Quinostatin

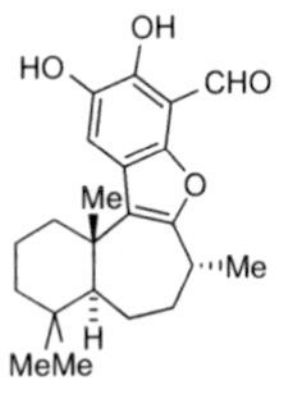

(69)
Liphagal

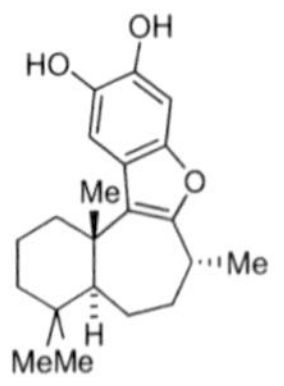

(70)
Desformylliphagel

A series of pteridinone-based PI3K inhibitors, exemplified by (60), were disclosed by Boehringer Ingelheim; no information on sub-type activity was disclosed for these compounds, however [162, 163].

Researchers at Bayer Schering Pharma have reported the discovery of a series of fused bicyclic pyrimidine-based PI3K inhibitors [166] and a related class of fused imidazole derivatives [165] for the treatment of cancer. Potent *in vitro* modulation of p-AKT and p-GSK3β levels, and anti-proliferative effects in A549 (lung adenocarcinoma) and OVCAR-3 (ovarian carcinoma) lines were reported for a number of compounds, including (61) and (62) [164, 165].

GlaxoSmithKline developed a series of thiazolidenediones with PI3K activity, for example (63) [166]. No data were disclosed, though GSK1059615, a thiazolidenedione with anti-cancer activity, was identified, and has been taken into Phase I clinical studies.

Bioimage disclosed the development of a series of isatin derivatives, exemplified by (64), which were reported to selectively inhibit mTOR pathway activation, though the precise mechanism of action is unknown. Potent inhibition of tumour cell proliferation was disclosed [167].

The compound S9, (65), was reported by researchers at the Chinese Academy of Sciences to down-regulate phosphorylation of AKT, mTOR, p70S6K and 4E-BP1 *in vitro* [168, 169]. Further studies into its mechanism of action revealed that S9 abrogated the EGF-activated PI3K-AKT-mTOR signalling cascade and AKT translocation, via dual inhibition of both PI3K and mTOR. S9 was further noted to inhibit the polymerisation of tubulin and induce M-phase cell cycle arrest. The *in vivo* anti-tumour activity of S9 was assessed in RH30 (rhabdomyosarcoma) and SMMC-7721 (liver cancer) mouse xenograft models, and was seen to confer a 50% tumour growth inhibition after 14 days of treatment at 50 mg/kg, i.p.

The discovery of several natural product-derived PI3K inhibitors has been reported. Researchers at NPIL developed furoquinoline-based inhibitors, derived from the natural product evolitrine, with anti-tumour activity [170–172]. Compound (66) was reported to selectively inhibit class I PI3K (67% inhibition at 10 μM) and compound (67) was disclosed as a dual class I PI3K/mTOR inhibitor. In a separate report, researchers at MIT reported the use of a high-throughput-cell-based assay to identify quinostatin (68) as a weak biochemical inhibitor of class IA PI3Ks (IC_{50}s for p110α = 15 μM and p110γ = 30 μM); (68) also inhibited tumour cell proliferation with IC_{50}s in the 2–8 μM concentration range [173]. Finally, liphagal (69), originally isolated from the marine sponge *Aka coralliphaga*, has been reported as a selective p110α inhibitor (IC_{50} = 100 nM), displaying 10-fold weaker potency against p110γ. Liphagal was cytotoxic to both LoVo (colorectal) and MDA468 (breast) tumour cell lines with IC_{50}s of

580 nM and 1.58 μM, respectively [174]. Derivatives of liphagal (including (70), desformylliphagel) have also been reported to be selective for p110α (IC$_{50}$ = 25 nM) over p110γ (inactive) [175].

CLASS I-TARGETED INHIBITORS FOR THE TREATMENT OF CANCER, INFLAMMATORY AND CARDIOVASCULAR DISEASES

Several examples of class I sub-type-selective inhibitors have been developed over recent years as putative therapeutics for the treatment of cancer and other chronic diseases. Cerylid (formerly Thrombogenix and Kinacia) have perhaps made the most significant progress in the development of selective class I small molecule inhibitors as anti-platelet and anti-coagulant drugs for the treatment of cardiovascular disorders. In one report, 2-morpholino-4*H*-pyrido[1,2-*a*]pyrimidin-4-ones, quinolin-4(1*H*)-ones, and chromen-4-one derivatives were disclosed, examples of which included (71), which had IC$_{50}$s for p110α and p110β of 50 nM, and of 5 μM for p110γ. Compound (72) displayed similar levels of class IA/IB selectivity to (71), whilst (73) displayed a profile showing more pan-class IA/IB activity (IC$_{50}$s for p110α and p110β = 100 nM; p110γ = 200 nM) [176]. In separate reports, the discovery of TGX-221 (74), a p110β-selective inhibitor and anti-thrombotic agent, was also disclosed. This compound displayed strong p110β activity (IC$_{50}$ = 5 nM) and high selectivity over the other class I PI3Ks (IC$_{50}$s for p110α = 5 μM, p110δ = 100 nM and p110γ > 10 μM). TGX-221 was also inactive against a panel of tyrosine kinases (IC$_{50}$ > 10 μM against Abl, EGFR and Fyn) and serine/threonine kinases (IC$_{50}$ > 10 μM against CK2, CDK2, ERK1, CaMK, PKA and PKC) [35, 36, 177–181]. The discovery and subsequent characterisation of TGX-221 represented a key development in the PI3K inhibitor field, and to date this compound remains the most extensively studied p110β-specific small molecule inhibitor.

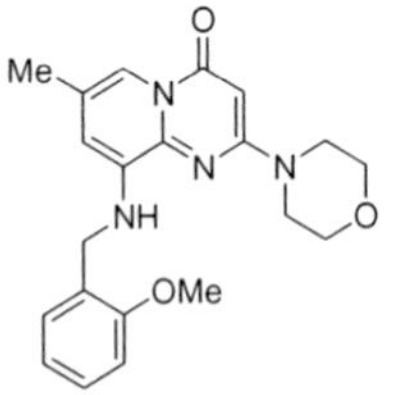

(71)

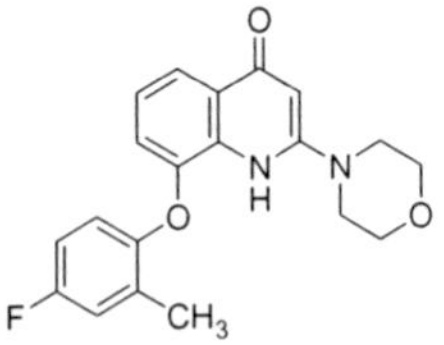

(72)

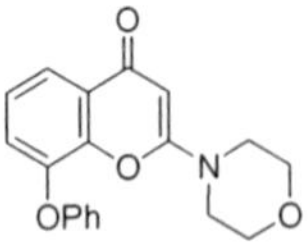

(73)

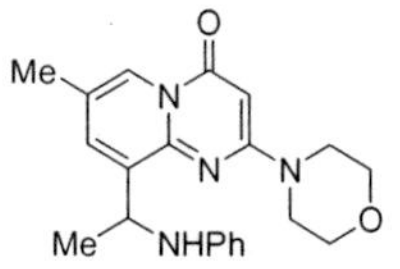

(74)
TGX-221

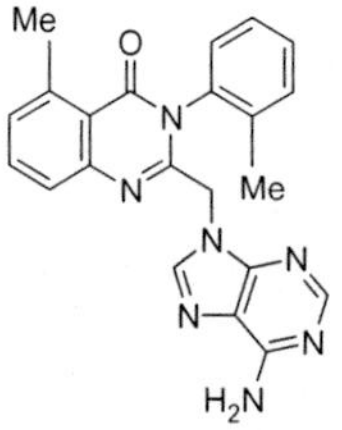

(75)
IC87114

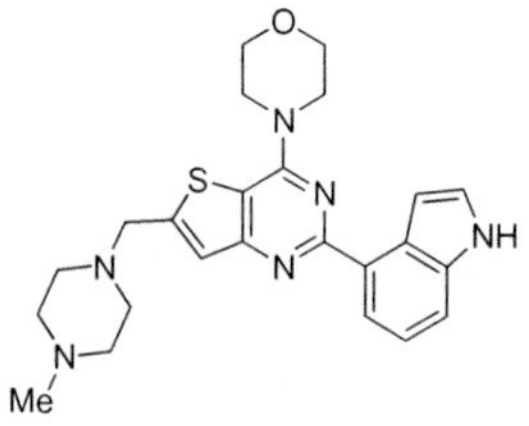

(76)

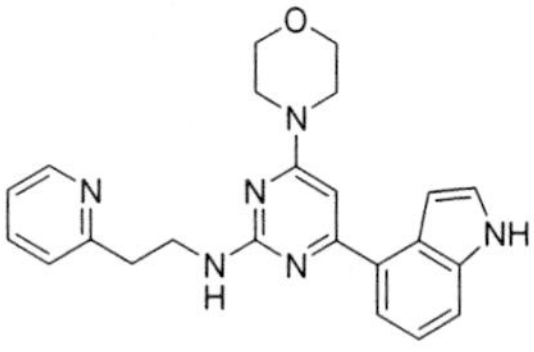

(77)

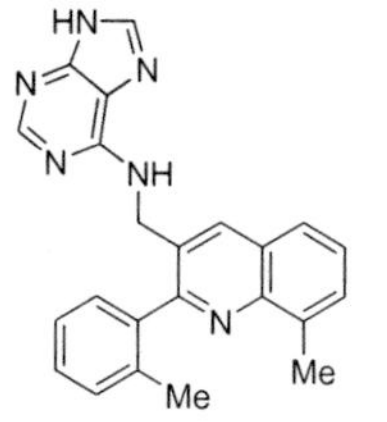

(78)

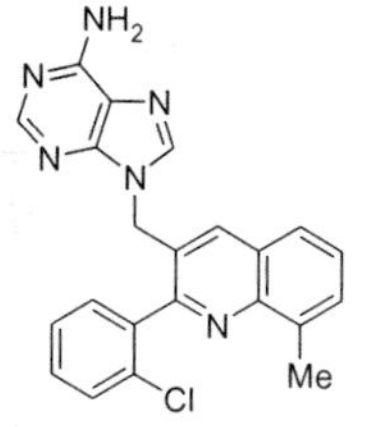

(79)

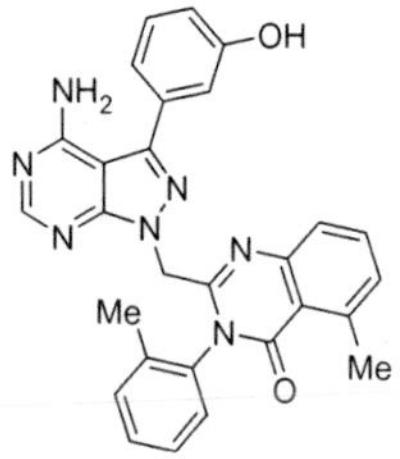

(80)

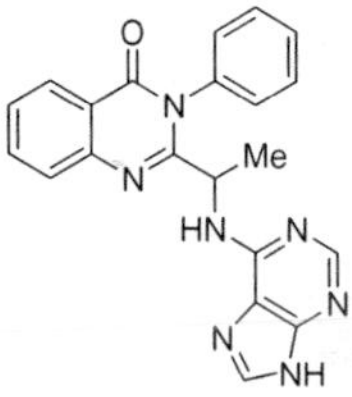

(81)

In light of the therapeutic potential of modulating p110δ activity to treat immune and inflammatory disorders, allergic diseases and haematological malignancies [29–33, 40–51]; a burgeoning level of attention has been focused over recent years on the design and development of small molecule inhibitors of this isoform [182, 183]. The potential for PI3K inhibitors to treat inflammation had previously been mooted by Lilly, with some key studies being performed with LY294002 [184]. However, as LY294002 lacks the specificity and possesses neither acceptable pharmacological nor toxicological properties to warrant its development as a clinical agent for this therapeutic area, the demand for class I sub-type-selective inhibitors has increased [185].

In 2005, ICOS disclosed a series of p110δ-selective quinazolinone derivatives that displayed anti-inflammatory and anti-cancer activities [186–189]. One such example was IC87114 (75), which displayed > 17-fold selectivity for p110δ over the other class I isoforms (IC$_{50}$s for p110α, β, δ, γ = > 100 μM, 1.82 μM, 70 nM, 1.24 μM respectively) [190]. ICOS revealed that (75) blocked fMLP- and TNFα-induced neutrophil superoxide generation and elastase exocytosis *in vitro*, and also suppressed TNFα-stimulated elastase exocytosis from neutrophils in pre-clinical inflammation models [43]. This suggests a key role for p110δ in neutrophil degranulation and migration, and potential for compounds of this class as anti-inflammatory agents [43, 44]. In a separate report, (75) was seen to inhibit ovalbumin-induced lung eosinophilia, airway mucus production and the inflammation score in an asthma mouse model [49]. Inhibition of p110δ activity by (75) also resulted in reduced neutrophil influx into inflamed tissues, as a direct consequence of the compound's ability to both impede chemoattractant-directed migration across the vascular endothelium, and to block adhesive interactions between neutrophils and the cytokine-stimulated endothelium [191].

The potential for the utility of p110δ-selective inhibitors in the treatment of cancer has also been explored [28–32]. In one report, IC87114 was shown to reduce the proliferation of primary AML cells, and had a similar effect on colony formation. In addition, IC87114 was seen to enhance the cytotoxic effects of etoposide and to augment apoptosis in primary AML cells, suggesting its potential both as a single agent therapy and also for use in combination therapy for this and other haematological cancers [30]. In a separate report, another compound developed by ICOS, IC486068 was reported to enhance radiation-induced apoptosis in endothelial cells and to reduce cell migration and tubule formation in endothelial cells in Matrigel following irradiation [192]. In the same report, it was revealed that the combined treatment of IC486068 and radiation significantly reduced tumour volume in mice with Lewis lung carcinoma and GL261 hind limb tumours, as compared with either treatment alone, suggesting that p110δ inhibitors have the potential to enhance radiation-induced tumour control.

A Phase I clinical trial of the orally active p110δ-selective inhibitor, CAL-101 (IC$_{50}$ p110δ = 2.5 nM), designed by ICOS, was initiated in 2008 in patients with haematological malignancies under the auspices of Calistoga Pharmaceuticals. A related compound, CAL-263, is currently in pre-clinical development at Calistoga for the treatment of inflammatory disorders.

The p110δ isoform has also been implicated in arterial hyper-responsiveness. IC87114 has been shown to confer anti-hypertensive efficacy *in vivo*, in the hypertensive deoxycorticosterone acetate salt rat model [193].

Piramed have reported the discovery of a series of compounds related to GDC-0941 (13), though possessing a 2-indolyl in place of the 2-indazolyl group as exemplified by (76), and eliciting p110δ-selectivity [194]. In a subsequent disclosure, Piramed reported the development of a series of p110δ-selective pyrimidine derivatives, such as (77) [195].

Amgen have reported the discovery of a class of bicyclic heteroaryl p110δ-selective inhibitors, exemplified by (78) and the structurally related purine derivative (79). Potent inhibition of B-cell proliferation was disclosed for several compounds, with an IC_{50} range of 1.9 nM – 0.81 μM being reported [196, 197].

Researchers at the University of California, San Francisco, have reported the discovery of the highly potent and selective p110δ inhibitor, 1*H*-pyrazolo[3,4-*d*]pyrimidin-4-amine, (80) (IC_{50} = 3 nM) [134]. This compound displayed >60-fold selectivity over p110γ, and >200-fold specificity over p110α, p110β, DNA-PK and mTOR. This compound was essentially inactive when screened against a panel of receptor tyrosine kinases.

Lilly have disclosed the discovery of thienopyrimidinone derivatives with p110δ selectivity [198]. Compound (81) was reported to have a K_i of 5 nM against p110δ, with selectivity over the other class I isoforms.

UCB have disclosed a class of novel morpholino-dihydrothiazolopyridinones that display dual p110δ/γ inhibition [199, 200], for example (82) (IC_{50}s for p110α, β, δ, γ = 738, 1325, 139 and 107 nM respectively) and (83) (IC_{50}s for p110α, β, δ, γ = 295, 540, 32 and 78 nM respectively). Optimisation led to (84), which displayed improved *in vitro* microsomal metabolic stability and solubility whilst maintaining PI3K activity (IC_{50}s for p110δ = 14 nM and p110γ = 52 nM). Efficacy studies in male Lewis rats showed that (84) inhibited CD3-induced IL-2 release with an ED_{50} of 5 mg/kg, compared with 25 mg/kg for (82).

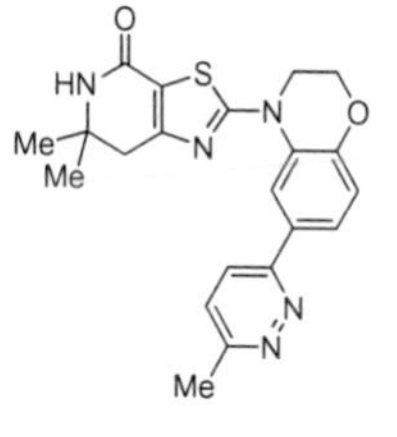

(82)

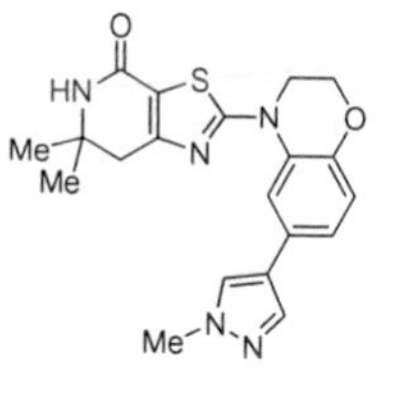

(83)

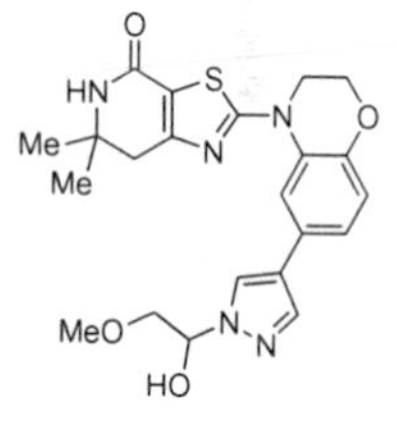

(84)

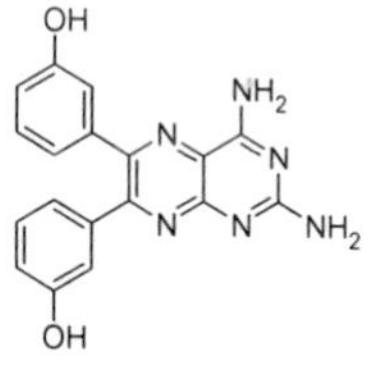

(85)
TG100115

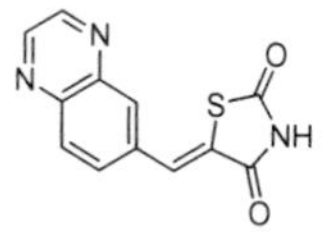

(86)
AS-605240

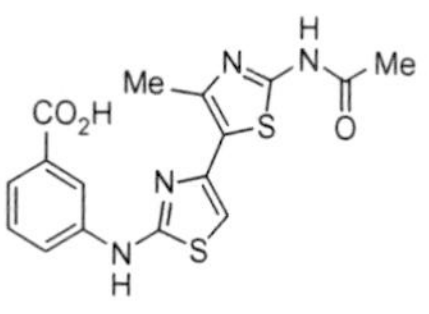

(87)

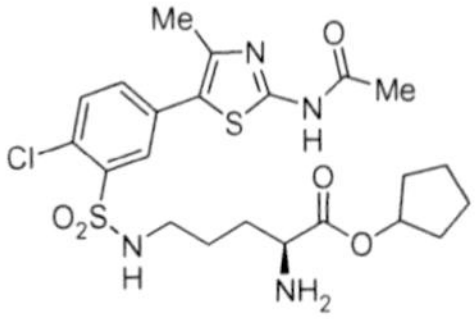

(88)

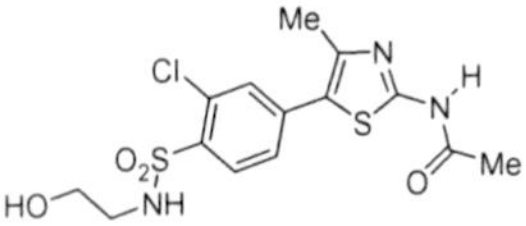

(89)

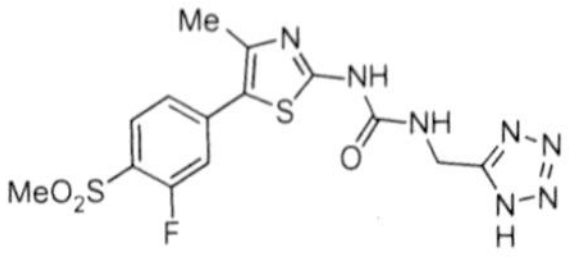

(90)

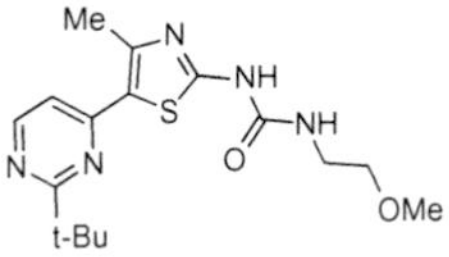

t-Bu

(91)

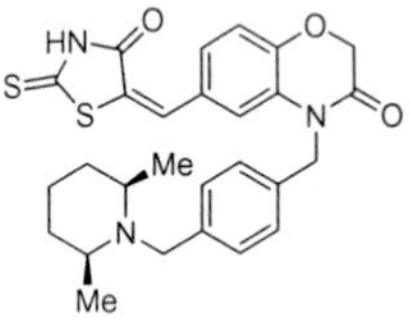

Me
Me

(92)

Cl

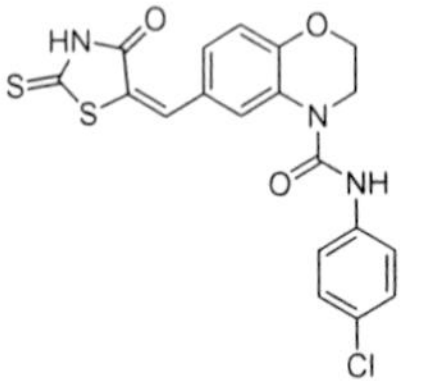

(93)

TargeGen have reported the design and development of TG100115 (85), a dual p110δ/γ inhibitor, with therapeutic potential for the treatment of inflammatory and cardiovascular disease. TG100115 showed selectivity over the other class I isoforms (IC_{50}s for p110α, β, δ, γ = 1.3 μM, 1.2 μM, 235 nM, and 83 nM respectively) and over a number of protein kinases [75]. In a murine asthma model, aerosolised (85) markedly reduced pulmonary eosinophilia and airway hyper-responsiveness; levels of IL-13 and mucin accumulation – key characteristics of the disease – were also diminished. In addition, the compound was also seen to reduce pulmonary neutrophilia in a mouse COPD model [200]. In a rodent model of myocardial infarction, intravenous administration of (85) at 0.5 mg/kg decreased total infarct area by at least 47% compared with vehicle control group, and provided significant cardioprotection [201–203]. TG100115 has since been taken into clinical trials for the treatment of myocardial infarction.

The first class I sub-family selective inhibitors to be developed were against p110γ, aided in part by the disclosure of a series of p110γ co-crystal structures in 2000. This shed light on the key structural elements required for ATP-competitive activity, thereby enabling the design and rapid development of a range of structurally diverse small molecule inhibitors [74]. Serono have made a significant contribution to the p110γ inhibitor field, most notably having developed the fused 1,3-thiazolidine-2,4-dione-based p110γ inhibitors, exemplified by AS-605240 (86) [204–207]. This compound displayed at least 7-fold greater selectivity for p110γ over the class IA isoforms, and has been assessed for its potential in the treatment of inflammatory diseases. In one disclosure, (86) was tested in a cultured mouse macrophage cell line, in which it was seen to inhibit the phosphorylation of

AKT. In addition, the compound was seen to temper migration of neutrophils and monocytes both *in vitro* and *in vivo*, and was efficacious in two mouse models of arthritis when administered orally: the antibody-induced model and the bovine type II collagen model [205]. AS-605240 treatment reduced the clinical disease score in this model after the mice had developed arthritis. In a further study, the potential of AS-605240 to inhibit a systemic lupus erythematosus-like disease in the MRL-lpr mouse model was confirmed: treatment of MRL-lpr mice with the compound led to increased survival of the mice by reducing immune complex deposition, levels of protein in the urine and levels of DNA-specific autoantibodies. This study demonstrated that AS-605240 could have therapeutic potential for the treatment of several autoimmune diseases [206].

In a separate report, Serono disclosed the development of thiazole-based $p110\gamma$ inhibitors, exemplified by (87), for the treatment of anaemia. Compound (87) displayed potent inhibition of $p110\gamma$ ($IC_{50} = 10\,nM$) and concomitant cellular down-regulation of p-AKT ($IC_{50} = 3.18\,\mu M$) [208].

Chroma has disclosed the development of a series of functionalised thiazoles as putative anti-inflammatory agents, exemplified by (88) [209]. Specific data were not disclosed, though (88) was reported to have activity against $p110\gamma$ at $< 1\,\mu M$ and inhibited $TNF\alpha$ release from THP-1 cells with an $IC_{50} < 10\,\mu M$.

A series of thiazole derivatives has been reported by Novartis for the treatment of inflammatory and allergic diseases. These compounds were shown to display potent and specific $p110\gamma$ activity, with (89) having an IC_{50} against this isoform of $2\,nM$ [210]. Scientists from Novartis have developed an additional series of functionalised thiazoles with $p110\gamma$-specific activity, exemplified by compound (90), which had an IC_{50} of $7\,nM$ [211], and have disclosed the development of a related class of pyrimidino-thiazoles, again displaying potent $p110\gamma$ inhibition: (91) had an IC_{50} against $p110\gamma$ of $5\,nM$ [212].

Pfizer have been prolific in the design and development of a number of small molecule $p110\gamma$ inhibitors during the last five years. In one report, a series of $p110\gamma$-active benzoxazin-3-ones, including (92), was disclosed for the treatment of inflammatory and cardiovascular diseases, and cancer. (92) was reported to have an IC_{50} of $3\,nM$ against $p110\gamma$ [213]. Pfizer have also disclosed the development of a class of benzoxazine derivatives as inhibitors of $p110\gamma$ for the treatment of inflammatory and cardiovascular diseases and cancer, with the compound (93) being reported to have an IC_{50} of $2\,nM$ against this isoform [214]. In a subsequent report, Pfizer provided further insight into the optimisation of this class of inhibitors, notably the SAR generated courtesy of modifications made to the substitution patterns on the phenethyl ring of the benzo[1,4]oxazine-3-one core [215].

Para-substitution was found to give rise to the most potent analogues in the series, with the 4-trifluromethyl derivative (94) exhibiting the highest level of *in vitro* inhibition (IC_{50} p110γ = 1.9 nM). Subsequent examination of functional cellular activity confirmed that (94) and derivatives displayed potent inhibition of superoxide production from neutrophils with an IC_{50} of 370 nM. Encouraging *in vivo* efficacy in the mouse peritonitis model of cell migration was also established for (93) and related compounds in this series [215].

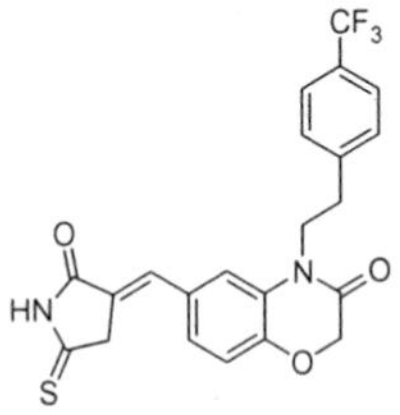

(94)

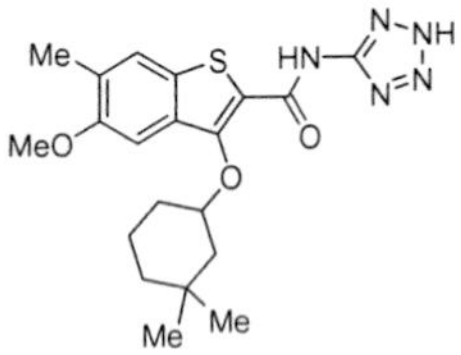

(95)

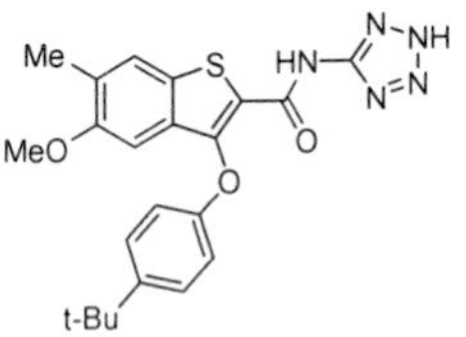

(96)

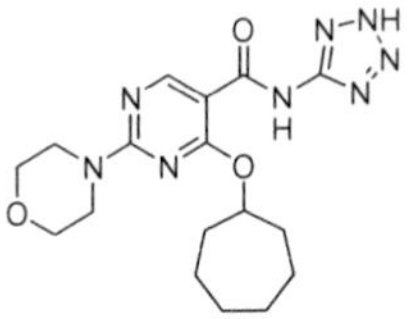

(97)

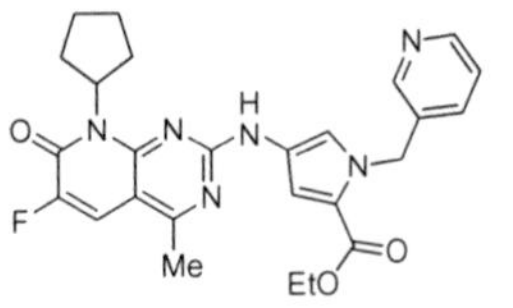

(98)

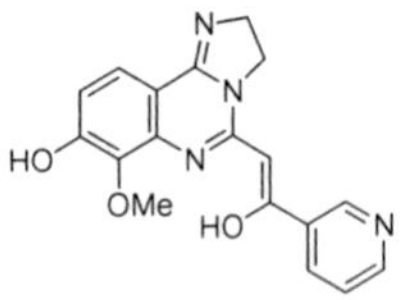

(99)

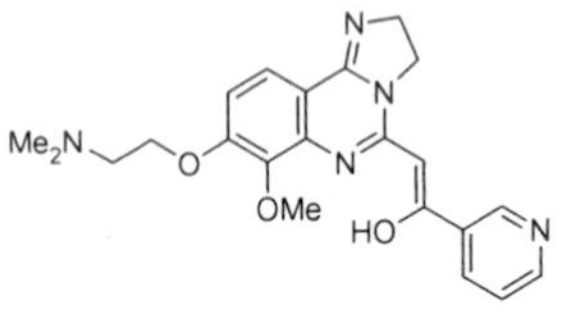

(100)

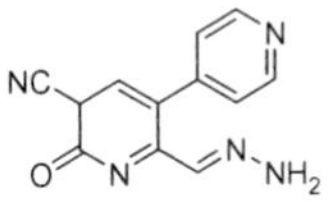

(101)
OK-1035

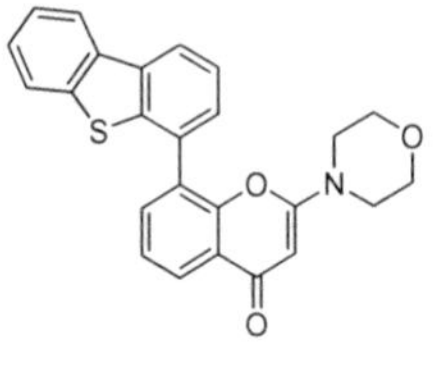

(102)
NU7441

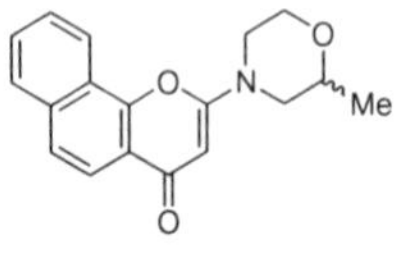

(103)
NU7163

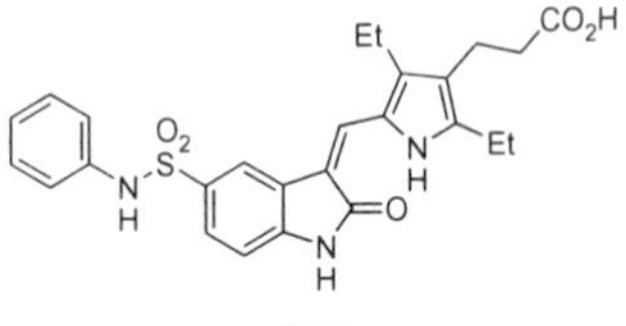

(104)
SU-11752

A series of benzo(b)thiophene-based p110γ inhibitors for the treatment of inflammatory and cardiovascular diseases and cancer has been developed by Pfizer, for example (95), which had an IC_{50} of 3 nM [216, 217] and (96), which had an IC_{50} of 9 nM [217]. Pyrimidine-based p110γ inhibitors [218, 219], exemplified by compound (97) (IC_{50} = 197 nM), and pyrido[2,3-d] pyrimidine-7-ones, exemplified by (98), have been disclosed by Pfizer. Compound (98) was reported to have an IC_{50} of 4 nM against p110γ, and efficacy in pre-clinical models of arthritis was reported for unspecified compounds in this series [220].

Bayer have reported the discovery of a series of 2,3-dihydroimidazo[1,2-c] quinazolines, exemplified by (99) and (100), which were reported to have activity for p110γ and p110β, respectively, though specific biochemical activities were not disclosed [221].

CLASS II-TARGETTED INHIBITORS

As mentioned above, the principal focus of PI3K small molecule drug discovery and development has centred on class I isoform modulators, and inhibitors of the class IV (PIKK) family isoforms; examples of the latter are outlined in the next section. To date, there have been no reported small molecule inhibitors of the class II enzymes, C2-α and -γ, or the sole class III member, Vps34. However, in the account of their discovery of PI-103 (12), Astellas reported that this compound also displayed potent inhibition of C2-β, with an IC_{50} of 10 nM [96, 97]. In addition, Astellas reported an IC_{50} of 220 nM against C2-β for the thienopyrimidine derivative (11) and of 2.1 μM for LY294002 (1) against the same enzyme [96, 97]. The imidazopyridine derivative (22) was also assessed for C2-β biochemical activity by Astellas, with an IC_{50} of 100 nM being reported [124, 125]. None of these compounds displayed selectivity for class II PI3Ks however, and owing to the emerging therapeutic potential of class II activity regulation, particularly in the context of cancer therapy, the development of ATP-competitive, class II-specific inhibitors represents a potentially attractive approach in this area.

CLASS IV-TARGETTED INHIBITORS

A number of ATP-competitive small molecule inhibitors of the PIKK family member DNA-PK have been developed. One of the first putative DNA-PK inhibitors to be disclosed was OK-1035 (101), which was originally reported to display moderate potency against DNA-PK, with an IC_{50} of 8 μM [222], though this biochemical activity was disputed in a subsequent report [223]. Since then, however, Kudos Pharmaceuticals, in collaboration with the Northern Institute of Cancer Research, have made several contributions to the development of inhibitors of this isoform, with the goal of identifying potent and selective inhibitors suitable for clinical evaluation as agents to augment the cytotoxicity of DNA-damaging chemotherapeutic agents [224–230]. Initial studies focused on utilising LY294002 (1) as a template for DNA-PK inhibitor design, since it was seen to have moderate activity against DNA-PK ($IC_{50} = 1.2\,\mu M$). Examination of substitution patterns at the 6-, 7- and 8-positions on the chromenone core using parallel synthesis led to the identification of NU7441 (102) as a potent and selective inhibitor, with IC_{50}s against DNA-PK, p110α and mTOR of 14 nM, 5 μM and 1.7 μM respectively; no activity against ATM or ATR was observed [224]. The high DNA-PK selectivity of NU7441 was further confirmed following screening against a panel of 60 diverse kinases, with no inhibitory activity being observed at 10 μM. The same group subsequently developed an improved synthetic route to NU7741 employing a Baker–Venkataraman rearrangement [225].

In a separate disclosure, Kudos reported the design and synthesis of the racemic chromone derivative NU7163 (103). This compound exhibited weaker DNA-PK activity than NU7741 ($IC_{50} = 0.19\,\mu M$), though it displayed a similar PI3K selectivity profile [226]. The cellular activity of NU7163 as a DNA-PK inhibitor and radio-sensitiser was assessed in a human tumour cell line *in vitro*, and at concentrations of 5 and 10 μM, NU7163 was seen to enhance the cytotoxicity of ionising radiation against HeLa B tumour cells [226].

Additional examples of DNA-PK inhibitors have been disclosed by Sugen, ICOS and Cancer Research, UK. Sugen reported the discovery of SU11752 (104), which has an *in vitro* IC_{50} of 100 nM. The DNA-PK cellular activity of SU11752 was confirmed through its inhibition of double-stranded break repair, and its ability to sensitise tumour cells to radiation. Sugen revealed that, in *in vitro* biochemical screens, SU11752 was approximately 10-fold weaker against p110γ and had negligible activity against the other class IV family member, ATM [231]. In a separate disclosure, ICOS reported the discovery of 9*H*-xanthen-9-one-based DNA-PK inhibitors, exemplified by (105), though specific data were not disclosed for these compounds [232]. Finally, researchers at Cancer Research, UK have reported that derivatives of vanillin – a tumour cell chemosensitiser – act as selective inhibitors of DNA-PK [233]. Examples disclosed included 4,5-dimethoxy-2-nitrobenzaldehyde (106) that had an IC_{50} of 15 μM, and 2-iodo-4,5-dimethoxybenzaldehyde (107) ($IC_{50} = 30\,\mu M$).

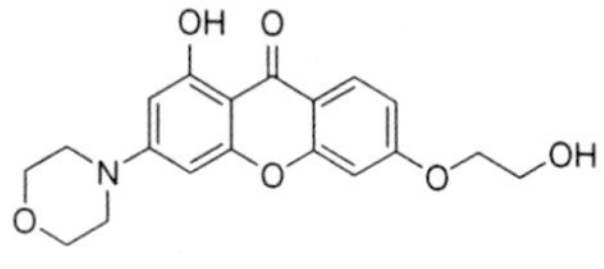

(105)

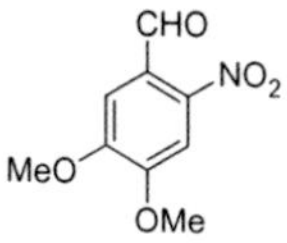

(106)

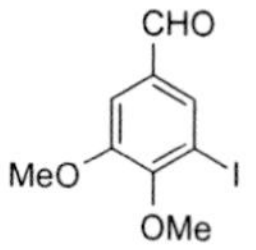

(107)

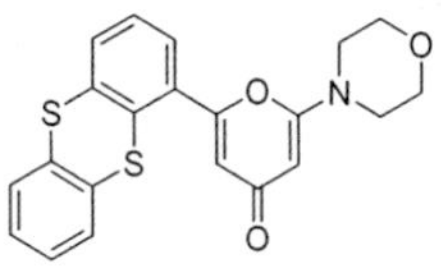

(108)
KU-55933

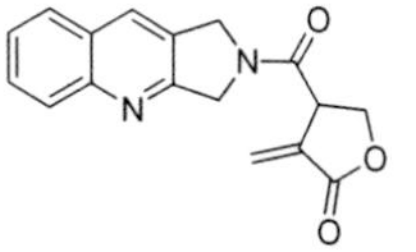

(109)

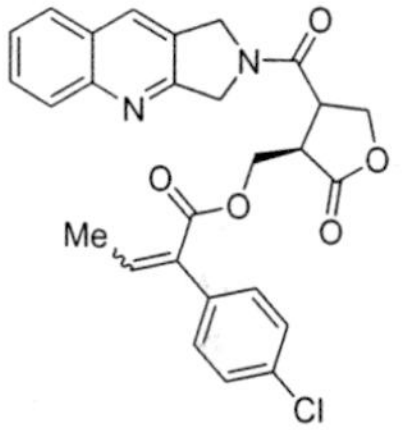

(110)

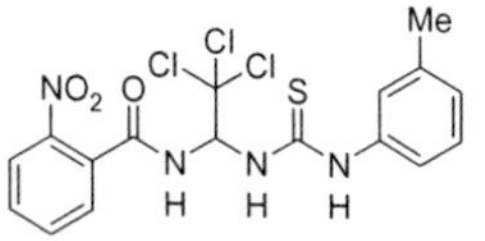

(111)

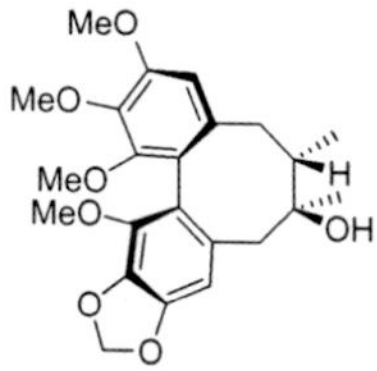

(112)

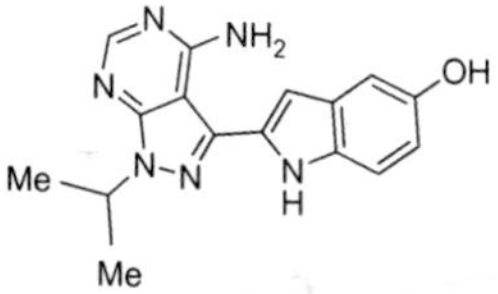

(113)

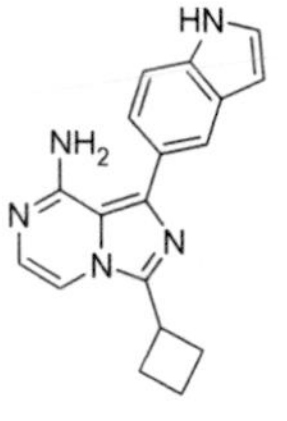

(114)

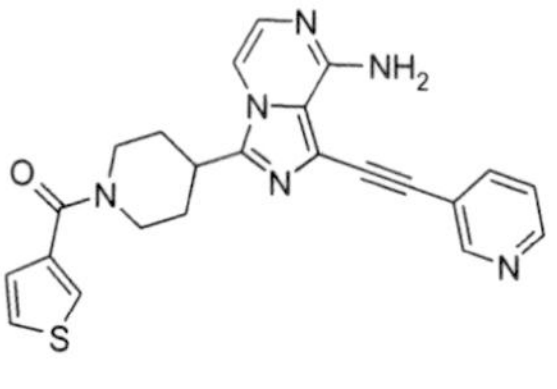

(115)

Kudos Pharmaceuticals have disclosed the ATM and DNA-PK SAR for a series of pyran-2-ones, pyran-4-ones, thiopyran-4-ones, and pyridin-4-ones [234, 235]. Lead optimisation led to the discovery of 6-thianthren-1-yl-pyran-4-one, KU-55933 (108), a potent ATP-competitive inhibitor of ATM. This compound inhibited ATM with an IC_{50} of 13 nM and a K_i of 2.2 nM, and showed good selectivity over other PI3K family members, including DNA-PK ($IC_{50} = 1.8\,\mu M$). No significant activity was observed for KU-55933 against a panel of 60 protein kinases at $10\,\mu M$. Cellular inhibition of ATM by KU-55933 was confirmed through the ablation of ionising radiation-dependent phosphorylation of a range of ATM substrates, including p53, γH2AX, NBS1 and SMC1. Exposure of cells to KU-55933 resulted in significant sensitisation to the cytotoxic effects of ionising radiation and to the DNA double-strand break-inducing chemotherapeutic agents, etoposide, doxorubicin and camptothecin. Inhibition of ATM by KU-55933 also caused a loss of ionising radiation-induced cell cycle arrest [236].

Researchers at the Georgia Institute of Technology have developed a series of pyrrolo-quinoline derived γ-lactones that display potent inhibition of ATM and mTOR [237]. Compound (109) exhibited selective ATM biochemical inhibition ($IC_{50} = 0.6\,\mu M$) over mTOR ($IC_{50} = 7\,\mu M$), together with potent NCI-H226 human tumour cell line growth inhibition ($IC_{50} = 0.38\,\mu M$). Additional examples in the series, such as (110) elicited selective inhibition of mTOR over ATM (IC_{50}s for mTOR, ATM = 500 nM and $6.5\,\mu M$ respectively).

Researchers at CKG have disclosed the synthesis of compounds with ATM and ATR activity, exemplified by compound (111) [238]. A new class of ATR inhibitors, including (112), has been disclosed by researchers at Niigata TLO [239]. No *in vitro* or *in vivo* data were provided for these compounds.

From a drug development perspective, mTOR is the most extensively studied of the PIKK family members. Rapamycin, and derivatives and formulations thereof, have been used clinically as immunosuppressants for organ transplant rejection, and for the treatment of cancer as both single-agent therapies and in combination with other molecular-targeted cyto-statics and cytotoxic chemotherapeutics. Of late, however, there has been an increasing level of attention geared towards the development of ATP-competitive inhibitors of mTOR [240]. Amongst the principal examples are small molecule heterocycles developed by researchers at the University of California, San Francisco, who disclosed the mTOR-selective 1H-pyrazolo [3,4-d]pyrimidin-4-amine (113) ($IC_{50} = 8$ nM) [134]. A series of fused bicyclic mTOR inhibitors, of the generic structure (114), has been developed by OSIP for the treatment of cancer. No data were disclosed for specific compounds, though mTOR biochemical activities in the range of $1\,nM - 11\,\mu M$, and cellular activities below $40\,\mu M$ were reported [241]. In a further disclosure, a series of fused bicyclic mTOR inhibitors, of the generic structure (115), were reported as anti-tumour agents by OSIP [242]. Researchers at Kudos have developed a series of pyrido[2,3-d]pyrimidines as mTOR inhibitors, illustrated by (116) [243], and compound (117) has been disclosed as a selective mTOR inhibitor with an IC_{50} of 5.3 μM. Compound (117) has been reported to be selective for mTOR over class I PI3Ks [244, 245].

(116)   (117)

CLINICAL LANDSCAPE

During the last three years, several of the compounds outlined above have successfully been approved for clinical development, and are currently in early stage clinical trials, details of which are presented in Table 3.1 and discussed below [246].

Scientists at Semafore Pharmaceuticals are developing SF-1126 (3) for the treatment of cancer. A Phase I trial of this compound in patients with solid tumours began in March 2007, for which positive interim data were

Table 3.1 PI3K INHIBITORS IN CLINICAL DEVELOPMENT

Drug	Primary target(s)	Therapeutic focus	Status	Organisation
SF-1126	p110α	Cancer (multiple myeloma, solid tumours)	Phase I	Semafore Pharmaceuticals
GDC-0941	p110α	Cancer (locally advanced or metastatic tumours, non-Hodgkin lymphoma)	Phase I	Genentech
BEZ-235	p110α/mTOR	Cancer (solid tumours)	Phase II	Novartis
XL-147	p110α	Cancer (solid tumours)	Phase II	Sanofi-Aventis/Exelixis
XL-765	p110α/mTOR	Cancer (solid tumours, glioblastoma, non-small cell lung cancer)	Phase II	Sanofi-Aventis/Exelixis
CAL-101	p110δ	Cancer; inflammatory disease	Phase I	Calistoga Pharmaceuticals
PX-866	p110α	Cancer (solid tumours)	Phase I	Oncothyreon/ProlX
GSK-1059615	p110α	Cancer (solid tumours, lymphoma)	Phase I	GlaxoSmithKline
TG100115	p110δ/γ	Myocardial infarction	Phase II	TargeGen

reported in June 2008. In this multicentre, dose-cohort escalating trial, patients received SF-1126 twice weekly as an intravenous infusion for four weeks in a 28-day cycle. Stable disease was observed in 25% of the patients who had completed one cycle of treatment. In May 2009, further clinical data were presented at the 45th ASCO meeting. In the study, 32 patients received 90, 140 180, 240, 320, 430 and 630 mg/m^2 intravenous infusions of SF-1126, twice weekly. Preliminary results showed that the compound was well tolerated, and that the maximum tolerated dose had not been reached. A total of 16 patients showed stable disease after a median duration of 12 weeks. SF-1126 was seen to be well tolerated and all adverse events were non-cumulative and reversible. An additional Phase I trial of SF-1126 for the treatment of multiple myeloma began in January 2008, data from which had not been disclosed at the time of writing.

Genentech, under licence from Piramed, is developing GDC-0941 (13) for the treatment of cancer. In October 2007, an open-labelled, uncontrolled, dose-escalation Phase I trial of GDC-0941 was initiated in patients in the USA with locally advanced or metastatic tumours, for which standard therapy either did not exist or had proven to be ineffective. The safety, tolerability, and pharmacokinetics of escalating oral doses of GDC-0941 administered daily or twice daily were assessed, with the primary endpoints being the occurrence of adverse events and dose-limiting toxicities. In June 2009, clinical data were presented at the 45th ASCO meeting. Results indicated that GDC-0941 displayed favourable pharmacokinetic and safety profiles, was well tolerated, and exhibited signs of anti-tumour activity. In March 2008, an open-labelled, uncontrolled, dose-escalation Phase I trial was initiated in the UK in patients with locally advanced or metastatic cancer, or non-Hodgkin's lymphoma, for which standard therapy either did not exist or had proven to be ineffective. Safety, tolerability and pharmacokinetic properties of escalating oral doses of GDC-0941 were assessed, with the primary endpoints being occurrence of adverse events and dose-limiting toxicities. Clinical studies were ongoing at the time of writing.

Novartis are developing NVP-BEZ235 (19) for the treatment of cancer. In December 2006, an open-label, multicentre Phase I/II trial was initiated in the USA patients with advanced solid malignancies, including breast cancer. In the Phase I portion, subjects received daily oral doses of NVP-BEZ235 on days 1–28. The safety and efficacy Phase II part of the trial began once the maximum tolerated dose of the compound had been reached. In September 2008, data from this trial were presented at the European Federation of Medicinal Chemistry conference in Austria. NVP-BEZ235 was reported to be well tolerated, with no evidence of disturbances to cardiac function, vital signs or glucose homeostasis. In October 2008, further data from this trial were presented at the NCRI meeting in the UK at which it was reported that

NVP-BEZ235 displayed significant anti-angiogenic properties. Prolonged stable disease was observed in a number of patients, including two patients with PIK3CA gene mutations. In January 2007, an additional Phase I trial was initiated in the USA in patients with solid tumours or Cowden Syndrome, and in February 2009, Japanese Phase I trials were expected to begin; it was reported that Novartis are also developing a follow-up compound, BGT-226 (structure undisclosed).

Sanofi-Aventis, under license from Exelixis, are developing XL-147 for the treatment of solid tumours. The Phase I trial was underway by August 2007. The purpose of this non-randomised, uncontrolled, dose-escalation study was to assess the safety and pharmacokinetic properties of XL-147 in patients with solid tumours. The primary endpoints were assessment of safety, tolerability and determination of the maximum tolerated dose; secondary endpoints included determination of pharmacokinetic and pharmacodynamic profiles. In October 2007, preliminary data from six patients in this trial were presented at the 19th AACR-NCI-EORTC conference in the USA. No dose-limiting toxicities or serious adverse events that were considered to be drug-related were reported. The median T_{max} was between 2 and 24 h post-dosing, and a terminal half-life of between 2.5 and 8 days was established; plasma concentrations of the XL-147 reached a steady state by day 20. In October 2008, additional Phase I clinical data were presented at the 20th EORTC-NCI-AACR in Switzerland. Dose-limiting toxicities were observed at both 600 and 900 mg; 900 mg was confirmed as being the maximum administered dose. Five serious adverse events were reported, though none were viewed to be treatment-related. In June 2009, further data were presented at the 45th ASCO meeting; at that time, 48 subjects had received treatment and the maximum tolerated dose was determined to be 600 mg on the 21-days-on/7-days-off dosing schedule. At the time of writing, dose escalation using a continuous dosing schedule was ongoing. At least 10% of the patients treated experienced adverse events of mostly grade 1 or 2; additionally, skin rash was observed in 12 patients, and four subjects experienced grade 3 rashes in the 21/7 dosing schedule.

Sanofi-Aventis are also developing the dual p110α/mTOR inhibitor XL-765 (structure undisclosed) under license from Exelixis for the treatment of solid tumours, including gliomas and non-small cell lung cancer. Previously, it had been reported that Exelixis were developing XL-765, which had reached Phase II in May 2009; however, Exelixis licensed worldwide development and commercialisation rights for the compound to Sanofi-Aventis. In June 2007, a non-randomised, open-label, uncontrolled, dose-escalating, Phase I trial was initiated in patients with solid tumours in the USA. The subjects received daily oral doses of XL-765, with the primary outcome measures including safety, tolerability and the maximum tolerated dose assessment of the compound. In May 2008, Phase I data were presented at the 44th annual ASCO meeting, at which XL-765 was reported to be well tolerated. In October 2008, further Phase I data were presented at the 20th EORTC-NCI-AACR meeting in Switzerland, when it was reported that five of the 28 evaluable patients achieved prolonged stable disease. The most common adverse events were reported to be grade 1 or 2. In June 2009, additional data was presented at the 45th ASCO meeting. At that time, a total of 51 patients had received either between 15 and 120 mg of XL-765 twice daily, or between 70 and 100 mg daily for 28 days. Dose-limiting toxicities were observed when the compound was dosed at 120 and 60 mg bi-daily. At that time, the preliminary maximum tolerated dose was reported to be between 50 mg bi-daily and 90 mg daily. In June 2008, an open-label, uncontrolled, single-group-assignment USA Phase I trial was initiated to determine the safety and tolerability of XL-765 in combination with temozolomide in adults with anaplastic glioma. Patients received a continuous daily dosing of XL-765 together with temozolomide (200 mg/m^2/day for five consecutive days, repeated every 28 days), with the primary outcome measures being safety, tolerability and assessment of the maximum tolerated dose. The trial was estimated to be completed in November 2009. By March 2009, an additional Phase Ib/II trial of XL-765 in combination with erlotinib in patients with non-small cell lung cancer was ongoing.

Scientists at Calistoga Pharmaceuticals, under license from ICOS, are developing CAL-101 (structure undisclosed), the lead compound from a series of orally active p110δ inhibitors that also includes IC-87114, for the treatment of inflammatory diseases and cancer. In March 2008, Calistoga began a European Phase I trial of CAL-101 in healthy volunteers. In June 2008, the company initiated a Phase I trial in the USA; this was a sequential dose-escalation study, in 60 patients with haematologic malignancies, including chronic lymphocytic leukemia, non-Hodgkin lymphoma or acute myeloid leukemia. The subjects received 50, 100, 200 or 375 mg of CAL-101 twice daily for 28 days. The primary endpoints were safety and determination of dose-limiting toxicity. Secondary endpoints included determination of pharmacokinetic parameters, pharmacodynamic effects and the clinical response rate. In May 2009, preliminary clinical data were presented at the 45th ASCO meeting, at which it was reported that CAL-101-treated patients had achieved a partial response, with no severe treatment-related adverse events having been observed. At that time, dose levels were still being evaluated in ongoing cohort-expansion studies, and subsequently, in June 2009, interim results were presented at the European Haematology Association 14th Congress in Germany. CAL-101 showed partial responses in 50% of the patients during the one-month study. The company reported that they planned to initiate a Phase II trial in 2010.

Oncothyreon, following their acquisition of ProlX, are developing PX-866 (9) for the treatment of cancer. In June 2008, a Phase I trial for PX-866 was initiated in 63 patients with advanced metastatic cancer, who had failed standard therapy. The subjects received once-daily doses of the agent on days 1–5 and days 8–12 of a four-week cycle. The primary endpoints included safety and maximum tolerated dose determination, and pharmacodynamic and pharmacokinetic analysis. In May 2009, data from the trial were presented at the 45th ASCO meeting. In advanced solid tumour patients, plasma fibroblast growth factor (FGF-2) levels were seen to decrease during treatment in patients with stable disease, but not in patients with progressive disease, after two cycles of treatment. Stabilisation of disease was observed in patients with progressive disease after administration of the drug.

GlaxoSmithKline are investigating GSK-1059615 (structure undisclosed) for the treatment of cancer. In July 2008, a Phase I open-label, dose-escalation study began in patients with lymphoma or solid tumours, including metastatic breast tumour or endometrial tumours. The primary outcome measures included adverse event assessment, and changes in laboratory values and vital signs. The trial was scheduled to be completed in January 2010.

TargeGen are developing TG100115 (85) for the treatment of acute myocardial infarction. In February 2005, a randomised, double-blind, placebo-controlled, single-group-assignment safety/efficacy Phase I/II trial was initiated in patients undergoing percutaneous coronary intervention for acute anterior ST elevation myocardial infarction in the USA. The study was completed in May 2008.

FUTURE PERSPECTIVE

The development of small molecule, ATP-competitive PI3K inhibitors has emerged as an important therapeutic strategy in the treatment of cancer, cardiovascular disease, and immune and inflammatory disorders. Significant progress has been made over recent years in the design and development of small molecule modulators of the PI3K/AKT/mTOR signalling axis, as evidenced by the growing level of clinical activity in this area. Based on the therapeutic promise shown to date – in particular by the agents highlighted in Table 3.1 – it is clear that the design of small molecule PI3K inhibitors will continue to play an important role in the future development of personalised medicines.

To date, much activity has been focused on the discovery and characterisation of anti-tumour agents that inhibit class IA PI3Ks, in particular p110α and, increasingly, p110δ, whilst p110β, p110δ and p110γ have emerged as important targets in the context of cardiovascular and inflammation therapy. It is anticipated that, over the coming years, we will see further progress in the discovery of novel therapeutics that target these isoforms. Additionally, with a developing knowledge of the structural features of the individual PI3K sub-types, gained through X-ray crystallography and *in-silico* modelling studies [74, 247–254], we should expect to see the development of an increasing number of inhibitors with augmented selectivities both for and within the PI3K superfamily. It is also anticipated that, through our growing understanding of the structural nuances of the PI3K family isoforms, and through the use of more detailed cellular selectivity analysis techniques [76], the design of finely tuned, multi-sub-type selective drugs for use in specific therapeutic areas will emerge as an important pursuit, including, for example, dual p110α/β inhibitors for the treatment of PIK3CA mutant and PTEN-null tumours. There is increasing evidence that several class II and III family members could become important targets in several disease areas, notably oncology, and a great opportunity exists to develop both classes II and III sub-type-specific inhibitors, as well as dual class I/II inhibitors for the treatment of specific tumour types. The use of classes I and IV inhibitors in combination with molecular-targeted cytostatic and cytotoxic agents, and with radiation co-therapy, is gaining momentum in the clinic, and we should expect to see this area grow substantially over the coming years for the treatment of a spectrum of solid tumours and haematological malignancies [254, 255].

REFERENCES

[1]   Stein, R.C. and Waterfield, M.D. (2000) *Mol. Med. Today* 6, 347–358.
[2]   Vivanco, I. and Sawyers, C.L. (2002) *Nat. Rev. Cancer* 2, 489–501.
[3]   Stephens, L., Williams, R. and Hawkins, P. (2005) *Curr. Opin. Pharmacol.* 5, 357–365.
[4]   Ihle, N.T. and Powis, G. (2009) *Mol. Cancer Ther.* 8, 1–9.
[5]   Drees, B.E., Mills, G.B., Rommel, C. and Prestwich, G.D. (2004) *Exp. Opin. Ther. Patents* 14, 703–732.
[6]   Kok, K., Geering, B. and Vanhaesebroeck, B. (2009) *Trends Biochem. Sci.* 648, 115–127.
[7]   Guillermet-Guibert, J., Bjorklof, K., Salpekar, S., Gonella, C., Ramadani, F., Bilancio, A., Meek, S., Smith, A.J.H., Okkenhaug, K. and Vanhaesebroeck, B. (2008) *Proc. Natl. Acad. Sci. USA* 105, 8292–8297.
[8]   Murga, C., Fukuhara, S. and Gutkind, J.S. (2000) *J. Biol. Chem.* 276, 12069–12073.
[9]   Rückle, T., Schwarz, M.K. and Rommel, C. (2006) *Nat. Rev. Drug Discov.* 5, 903–918.
[10] Gaidarov, I., Smith, M.E., Domin, J. and Keen, J.H. (2001) *Mol. Cell* 7, 443–449.
[11] Traer, C.J., Foster, F.M., Abraham, S.M. and Fry, M.J. (2006) *Bull. Cancer* 93, E53–E58.
[12] Fruman, D.A., Meyers, R.E. and Cantley, L.C. (1998) *Annu. Rev. Biochem.* 67, 481–507.
[13] Wurmser, A.E. and Emr, S.D. (2002) *J. Cell Biol.* 158, 761–772.
[14] Backer, J.M. (2008) *Biochem. J.* 410, 1–17.

[15] Nobukuni, T., Joaquin, M., Roccio, M., Dann, S.G., Kim, S.Y., Gulati, P., Byfield, M.
      P., Backer, J.M., Natt, F., Bos, J.L., Zwartkruis, F.J. and Thomas, G. (2005) *Proc. Natl.
      Acad. Sci. USA* **102**, 14238–14243.
[16] Kuruvilla, F.G. and Schreiber, S.L. (1999) *Chem. Biol.* **6**, R129–R136.
[17] Guertin, D.A. and Sabatini, D.M. (2007) *Cancer Cell* **12**, 9–22.
[18] Strimpakos, A.S., Karapanagiotou, E.M., Saif, W.M. and Syrigos, K.N. (2009) *Cancer
      Treat. Rev.* **35**, 148–159.
[19] Guertin, D.A. and Sabatini, D.M. (2009) *Sci. Signal* **2**, pe24.
[20] Samuels, Y., Wang, A., Bardelli, A., Silliman, N., Ptak, J., Szabo, A., Yan, H., Gazdar,
      A., Powell, S.M., Riggins, G.J., Willison, J.K.V., Markowitz, A., Kinzler, K.W.,
      Vogelstein, B. and Velculescu, V.E. (2004) *Science* **304**, 554.
[21] Zhao, J.J., Cheng, H., Jia, S., Wang, L., Gjoerup, O.V., Mikami, A. and Roberts, T.M.
      (2006) *Proc. Nat. Acad. Sci. USA* **103**, 16296–16300.
[22] Gymnopolous, M., Elsliger, M.A. and Vogt, P. (2007) *Proc. Natl. Acad. Sci. USA* **104**,
      5569–5574.
[23] Miled, N., Yan, Y., Hon, W.C., Perisic, O., Zvelebil, M., Inbar, Y., Scheidman-Duhovny,
      D., Wolfson, H.J., Backer, J.M. and Williams, R.L. (2007) *Science* **317**, 239–242.
[24] Vogt, P.K., Kang, S., Elsliger, M.A. and Gymnopoulos, M. (2007) *Trends Biochem. Sci.*
      32, 342–349.
[25] Sun, X., Huang, J., Homma, T., Kita, D., Klocker, H., Schafer, G., Boyle, P. and
      Ohgaki, H. (2009) *Anticancer Res.* **29**, 1739–1744.
[26] Wee, S., Wiederschain, D., Maira, S.M., Loo, A., Miller, C., DeBeaumont, R.,
      Stegmeier, F., Yao, Y.M. and Lengauer, C. (2008) *Proc. Natl. Acad. Sci. USA* **105**,
      13057–13062.
[27] Jia, S., Roberts, T.M. and Zhao, J.J. (2009) *Curr. Opin. Cell Biol.* **21**, 199–208.
[28] Vogt, P.K., Gymnopoulos, M. and Hart, J.R. (2009) *Curr. Opin. Genet. Dev.* **19**, 12–17.
[29] Sujobert, P., Bardet, V., Cornillet-Lefebvre, P., Hayflick, J.S., Prie, N., Verdier, F.,
      Vanhaesebroeck, B., Muller, O., Pesce, F., Ifrah, N., Hunault-Berger, M., Berthou, C.,
      Villemagne, B., Jourdan, E., Audhuy, B., Solary, E., Witz, B., Harousseau, J.L.,
      Himberlin, C., Lamy, T., Lioure, B., Cahn, J.Y., Dreyfus, F., Mayeux, P., Lacombe, C.
      and Bouscary, D. (2005) *Blood* **106**, 1063–1066.
[30] Billottet, C., Banerjee, L., Vanhaesebroeck, B. and Khwaja, A. (2009) *Cancer Res.* **69**,
      1027–1036.
[31] Billottet, C., Grandage, V.L., Gale, R.E., Quattropani, A., Rommel, C., Vanhaeseb-
      roeck, B. and Khwaja, A. (2006) *Oncogene* **25**, 6648–6659.
[32] Cornillet-Lefebvre, P., Cuccuini, W., Bardet, V., Tamburini, J., Gillot, L., Ifrah, N.,
      Nguyen, P., Dreyfus, F., Mayeux, P., Lacombe, C. and Bouscary, D. (2006) *Leukemia*
      20, 374–376.
[33] Sawyer, C., Sturge, J., Bennett, D.C., O'Hare, M.J., Allen, W.E., Bain, J., Jones, G.E.
      and Vanhaesebroeck, B. (2003) *Cancer Res.* **63**, 1667–1675.
[34] Boller, D., Schramm, A., Doepfner, K.T., Shalaby, T., von Bueren, A.O., Eggert, A.,
      Grotzer, M.A. and Arcaro, A. (2008) *Clin. Cancer Res.* **14**, 1172–1181.
[35] Jackson, S.P., Schoenwaelder, S.M., Goncalves, I., Nesbitt, W.S., Yap, C.L., Wright, C.
      E., Kenche, V., Anderson, K.E., Dopheide, S.M., Yuan, Y., Sturgeon, S.A.,
      Prabaharan, H., Thompson, P.E., Smith, G.D., Shepherd, P.R., Daniele, N., Kulkarni,
      S., Abbott, B., Saylik, D., Jones, C., Lu, L., Giuliano, S., Hughan, S.C., Angus, J.A.,
      Robertson, A.D. and Salem, H.H. (2005) *Nat. Med.* **11**, 507–514.
[36] Jackson, S.P., Yap, C.L. and Anderson, K.E. (2004) *Biochem. Soc. Trans.* **32**, 387–392.

[37] Pretorius, L., Owen, K.L. and McMullen, J.R. (2009) *Front. Biosci.* 14, 2221–2229.
[38] Utsugi, M., Dobashi, K., Ono, A., Ishizuka, T., Matsuzaki, S., Hisada, T., Shimizu, Y., Kawata, T., Aoki, H., Kamide, Y. and Mori, M. (2009) *J. Immunol.* 182, 5225–5231.
[39] Ciraolo, E., Iezzi, M., Marone, R., Marengo, S., Curcio, C., Costa, C., Azzolino, O., Gonella, C., Rubinetto, C., Wu, H., Dastrù, W., Martin, E.L., Silengo, L., Altruda, F., Turco, E., Lanzetti, L., Musiani, P., Rückle, T., Rommel, C., Backer, J.M., Forni, G., Wymann, M.P. and Hirsch, E. (2008) *Sci. Signal* 1, ra3.
[40] Vanhaesebroeck, B., Welham, M.J., Kotani, K., Stein, R., Warne, P.H., Zvelebil, M.J., Higashi, K., Volinia, S., Downward, J. and Waterfield, M.D. (1997) *Proc. Natl. Acad. Sci. USA* 94, 4330–4335.
[41] Okkenhaug, K., Bilancio, A., Farjot, G., Priddle, H., Sancho, S., Peskett, E., Pearce, W., Meek, S.E., Salpekar, A., Waterfield, M.D., Smith, A.J.H. and Vanhaesebroeck, B. (2002) *Science* 297, 1031–1034.
[42] Ali, K., Bilancio, A., Thomas, M., Pearce, W., Gilfillan, A.M., Tkaczyk, C., Kuehn, N., Gray, A., Giddings, J., Peskett, E., Fox, R., Bruce, I., Walker, C., Sawyer, C., Okkenhaug, K., Finan, P. and Vanhaesebroeck, B. (2004) *Nature (London)* 431, 1007–1011.
[43] Sadhu, C., Dick, K., Tino, W.T. and Staunton, D.E. (2003) *Biochem. Biophys. Res.* 308, 764–769.
[44] Sadhu, C., Masinovsky, B., Dick, K., Sowell, C.G. and Staunton, D.E. (2003) *J. Immunol.* 170, 2647–2654.
[45] Papakonstanti, E.A., Zwaenepoel, O., Bilancio, A., Burns, E., Nock, G.E., Houseman, B., Shokat, K., Ridley, A.J. and Vanhaesebroeck, B. (2008) *J. Cell Sci.* 121, 4124–4133.
[46] Bilancio, A., Okkenhaug, K., Camps, M., Emery, J.L., Ruckle, T., Rommel, C. and Vanhaesebroeck, B. (2006) *Blood* 107, 642–650.
[47] Hebels, B.J., Vigorito, E. and Turner, M. (2004) *Biochem. Soc. Trans.* 32, 789–791.
[48] Kim, N., Saudemont, A., Webb, L., Camps, M., Ruckle, T., Hirsch, E., Turner, M. and Francesco Colucci, F. (2007) *Blood* 110, 3202–3208.
[49] Lee, K.S., Lee, H.K., Hayflick, J.S., Lee, Y.C. and Puri, K.D. (2006) *FASEB J.* 20, 433–465.
[50] Kwak, Y.G., Song, C.H., Yi, H.K., Hwang, P.H., Kim, J.S., Lee, K.S. and Lee, Y.C. (2003) *J. Clin. Invest.* 111, 1083–1092.
[51] Randis, T.M., Puri, K.D., Zhou, H. and Diacovo, T.G. (2008) *Eur. J. Immunol.* 38, 1215–1224.
[52] Laffargue, M., Calvez, R., Finan, P., Trifilieff, A., Barbier, M., Altruda, F., Hirsch, E. and Wymann, M.P. (2002) *Immunity* 16, 441–451.
[53] Pinho, V., Souza, D.G., Barsante, M.M., Hamer, F.P., De Freitas, M.S., Rossi, A.G. and Teixeir, M.M. (2005) *J. Leukoc. Biol.* 77, 800–810.
[54] Del Prete, A., Vermi, W., Dander, E., Otero, K., Barberis, L., Luini, W., Bernasconi, S., Sironi, M., Santoro, A., Garlanda, C., Facchetti, F., Wymann, M.P., Vecchi, A., Hirsch, E., Mantovani, A. and Sozzani, S. (2004) *EMBO J.* 23, 3505–3515.
[55] Reif, K., Okkenhaug, K., Sasaki, T., Penninger, J.M., Vanhaesebroeck, B. and Cyster, J.G. (2004) *J. Immunol.* 173, 2236–2240.
[56] Condliffe, A.M., et al. (2005) *Blood* 106, 150–157.
[57] Rommel, C., Camps, M. and Ji, H. (2007) *Nat. Rev. Immunol.* 7, 191–201.
[58] Kang, S., Song, J., Kang, J., Kang, H., Lee, D., Lee, Y. and Park, D. (2005) *Biochem. Biophys. Res. Commun.* 329, 6–10.

[59] Elis, W., Triantafellow, E., WOlters, N.M., Sian, K.R., Caponigro, G., Borawski, J., Gaither, L.A., Murphy, L.O., FInan, P.M. and MacKeigan, J.P. (2008) *Mol. Cancer Res.* **6**, 614–623.
[60] Wen, P.J., Osborne, S.L., Morrow, I.C., Parton, R.G., Domin, J. and Meunier, F.A. (2008) *Mol. Biol. Cell* **19**, 5593–5603.
[61] Arcaro, A., Khanzada, U.K., Vanhaesebroeck, B., Tetley, T.D., Waterfield, M.D. and Seckl, M.J. (2002) *EMBO J.* **21**, 5097–5108.
[62] Maffuci, T., Cooke, F.T., Foster, F.M., Traer, C.J. and Fry, M.J. (2006) *J. Cell Biol.* **169**, 789–799.
[63] Parkinson, G., Vines, D., Driscoll, P.C. and Djordjevic, S. (2008) *BMC Struct. Biol.* **8**, 1–9.
[64] Dann, S.G. and Thomas, G. (2006) *FEBS Lett.* **580**, 2821–2829.
[65] Gulati, P. and Thomas, G. (2007) *Biochem. Soc. Trans.* **35**, 236–238.
[66] Nobukini, T., Kozma, S.C. and Thomas, G. (2007) *Curr. Opin. Cell Biol.* **19**, 135–141.
[67] Hipper, M.M., O'Toole, P.S. and Thorburn, A. (2006) *Cancer Res.* **66**, 9349–9351.
[68] Lane, H.A. and Breuleux, M. (2009) *Curr. Opin. Cell Biol.* **21**, 1–11.
[69] Ma, W.W. and Hidalgo, M. (2007) *World J. Gastroenterol.* **13**, 5845–5856.
[70] Jimeno, A., Tan, A.C., Coffa, J., Rajeshkumar, N.V., Kulesza, P., Rubio-Viquira, B., Wheelhouse, J., Diosdado, B., Messersmith, W.A., Maitra, A., Iacobuzio-Donahue, C., Meijer, G.A., Garcia, M.L., Tan, A.C., Varela-Garcia, M., Hirsch, F., Gerrit, M. and Hidalgo, M. (2008) *Cancer Res.* **68**, 2841–2849.
[71] Ma, W.W. and Adjei, A.A. (2009) *CA Cancer J. Clin* **59**, 113–137.
[72] Smith, G.C.M. and Jackson, S.P. (2004) *Handbook of Cell Signaling* **1**, 557–561.
[73] Vlahos, C.J., Matter, W.F., Hui, K.Y. and Brown, R.F. (1994) *J. Biol. Chem.* **269**, 5241–5248.
[74] Walker, E.H., Pacold, M.E., Perisic, O., Stephens, L., Hawkins, P., Wymann, M. and Williams, R.L. (2000) *Mol. Cell* **6**, 909–919.
[75] Kong, D. and Yamori, T. (2008) *Cancer Sci.* **99**, 1734–1740.
[76] Gharbi, S., Zvelebil, M.J., Shuttleworth, S.J., Hancox, T., Saghir, N., Timms, J.F. and Waterfield, M.D. (2007) *Biochem. J.* **404**, 15–21.
[77] Davies, S.P., Reddy, H., Calvano, M. and Cohen, P. (2000) *Biochem. J.* **351**, 95–105.
[78] Abbott, B.M. and Thompson, P.E. (2004) *Bioorg. Med. Chem. Lett.* **14**, 2847–2851.
[79] Jacobs, M.D., Black, J., Futer, O., Swenson, L., Hare, B., Fleming, M. and Saxena, K. (2005) *J. Biol. Chem.* **280**, 13728–13734.
[80] Abbott, B. and Thompson, P. (2003) *Aust. J. Chem.* **56**, 1099–1106.
[81] Chiosis, G., Rosen, N. and Sepp-Lorenzino, L. (2001) *Bioorg. Med. Chem. Lett.* **11**, 909–913.
[82] Garlich, J.R., Durden, D.L., Patterson, M., Su, J. and Suhr, R.G. (2004). *PCT Int. Appl.*, WO 04 089925.
[83] Garlich, J.R., De, P., Dey, N., Su, J., Peng, X., Miller, A., Murali, R., Lu, Y., Mills, G. B., Kundra, V., Shu, H.K., Peng, Q. and Durden, D.L. (2008) *Cancer Res.* **68**, 206–215.
[84] Berrie, C.P. (2001) *Expert Opin. Investig. Drugs* **10**, 1085–1098.
[85] Wymann, M.P., Bulgarelli-Leva, G., Zvelebil, M.J., Pirola, L., Vanhaesebroeck, B., Waterfield, M.D. and Panayotou, G. (1996) *Mol. Cell Biol.* **16**, 1722–1733.
[86] Zask, A., Kaplan, J., Toral-Barza, L., Hollander, I., Young, M., Tischler, M., Gaydos, C., Cinque, M., Lucas, J. and Yu, K. (2008) *J. Med. Chem.* **51**, 1319–1323.
[87] Yu, K., Lucas, J., Zhu, T.M., Zask, A., Gaydos, C., Toral-Barza, L., Gu, J.X., Li, F.B., Chaudhary, I., Cai, P., Lotvin, J., Petersen, R., Ruppen, M., Fawzi, M., Ayral-

Kaloustian, J., Skotnicki, J., Mansour, T., Frost, P. and Gibbons, J. (2005) *Cancer Biol. Ther.* **4**, 538–545.

[88]   Zask, A., Gu, J., Cai, P., Yu, K., Kaplan, J., Gilbert, A.M. and Bursavich, M.G. (2006) *PCT Int. Appl.* WO 06 044453.

[89]   Zhu, T., Gu, J., Yu, K., Lucas, J., Cai, P., Tsao, R., Gong, Y., Li, F., Chaudhary, I., Desai, P., Ruppen, M., Fawzi, M., Gibbons, J., Ayral-Kaloustian, S., Skotnicki, J., Mansour, T. and Zask, A. (2006) *J. Med. Chem.* **49**, 1373–1378.

[90]   Ihle, N.T., Williams, R., Chow, S., Chew, W., Berggre, M.I., Paine-Murrieta, G., Minion, D.J., Halter, R.J., Wipf, P., Abraham, R., Kirkpatrick, L. and Powis, G. (2004) *Mol. Cancer Ther.* **3**, 763–772.

[91]   Wipf, P., Minion, D.J., Halter, R.J., Berggren, M.I., Ho, C.B., Chiang, G.G., Kirkpatrick, L., Abraham, R. and Powis, G. (2004) *Org. Biomol. Chem.* **2**, 1911–1920.

[92]   Kirkpatrick, L., Powis, G. and Wipf, P. (2007) *PCT Int. Appl.* WO 07 008200.

[93]   Yuan, H., Luo, J., Weissleder, R., Cantley, L. and Josephson, L. (2006) *J. Med. Chem.* **49**, 740–747.

[94]   Yuan, H., Luo, J., Weissleder, R., Cantley, L. and Josephson, L. (2007) *PCT Int. Appl.* WO 07 086943.

[95]   Hayakawa, M., Kaizawa, H., Moritomo, H., Kawaguchi, K., Koizumi, T., Yamano, M., Matsuda, K., Okada, M. and Ohta, M. (2001) *PCT Int. Appl.* WO 01 083456.

[96]   Hayakawa, M., Kaizawa, H., Moritomo, H., Koizumi, T., Ohishi, T., Okada, M., Ohta, M., Tsukamoto, S., Parker, P., Workman, P. and Waterfield, M. (2006) *Bioorg. Med. Chem.* **14**, 6847–6858.

[97]   Hayakawa, M., Kaizawa, H., Moritomo, H., Koizumi, T., Ohishi, T., Yamano, M., Okada, M., Ohta, M., Tsukamoto, S., Raynaud, F.I., Workman, P., Waterfield, M. and Parker, P. (2007) *Bioorg. Med. Chem. Lett.* **17**, 2438–2442.

[98]   Knight, Z.A., Gonzalez, B., Feldman, M.E., Zunder, E.R., Goldenberg, D.D., Williams, O., Loewith, R., Stokoe, D., Balla, A., Toth, B., Bala, T., Weiss, W.A., Williams, R.L. and Shokat, K.M. (2006) *Cell* **125**, 5840–5850.

[99]   Park, S., Chapuis, N., Bardet, V., Tamburini, J., Gallay, N., Willems, L., Knight, Z.A., Shokat, K.M., Azar, N., Viguie, F., Ifrah, N., Dreyfus, F., Mayeux, P., Lacombe, C. and Bouscary, D. (2008) *Leukemia* **22**, 1698–1706.

[100] Fan, Q.W., Cheng, C.K., Nicolaides, T.P., Hackett, C.S., Knight, Z.A., Shokat, K.M. and Weiss, W.A. (2007) *Cancer Res.* **67**, 7960–7965.

[101] Chen, J.S., Zhou, L.J., Entin-Meer, M., Yang, X., Donker, M., Knight, Z.A., Weiss, W., Shokat, K.M., Haas-Kogan, D. and Stokoe, D. (2008) *Mol. Cancer Ther.* **7**, 841–850.

[102] Chaisuparat, R., Hu, J., Jham, B.C., Knight, Z.A., Shokat, K.M. and Montaner, S. (2008) *Cancer Res.* **68**, 8361–8368.

[103] Shuttleworth, S.J., Folkes, A.J., Chuckowree, I.S., Wan, N.C., Hancox, T.C., Baker, S.J., Sohal, S. and Latif, M.A. (2006) *PCT Int. Appl.* WO 06 046031.

[104] Folkes, A.J., Ahmadi, K., Alderton, W.K., Alix, S., Baker, S.J., Box, G., Chuckowree, I.S., Clarke, P.A., Depledge, P., Eccles, S.A., Friedman, L.S., Hayes, A., Hancox, T.C., Kugendradas, A., Lensun, L., Moore, P., Olivero, A.G., Pang, J., Patel, S., Pergl-Wilson, G.H., Raynaud, F.I., Robson, A., Saghir, N., Salphati, L., Sohal, S., Ultsch, M.H., Valenti, M., Wallweber, H.J.A., Wan, N.C., Wiesmann, C., Workman, P., Zhyvoloup, A., Zvelebil, M.J. and Shuttleworth, S.J. (2008) *J. Med. Chem.* **51**, 5522–5532.

[105] Castanedo, G., Goldsmith, R., Gunzner, J., Heffron, T., Malesky, K., Mathieu, S., Olivero, A., Sutherlin, D. P., Tsui, V., Wang, S., Wiesman, C., Zhu, B. Y. Shuttleworth,

S.J., Folkes, A.J., Oxenford, S., Hancox, T. and Bayliss, T. (2007) *PCT Int. Appl.* WO 07 127183.

[106]   Shuttleworth, S.J., Folkes, A.J., Chuckowree, I.S., Wan, N.C., Hancox, T.C., Baker, S. J., Sohal, S. and Latif, M.A. (2006) *PCT Int. Appl.* WO 06 046040.

[107]   Raynaud, F.I., Eccles, S.A., Patel, S., Alix, S., Box, G., Chuckowree, I., Folkes, A., Gowan, S., De Haven-Brandon, A., Di Stefano, F., Hayes, A., Henley, A.T., Lensun, L., Pergl-Wilson, G., Robson, A., Saghir, N., Zhyvoloup, A., McDonald, E., Sheldrake, P., Shuttleworth, S., Valenti, M., Wan, N.C., Clarke, P.A. and Workman, P. (2009) *Mol. Can. Ther.* **8**, 1725–1738.

[108]   Chuckowree, I.S., Folkes, A.J., Goldsmith, P., Hancox, T.C. and Shuttleworth, S.J. (2007) *PCT Int. Appl.* WO 07 132171.

[109]   Stauffer, F., Maira, S.M., Furet, P. and Garcia-Echeverria, C. (2008) *Bioorg. Med. Chem. Lett.* **18**, 1027–1030.

[110]   Brachmann, S.M., Finan, P., Fritsch, C., Garcia-Echeverria, C., Maira, S.M., Murphy, L. and Nicklin, P.L. (2008) *PCT Int. Appl.* WO 08 103636.

[111]   Maira, S.M., Stauffer, F., Brueggen, J., Furet, P., Schnell, C., Fritsch, C., Brachmann, S., Chene, P., De Pover, A., Schoemaker, K., Fabbro, D., Gabriel, D., Simonen, M., Murphy, L., Finan, P., Sellers, W. and Garcia-Echeverria, C. (2008) *Mol. Cancer Ther.* **7**, 1851–1863.

[112]   Serra, V., Markman, B., Scaltriti, M., Eichhorn, P.J.A., Valero, V., Guzman, M., Botero, M.L., Llonch, E., Atzori, F., Di Cosimo, S., Maira, M., Garcia-Echeverria, C., Parra, J.L., Arribas, J. and Baselga, J. (2008) *Cancer Res.* **68**, 8022–8030.

[113]   Cao, P., Maira, S.M., Garcia-Echeverria, C. and Hedley, D.W. (2009) *Br. J. Cancer* **100**, 1267–1276.

[114]   Yamori, T., Matsunaga, A., Sato, S., Yamazaki, K., Komi, A., Ishizu, K., Mita, I., Edatsugi, H., Matsuba, Y., Takezawa, K., Nakanishi, O., Kohno, H., Nakajima, Y., Komatsu, H., Andoh, T. and Tsuruo, T. (1999) *Cancer Res.* **59**, 4042–4049.

[115]   Dan, S., Tsunoda, T., Kitahara, O., Yanagawa, R., Zembutsu, H., Katagiri, T., Yamazaki, K., Nakamura, Y. and Yamori, T. (2002) *Cancer Res.* **62**, 1139–1147.

[116]   Kong, D., Okamura, M., Yoshimi, H. and Yamori, T. (2009) *Eur. J. Cancer* **45**, 857–865.

[117]   Yaguchi, S.I., Fukui, L., Koshimizu, K., Yoshimi, H., Matsuno, T., Gouda, H., Hirono, S., Yamazaki, K. and Yamori, T. (2006) *J. Natl. Cancer Inst.* **98**, 545–556.

[118]   Kong, D. and Yamori, T. (2007) *Cancer Sci.* **98**, 1638–1642.

[119]   Nakatsu, N., Nakamura, T., Yamazaki, K., Sadahiro, S., Makuuchi, H., Kanno, J. and Yamori, T. (2007) *Mol. Pharmacol.* **72**, 1171–1180.

[120]   Kawashima, S., Matsuno, T., Yaguchi, S., Tsuchida, Y., Saitoh, K. and Watanabe, T. (2005) *PCT Int. Appl.* WO 05 095389.

[121]   Kong, D., Yaguchi, S. and Yamori, T. (2009) *Biol. Pharm. Bull.* **32**, 297–300.

[122]   Akashi, T. and Yamori, T. (2009) *Proteome Sci.* **7**, 14.

[123]   Hayakawa, M., Kaizawa, H., Moritomo, H., Kawaguchi, K., Koizumi, T., Yamano, M., Matsuda, K., Okada, M. and Ohta, M. (2002) *PCT Int. Appl.* US151544.

[124]   Hayakawa, M., Kaizawa, H., Kawaguchi, K., Ishikawa, N., Koizumi, T., Ohishi, T., Yamano, M., Okada, M., Ohta, M., Tsukamoto, S., Raynaud, F.I., Waterfield, M.D., Parker, P. and Workman, P. (2007) *Bioorg. Med. Chem.* **15**, 403–412.

[125]   Hayakawa, M., Kawaguchi, K.I., Kaizawa, H., Tomonobu, K., Ohishi, T., Yamano, M., Okada, M., Ohta, M., Tsukamoto, S., Raynaud, F.I., Parker, P., Workman, P. and Waterfield, M.D. (2007) *Biorg. Med. Chem.* **15**, 5837–5844.

[126] Kendall, J.D., Rewcastle, G.W., Frederick, R., Mawson, C., Denny, W.A., Marshall, E. S., Baguley, B.C., Chaussade, C., Jackson, S.P. and Shepherd, P.R. (2007) *Biorg. Med. Chem.* **15**, 7677–7687.

[127] Alexander, R.P., Bailey, S., Brand, S., Brookings, D.C., Brown, J.A., Haughan, A.F., Kinsella, N., Lowe, C., Mack, S.R., Pitt, W.R., Richard, M.D., Sharpe, A. and Tait, L.J. (2007) *PCT Int. Appl.* WO 07 141504.

[128] Buckley, G.M., Morgan, T. and Sabin, V.M. (2008) *PCT Int. Appl.* WO 08 044022.

[129] Alexander, R.P., Aujla, P.S., Crépy, K.V.L., Foley, A.M. and Franklin, R.J. (2009) *PCT Int. Appl.* WO 09 001089.

[130] Alexander, R.P., Aujla, P.S., Crépy, K.V.L., Foley, A.M., Franklin, R.J., Haughan, A. F., Horlsey, H.T., Jones, W.M., Lallemand, B.I.L.F., Mack, S.R., Morgan, T., Pasau, M.G., Phillips, D.J., Sabin, V., Buckley, G.M., Jenkins, K. and Perry, B.J. (2008) *PCT Int. Appl.* WO 08 001076.

[131] Alexander, R., Balasundaram, A., Batchelor, M., Brookings, D., Crepy, K., Crabbe, T., Deltent, M.F., Driessens, F., Gill, A., Harris, S., Hutchinson, G., Kulisa, C., Merriman, M., Mistry, P., Parton, T., Turner, J., Whitcombe, I. and Wright, S. (2008) *Bioorg. Med. Chem. Lett.* **18**, 4316–4320.

[132] Capraro, H.-G., Carvatti, G., Furet, P., Imbach, P., Lan, J., Pecchi, S. and Schoepfer, J. (2008) *PCT Int. Appl.* WO 08 138889.

[133] Shokat, K.M., Knight, Z. and Aspel, B. (2007) *PCT Int. Appl.* WO 07 114926.

[134] Aspel, B., Blair, J.A., Gonzalez, B., Nazif, T.M., Feldman, M.E., Aizenstein, B., Hoffman, R., Wiliams, R.L., Shokat, K.M. and Knight, Z.A. (2008) *Nat. Chem. Biol.* **4**, 691–699.

[135] Zask, A., Nowak, P.W., Verheijen, J., Curran, K.J., Kaplan, J., Malwitz, D., Bursavich, M.G., Cole, D.C., Ayral-Kaloustian, S., Yu, K., Richard, D.J. and Lefever, M. (2008) *PCT Int. Appl.* WO 08 115974.

[136] Ren, P., Liu, Y. and Wilson, T.E. (2009) *PCT Int. Appl.* WO 09 046448.

[137] Booker, S., D'Angelo, N., D'Amico, D.C., Kim, T., Liu, L., Meagher, K., Norman, M. H., Panter, K., Schenkel, L.B., Smith, A., Tamayo, N., Whittington, D.A., Xi, N. and Yang, K. (2009) *PCT Int. Appl.* WO 09 017822.

[138] McDonald, E., Large, J., Folkes, A., Shuttleworth, S.J. and Wan, N.C. (2007) *PCT Int. Appl.* WO 07 042806.

[139] Nuss, J.M., Pecchi, S. and Renhowe, P.A. (2004) *PCT Int. Appl.* WO 04 048365.

[140] Bailey, J.P., Giles, M.B. and Pass, M. (2006) *PCT Int. Appl.* WO 06 005914.

[141] Pike, K.G., Finlay, M.R.V., Fillery, S.M. and Dishington, A.P. (2007) *PCT Int. Appl.* WO 07 080382.

[142] Finlay, M.R.V., Morris, J. and Pike, K.G. (2008) *PCT Int. Appl.* WO 08 023159.

[143] Finlay, M.R.V. (2008) *PCT Int. Appl.* WO 08 023180.

[144] Morris, J.J. and Pike, K.G. (2009) *PCT Int. Appl.* WO 09 007748.

[145] Finlay, M.R.V. (2009) PCT Int. Appl. WO 09 007749.

[146] Pike, K.G. (2009) *PCT Int. Appl.* WO 09 007750.

[147] Finlay, M.R.V. and Pike, K.G. (2009) *PCT Int. Appl.* WO 09 007751.

[148] Pass, M. (2006) *PCT Int. Appl.* WO 06 005915.

[149] Pass, M. (2006) *PCT Int. Appl.* WO 06 005918.

[150] Bajjalieh, W., Bannen, L.C., Brown, D., Kearney, P., Mac, M., Marlowe, C.K., Nuss, J. M., Tesfai, Z., Wang, Y. and Xu, Wei. (2007) *PCT Int. Appl.* WO 07 044729.

[151] Buhr, C.A., Bajjalieh, W., Joshi, A.A., Lara, K., Ma, S., Marlowe, C.K., Wang, L. and Yeung, B.K.S. (2008) *PCT Int. Appl.* WO 08 127678.

[152] Kumar, S., Joshi, K.S., Deore, V., Bhonde, M.R., Yewalkar, N.N., Padgaonkar, A.A., Rathos, M.J., Kulkarni-Almeida, A.A., Parikh, S. and Dagia, N.M. (2009) *PCT Int. Appl.* WO 09 019656.

[153] Cheng, H., Bhumralkar, D., Dress, K.R., Hoffman, J.E., Johnson, M.C., Kania, R.S., Le, P.T.Q, Nambu, M.D., Pairish, M.A., Plewe, M.B. and Tran, K.T. (2008) *PCT Int. Appl.* WO 08 032162.

[154] Bruendl, M.M., Gogliotti, R.D., Goodman, A.P. and Reichard, G. (2005) *PCT Int. Appl.* WO 05 105801.

[155] Bengtsson, M., Larsson, J., Nikitidis, G., Storm, P., Bailey, J.P., Griffen, E.J., Arnould, J.C. and Bird, T.G.C. (2006) *PCT Int. Appl.* WO 06 051270.

[156] Arnould, J.C., Foote, K.M. and Griffen, E.J. (2007) *PCT Int. Appl.* WO 07 129044.

[157] David, L., Foote, K.M. and Lisius, A. (2007) *PCT Int. Appl.* WO 07 129052.

[158] Foote, K.M. and Griffen, E.J. (2007) *PCT Int. Appl.* WO 07 135398.

[159] Melese, T., Perkins, E.L., Nguyen, A.T.Q. and Sun, D. (2003) *PCT Int. Appl.* WO 034997.

[160] Melese, T., Perkins, E.L., Nguyen, A.T.Q. and Sun, D. (2003) *PCT Int. Appl.* WO 03 035618.

[161] Drees, B., Chakravarty, L., Prestwich, G.D., Dorman, G., Kavecz, M., Lukacs, A., Urge, L. and Darvas, F. (2005) *PCT Int. Appl.* WO 05 002514.

[162] Breitfelder, S., Maier, U., Brandl, T., Hoenke, C., Grauert, M., Pautsch, A., Hoffmann, M., Kalkbrenner, F., Joergensen, A., Schaenzle, G., Peters, S., Büttner, F. and Bauer, E. (2006) *PCT Int. Appl.* WO 06 040279.

[163] Giovannini, R., Frattini, S., Brandl, T., Breitfelder, S., Cereda, E., Grauert, M., Hoffmann, M., Joergensen, A., Maier, U., Pautsch, A., Quai, M. and Scheuerer, S. (2008) *PCT Int. Appl.* WO 08 092831.

[164] Hölder, S., Vennemann, M., Beneke, G., Zülch, A., Gekeler, V., Beckers, T., Zimmerman, A. And Joshi, H. (2009) *PCT Int. Appl.* WO 09 021992.

[165] Hölder, S., Zülch, A., Bär, T., Maier, T., Mermann, A., Beckers, T., Gekeler, V., Joshi, H. Munot, Y., Bhise, U., Chavan, S., Shivatare, S., Patel, S. and Gore, V. (2009) *PCT Int. Appl.* WO 09 021990.

[166] Darcy, M.G., Knight, S.D., Adams, N.D. and Schmidt, S.J. (2007) *PCT Int. Appl.* WO 07 136940.

[167] Felding, J., Pedersen, H. C., Krog-Jensen, C., Praestergaard, M., Butcher, S.P., Linde, V., Coulter, T.S., Montalbetti, C., Uddin, M. and Reigner, S. (2005) *PCT Int. Appl.* WO 05 097107.

[168] Ding, H.S., Zhang, C., Wu, X.H., Yang, C.H., Zhang, X.W., Ding, J. and Xie, Y.Y. (2005) *Bioorg. Med. Chem. Lett.* **15**, 4799–4802.

[169] Zhang, C., Yang, N., Yang, C.-H., Ding, H.-S., Luo, C., Zhang, Y., Wu, M.J., Zhang, X.-w., Shen, X., Jiang, H.-L., Meng, L.H. and Ding, J. (2009) *PLoS One* **4**(3), e4881.

[170] Chen, I.L., Chen, Y.L., Tzeng, C.C. and Chen, I.S. (2002) *Helv. Chim. Acta* **85**, 2214–2221.

[171] Lal, B., Bhise, N.B., Gidwani, R.M., Lakdawala, A.D., Joshi, K. and Patvardhan, S. (2005) *Arkivoc* 77–97.

[172] Lohar, M.V., Mundada, R., Bhonde, M., Padgaonkar, A., Deore, V., Yewalkar, N., Bhatia, D., Rathos, M., Joshi, K., Vishwakarma, R.A. and Kumar, S. (2008) *Bioorg. Med. Chem. Lett.* **18**, 3603–3606.

[173] Yang, J., Shamji, A., Matchacheep, S. and Schreiber, S.L. (2007) *Chem. Biol.* **14**, 371–377.

[174] Marion, F., Williams, D.E., Patrick, B.O., Hollander, I., Mallon, R., Kim, S.C.,
       Roll, D.M., Feldberg, L., Van Soest, R. and Andersen, R.J. (2006) *Org. Lett.* **8**,
       321–324.
[175] Andersen, R., Hollander, I., Roll, D.M., Kim, S.C., Mallon, R.G., Williams, D.E. and
       Marion, F. (2006) *PCT Int. Appl.* WO 06 081659.
[176] Robertson, A.D., Jackson, S., Kenche, A., Yaip, C., Parbaharan, H. and Thompson, P.
       (2001) *PCT Int. Appl.* WO 01 53266.
[177] Jackson, S., Robertson, A.D., Kenche, V., Thompson, P., Parbaharan, H., Andersen,
       K., Abbott, B., Goncalves, I., Nesbitt, W., Schoenwaelder, S., Saylik, D. and Yaip, C.
       (2004) *PCT Int. Appl.* WO 04 016607.
[178] Chaussade, C., Rewcastle, G.W., Kendall, J.D., Denny, W.A., Cho, K., Grønning, L.
       M., Chong, M.L., Anagnostou, S.H., Jackson, S.P., Daniele, N. and Shepherd, P.R.
       (2007) *Biochem J.* **404**, 449–458.
[179] Straub, A., Wendel, H.P., Dietz, K., Schiebold, D., Peter, K., Schoenwaelder, S.M. and
       Ziemer, G. (2008) *Thromb. Haemost.* **99**, 609–615.
[180] Sturgeon, S.A., Jones, C., Angus, J.A. and Wright, C.E. (2008) *Eur. J. Pharmacol.* **587**,
       209–215.
[181] Frazzetto, M., Suphioglu, C., Zhu, J., Schmidt-Kittler, O., Jennings, I.G., Cranmer, S.
       L., Jackson, S.P., Kinzler, K.W., Vogelstein, B. and Thompson, P.E. (2008) *Biochem J.*
       414, 383–390.
[182] Puri, K.D. (2006) *Curr. Enzyme Inhib.* **2**, 147–161.
[183] Okkenhaug, K. and Vanhaesebroeck, B. (2003) *Nat. Rev. Immunol.* **3**, 317–330.
[184] Duan, W., Aguinaldo Datiles, A.M.K., Leung, B.P., Vlahos, C.J. and Wong, W.S.F.
       (2005) *Int. Immunopharmacol.* **5**, 495–502.
[185] Ito, K., Caramori, G. and Adcock, I.M. (2007) *J. Pharmacol. Exp. Ther.* **321**, 1–8.
[186] Douangpanya, J., Hayflick, J.S. and Puri, K.D. (2005) *PCT Int. Appl.* WO 05 016348.
[187] Diacovo, T.G., Haflick, J.S. and Puri, K.D (2005) *PCT Int. Appl.* WO 05 016349.
[188] Hallahan, D., Hayflick, J.S. and Sadhu, C. (2005) *PCT Int. Appl.* WO 05 112935.
[189] Fowler, K.W., Huang, D., Kesicki, E.A., Ooi, H.C., Oliver, A.R., Ruan, F. and
       Treiberg, J. (2005) *PCT Int. Appl.* WO 05 113556.
[190] Ali, K., Camps, M., Pearce, W.P., Ji, H., Rückle, T., Kuehn, N., Pasquali, C., Chabert,
       C., Rommel, C. and Vanhaesebroeck, B. (2009) *J. Immunol.* **180**, 2538–2544.
[191] Puri, K.D., Doggett, T.A., Douangpanya, J., Hou, Y., Tino, W.T., Wilson, T., Graf, T.,
       Clayton, E., Turner, M., Hayflick, J.S. and Diacovo, T.G. (2004) *Blood* **103**, 3448–3456.
[192] Geng, L., Tan, J., Himmelfarb, E., Schueneman, A., Niermann, K., Fu, A., Cuneo, K.,
       Kesicki, E.A., Treiberg, J., Hayflick, J.S. and Hallahan, D.E. (2004) *Cancer Res.* **64**,
       4893–4899.
[193] Northcott, C.A., Hayflick, J.S. and Watts, S.W. (2004) *Hypertension* **43**, 885–890.
[194] Shuttleworth, S.J., Folkes, A.J., Chuckowree, I.S., Wan, N.C., Hancox, T.C., Baker, S.
       J., Sohal, S. and Latif, M.A. (2006) *PCT Int. Appl.* WO 06 046035.
[195] McDonald, E., Large, J. and Shuttleworth, S.J. (2007) *PCT Int. Appl.* WO 07 042810.
[196] Cushing, T.D., Hao, X., He, X., Reichelt, A., Rzasa, R.M., Seganish, J., Shin, Y. and
       Zhang, D. (2008) *PCT Int. Appl.* WO 08 118454.
[197] Cushing, T.D., Hao, X., He, X., Reichelt, A., Rzasa, R.M., Seganish, J., Shin, Y. and
       Zhang, D. (2008) *PCT Int. Appl.* WO 08 118454.
[198] White, S.L. (2008) *PCT Int. Appl.* WO 08 064018.
[199] Perry, B., Alexander, R., Bennett, G., Buckley, G., Ceska, T., Crabbe, T., Dale, V.,
       Gowers, L., Horsley, H., James, L., Jenkins, K., Crepy, K., Kulisa, C., Lightfoot, H.,

STEPHEN SHUTTLEWORTH ET AL. 129

Lock, S., Mack, S., Morgan, T., Nicolas, A.L., Pitt, W., Sabin, V. and Wright, S. (2008) *Bioorg. Med. Chem. Lett.* **18**, 4700–4704.
[200] Perry, B., Beevers, R., Bennett, G., Buckley, G., Crabbe, T., Gowers, L., James, L., Jenkins, K., Lock, C., Sabin, V. and Wright, S. (2008) *Bioorg. Med. Chem. Lett.* **18**, 5299–5302.
[201] Doukas, J., Eide, L., Stebbins, K., Racanelli-Layton, A., Dellamary, L., Martin, M., Dneprovskaia, E., Noronha, G., Soll, R., Wrasidlo, W., M. Acevedo, L.M. and Cheresh, D.A. (2009) *J. Pharmacol. Exp. Ther.* **328**, 758–765.
[202] Doukas, J., Wrasidlo, W., Noronha, G., Dneprovskaia, E., Fine, R., Weis, S., Hood, J., DeMaria, A., Soll, R. and Cheresh, D. (2006) *Proc. Natl. Acad. Sci. USA* **103**, 19866–19871.
[203] Doukas, J., Wrasidlo, W., Noronha, G., Dneprovskaia, E., Hood, J. and Soll, R. (2007) *Biochem. Soc. Trans.* **35**, 2041–2206.
[204] Wetzker, R., Mueller, A. and Rommel, C. (2006) *PCT Int. Appl.* WO 06 040318.
[205] Pomel, V., Klicic, J., Covini, D., Church, D.D., Shaw, J.P., Roulin, K., Burgat-Charvillon, F., Valognes, D., Camps, M., Chabert, C., Gillieron, C., Françon, B., Perrin, D., Leroy, D., Gretener, D., Nichols, A., Vitte, P.A., Carboni, S., Rommel, C., Schwarz, M.K. and Rückle, T. (2006) *J. Med. Chem.* **49**, 3857–3871.
[206] Camps, M., Ruckle, T., Ji, H., Ardissone, V., Rintelen, F., Shaw, J., Ferrandi, C., Chabert, C., Gillieron, C., Francon, B., Martin, T., Gretener, D., Perrin, D., Leroy, D., Vitte, P.A., Hirsch, E., Wymann, M.P., Cirillo, R., Schwarz, M.K. and Rommel, C. (2005) *Nat. Med.* **11**, 936–943.
[207] Barber, D.F., Bartolome, A., Hernandez, C., Flores, J.M., Redondo, C., Fernandez-Arias, C., Camps, M., Rückle, T., Schwarz, M.K., Rodriguez, S., Martinez-Alonso, C., Balomenos, D., Rommel, C. and Carrera, A.C. (2005) *Nat. Med.* **11**, 933–935.
[208] Quattropani, A., Rueckle, T., Schwarz, M., Dorbais, J., Sauer, W., Cleva, C. and Desforges, G. (2005) *PCT Int. Appl.* WO 05 068444.
[209] Moffat, D.F.C., Davies, S., Alesso, S.M. and Launary, D.F.M. (2007) *PCT Int. Appl.* WO 07 129005.
[210] Bruce, I., Finan, P., Leblanc, C., McCarthy, C., Whitehead, L., Blair, N.E., Bloomfield, F.C., Hayler, J., Kirman, L., Oza, M.S. and Shukla, L. (2003) *PCT Int. Appl.* WO 03 072557.
[211] Bloomfield, G.C., Bruce, I., Hayler, J.F., Leblanc, C., Le Grand, D.M. and McCarthy, C. (2005) *PCT Int. Appl.* WO 05 021519.
[212] Bruce, I., Cuenoud, B., Keller, T.H., Pilgrim, G.E., Press, N., Le Grand, D.M., Ritchie, C., Valade, B., Hayler, J. and Budd, E. (2004) *PCT Int. Appl.* WO 04 096797.
[213] Barvian, N.C., Kolz, C.N., Para, K.S., Patt, W.C. and Visnick, M. (2004) *PCT Int. Appl.* WO 04 052373.
[214] Gogliotti, R.D., Muccioli, K., Para, K.S. and Visnick, M. (2004) *PCT Int. Appl.* WO 04 056830.
[215] Lanni, T.B., Greene, K.L., Kolz, C.N., Para, K.S., Visnick, M., Mobley, J.L., Dudley, D.T., Baginski, T.J. and Liimatta, M.B. (2007) *Bioorg. Med. Chem. Lett.* **17**, 756–760.
[216] Connolly, M.K., Gogliotti, R.D., Lee, H.T., Plummer, M.S., Sexton, K.E. and Visnick, M. (2004) *PCT Int. Appl.* WO 04 108713.
[217] Connolly, M.K., Gogliotti, R.D., Hurt, C.R., Reichard, G.A. and Visnick, M. (2005) *PCT Int. Appl.* WO 05 023800.
[218] Bruendl, M.M., Connolly, M.K., Goodman, A.P., Gogliotti, R.D., Lee, H.T., Plummer, M.S., Sexton, K.E., Reichard, G.A., Visnick, M. and Wilson, M.W. (2004) *PCT Int. Appl.* WO 04 108715.

[219] Connolly, M.K., Gogliotti, R.D., Plummer, M.S. and Visnick, M. (2005) *PCT Int. Appl.* WO 05 042519.
[220] Bruendl, M.L., Gogliotti, R.D., Goodman, A.P. and Reichard, G. (2005) *PCT Int. Appl.* WO 05 105801.
[221] Shimada, M., Murata, T., Fuchikami, K., Tsujishita, H., Omori, N., Kato, I., Miura, M., Urbahns, K., Ganter, F. and Bacon, K. (2004) *PCT Int. Appl.* WO 04 029055.
[222] Take, Y., Kumano, M., Hamano Teraoka, H.Y., Fukatsu, H., Nishimura, S. and Okuyama, A. (1995) *Biochem. Biophys. Res. Commun.* 215, 41–47.
[223] Stockley, M., Clegg, W., Fontana, G., Golding, B.T., Martin, N., Rigoreau, L.J.M., Smith, G.C.M. and Griffin, R.J. (2001) *Bioorg. Med. Chem. Lett.* 11, 2837–2841.
[224] Leahy, J.J.J., Golding, B.T., Griffin, R.J., Hardcastle, I.R., Richardson, C., Rigoreau, L. and Smith, G.C.M. (2004) *Bioorg. Med. Chem. Lett.* 14, 6083–6087.
[225] Rodriguez Aristegui, S., Desage El-Murr, M., Golding, B.T., Griffin, R.J. and Hardcastle, I.R. (2006) *Org. Lett.* 8, 5927–5929.
[226] Griffin, R.J., Fontana, G., Golding, B.T., Guiard, S., Hardcastle, I.R., Leahy, J.J.J., Martin, N., Richardson, C., Rigoreau, L., Stockley, M. and Smith, G.C.M. (2005) *J. Med. Chem.* 48, 569–585.
[227] Frigerio, M., Hummersone, M.G., Menear, K.A., Bailey, C.S., Duggan, H.M.E., Gomez, S. and Martin, N.M.B. (2009) *PCT Int. Appl.* WO 09 010761.
[228] Smith, G.C.M., Martin, N.M.B., Cockcroft, X.L.F., Menear, K.A., Hummersone, M.G., Griffin, R.J., Frigerio, M., Golding, B.T., Hardcastle, I.R., Newell, D.R., Calvert, H.A., Curtin, N.J. and Desage-El Murr, M. (2006) *PCT Int. Appl.* WO 06 109084.
[229] Griffin, R.J., Golding, B.T., Newell, D.R., Calvert, H.A., Curtin, N.J., Hardcastle, I.R., Martin, N.M.B., Smith, G.C.M. and Rigoreau, L.J.M. (2003) *PCT Int. Appl.* WO 03 024949.
[230] Griffin, R.J., Golding, B.T., Newell, D.R., Calvert, H.A., Curtin, N.J. and Hardcastle, I.R., (2003) *PCT Int. Appl.* WO 03 015790.
[231] Ismail, I.H., Mårtensson, S., Moshinsky, D., Rice, A., Tang, C., Howlett, A., McMahon, G. and Hammarsten, O. (2004) *Oncogene* 23, 873–882.
[232] Halbrook, J.W., Kesicki, E.A., Burgess, L.E., Schlachter, S.T., Eary, C.T. and Schiro, J.G. (2004) *PCT Int. Appl.* WO 04 085418.
[233] Durant, S. and Karran, P. (2003) *Nucl. Acids Res.* 31, 5501–5512.
[234] Smith, G.C.M., Martin, N.M.B., Cockcroft, X.L.F., Matthews, I.T.W., Menear, K.A., Rigoreau, L.J.M., Hummersone, M.G. and Griffin, R.J. (2005) *PCT Int. Appl.* WO 05 016919.
[235] Hollick, J.J., Rigoreau, L.J.M., Cano-Soumillac, C., Cockcroft, X., Curtin, N.J., Frigerio, M., Golding, B.T., Guiard, S., Hardcastle, I.R., Hickson, I., Hummersone, M.G., Menear, K.A., Martin, N.M.B., Matthews, I., Newell, D.R., Ord, R., Richardson, C.J., Smith, G.C.M. and Griffin, R.J. (2007) *J. Med. Chem.* 50, 1958–1972.
[236] Hickson, I., Zhao, Y., Richardson, C.J., Green, S.J., Martin, N.M.B., Orr, A.I., Reaper, P.M., Jackson, S.P., Curtin, N.J. and Smith, G.C.M. (2004) *Cancer Res.* 64, 9152–9159.
[237] Peng, H., Kima, D.I., Sarkariab, J.N., Choa, Y.S., Abraham, R.T. and Zalkowa, L.H. (2002) *Bioorg. Med. Chem.* 10, 167–174.
[238] Kim, T.K., Won, J.J. and Yi, Y.W. (2007) *PCT Int. Appl.* WO 07 015632.
[239] Nishida, H., Hammamori, Y., Konishi, T. and Yamashita, J. (2007) *PCT Int. Appl.* WO 046246.
[240] Thoreen, C.C., Kang, S.A., Chang, J.W., Liu, Q., Zhang, J., Gao, Y., Reichling|, L.J., Sim, T., Sabatini, D.M. and Gray, N.S. (2009) *J. Biol. Chem.* 284, 8023–8032.

[241] Chen, X., Coate, H., Crew, A.-P., Dong, H.-Q, Honda, A., Mulvihill, M.J., Tavares, P. A.R., Wang, J., Werner, D.S., Mulvihill, K.M., Siu, K.W., Panicker, B., Bharadwaj, A., Arnold, L.D., Jin, M., Volk, B., Weng, Q. and Beard, J.D. (2007) *PCT Int. Appl.* WO 07 061737.

[242] Crew, A.-P., Werner, D.S. and Tavares, P.A.R. (2007) *PCT Int. Appl.* WO 07 087395.

[243] Hummersone, M.G., Gomez, S., Menear, K.A., Cockcroft, X.L.F., Edwards, P., Loh, V.J.M.L. and Smith, G.C.M. (2006) *PCT Int. Appl.* WO 06 090169.

[244] Ballou, L.M., Selinger, E.S., et al. (2007) *J. Biol. Chem.* **282**, 24463–24470.

[245] Lin, R., Dreuckhammer, D., Choi, J.Y. and Ballou, L. (2008) *PCT Int. Appl.* WO 08 148074.

[246] Data courtesy of the Thomson Pharma database.

[247] Walker, E.H., Perisic, O., Ried, C., Stephens, L. and Williams, R.L. (1999) *Nature (London)* **402**, 313–320.

[248] Zvelebil, M.J., Waterfield, M.D. and Shuttleworth, S.J. (2008) *Arch. Biochem. Biophys.* **15**, 404–410.

[249] Eccleston, J., Pacold, M., Stephens, L. and Williams, R. (2002) *PCT Int. Appl.* WO 02 016427.

[250] Alaimo, P.J., Knight, Z.A. and Shokat, K.M. (2005) *Bioorg. Med. Chem.* **13**, 2825–2836.

[251] Sundstrom, T.J., Anderson, A.C. and Wright, D.L. (2009) *Org. Biomol. Chem.* **7**, 840–850.

[252] Wymann, M.P., Zvelebil, M. and Laffargue, M. (2003) *Trends. Pharm. Sci.* **24**, 366–376.

[253] Frédérick, R. and Denny, W.A. (2008) *J. Chem. Inf. Model.* **48**, 629–638.

[254] Kuang, R.R., Qian, F., Li, Z. and Wei, D.Z. (2006) *J. Mol. Model.* **12**, 445–452.

[255] LoPiccolo, J., Blumenthal, G.M., Bernstein, W.B. and Dennis, P.A. (2008) *Drug Res. Updat.* **11**, 32–50.

4 Progress in the Development of β-secretase Inhibitors for Alzheimer's Disease

JEFFREY S. ALBERT

CNS Discovery Research, AstraZeneca Pharmaceuticals, 1800 Concord Pike, PO Box 15437, Wilmington, DE 19850-5437, USA

Progress in Medicinal Chemistry – Vol. 48   133
Edited by G. Lawton and D.R. Witty
DOI: 10.1016/S0079-6468(09)04804-8

INTRODUCTION

A leading hypothesis to explain the pathophysiology of Alzheimer's disease centres on the abnormal accumulation of plaques in the brain that are composed of the 40/42 amino acid polypeptide called β-amyloid peptide ($A\beta_{40/42}$). According to the amyloid cascade hypothesis, β-amyloid is produced in the brain as the product of two proteolytic processing events. β-amyloid precursor protein (APP) is first cleaved by β-secretase (BACE; also called BACE1), and second by γ-secretase (Figure 4.1). Thus formed, the β-amyloid peptide accumulates and is deposited as insoluble plaques that are associated, perhaps causatively, with Alzheimer's progression [1, 2].

Mutations in APP and γ-secretase are known to alter $A\beta_{40/42}$ processing and are associated with familial forms of Alzheimer's disease that lead to onset at an early age [3]. Although this is consistent with the role of $A\beta_{40/42}$ plaque formation in disease progression, it is apparent that severity does not correlate with the extent of plaque deposition. Thus, while plaque deposition (as well as τ-hyperphosphorylation) is a hallmark of the disease, it is not clear that plaque formation is actually causative. Soluble oligomers of $A\beta_{40/42}$ are associated with neurotoxic effects in cell culture [4] and with memory impairment effects in transgenic mice that express human APP [5]. Altering APP processing via inhibition or modulation of γ-secretase activity [6–9] shows considerable promise as a mechanism for disease therapy;

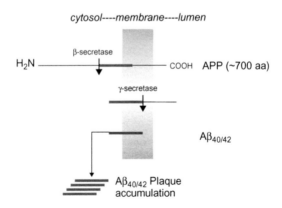

Fig. 4.1   The amyloid hypothesis in Alzheimer's disease. The transmembrane protein amyloid precursor protein (APP) is cleaved first by β-secretase (BACE) then by γ-secretase. The resulting peptidic fragment, $A\beta_{40/42}$ is liberated leading to plaque accumulation. These plaques are associated with neuronal degeneration.

however, this review will be restricted to consideration of inhibitors of BACE.

BACE is a particularly attractive target for Alzheimer's disease therapy because inhibition should directly reduce β-amyloid production. The safety of BACE inhibitors is supported by the observation that BACE-knockout mice appear normal except for their inability to produce the β-amyloid peptide [10–14]. Five research groups independently identified BACE in 1999 [15–19]. Since then, intense effort has been invested in discovering inhibitors for this enzyme. This area has been extensively reviewed [1, 2, 6, 20–26].

BACE is an aspartyl protease and is in the same family of proteins as HIV protease, pepsin, cathepsin D (Cat-D), and renin [27]. In addition, BACE-2 is a homologue of BACE, sharing 51% homology. BACE-2 proteolysis has similar substrate specificity to that of BACE and is also expressed in the CNS. However, it is expressed at higher levels in the periphery than BACE. The physiologic role for BACE-2 in the processing of APP and contribution to Alzheimer's progression remains uncertain.

BACE is among the most intensely investigated therapeutic targets and progress continues to accelerate. This review will focus primarily on discovery and research developments toward identifying, and optimising, inhibitors since 2007.

According to typical conventions in the protease literature, residues of the cleaved peptide or protein are designated relative to the cleaved amide bond (Figure 4.2). The residues to the amino- and carboxy-termini are referred to as P1 and P1', respectively. Regions of the enzyme get corresponding designations according to which particular residue is bound in that region for the natural substrate. For example, the P1' residue binds in the S1' pocket.

Like other aspartyl proteases, peptide amide bond hydrolysis by BACE is understood to involve activation of an enzyme bound water molecule by one of two catalytic aspartates (Figure 4.3). This activated water molecule undergoes nucleophilic addition to the carbonyl group of the scissile amide

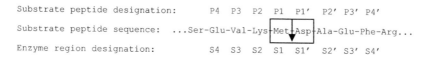

Substrate peptide designation: P4 P3 P2 P1 P1' P2' P3' P4'

Substrate peptide sequence: ...Ser-Glu-Val-Lys⌐Met┬Asp┬Ala-Glu-Phe-Arg...

Enzyme region designation: S4 S3 S2 S1 S1' S2' S3' S4'

Fig. 4.2 Terminology used to describe regions of the substrate peptide and the corresponding binding regions in the enzyme. Peptide cleavage of the native substrate by BACE occurs between Met and Asp; this is the active site and is shown by the box.

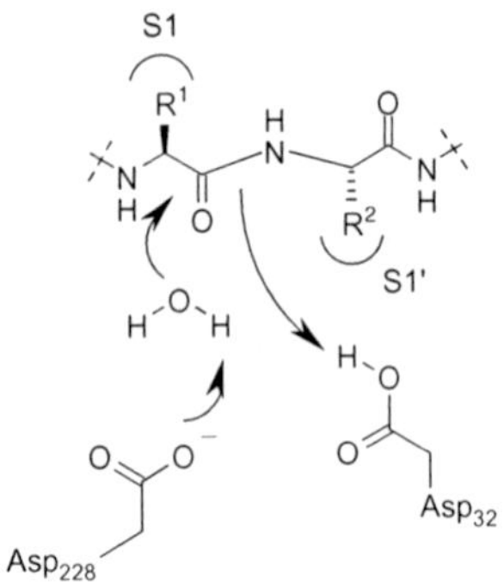

Fig. 4.3   Mechanism of peptide bond cleavage by BACE involving Asp32 and Asp228 as key residues at the catalytic site.

bond to form a hemiketal intermediate. Collapse of the tetrahedral intermediate then releases the products of the peptide cleavage.

ASTEX

Astex employed fragment-based approaches [28] to identify several chemotypes as BACE inhibitors. A set of just 347 small molecules (MW typically 100–250 g/mol) was screened in pools of six by direct soaking with preformed BACE crystals [29]. Hits were detected by crystallographic analysis of the resulting complexes. In this manner, hits (1) and (2) were identified. Both were quinoline isomers and may be binding in the protonated state with direct contacts to the catalytic aspartates. However, they do differ in their binding orientation and thus offer different possibilities for design and optimization.

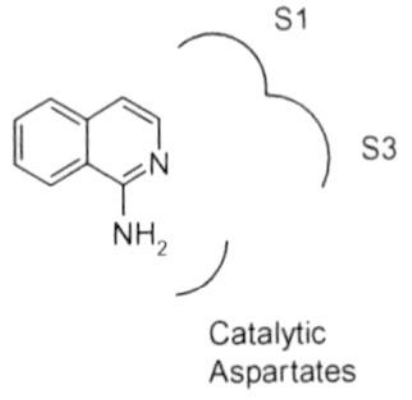

(1)
BACE 30–40% inhib. at 1 mM
(estimated IC_{50} 2 mM)

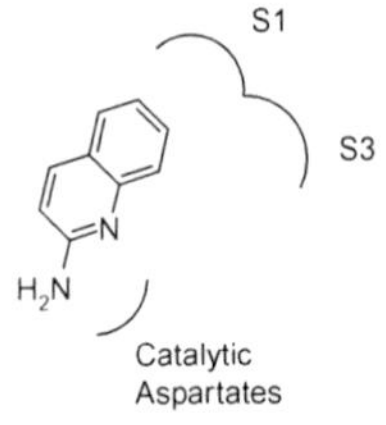

(2)
BACE 30–40% inhib. at 1 mM
(estimated IC_{50} 2 mM)

Virtual screening was employed to identify an additional set of 65 candidates for crystallographic screening. This led to the identification of compounds (3), (4), and (5). While structurally quite distinct, they make similar contacts with the catalytic aspartates and were oriented so as to permit further optimization into the S1/S3 regions. It is noteworthy that the binding affinity of (5) was so weak that enzyme inhibitory activity could not be detected at concentrations as high as 1 mM; nevertheless, the crystallographic screening system was sensitive enough to detect this binding and to provide details of the binding mechanism. Together, the fragment approaches enabled identification of several additional BACE inhibitor chemotypes with high ligand efficiency (lower molecular weight relative to comparable binding affinity [30, 31]) and thus greater potential for further optimization.

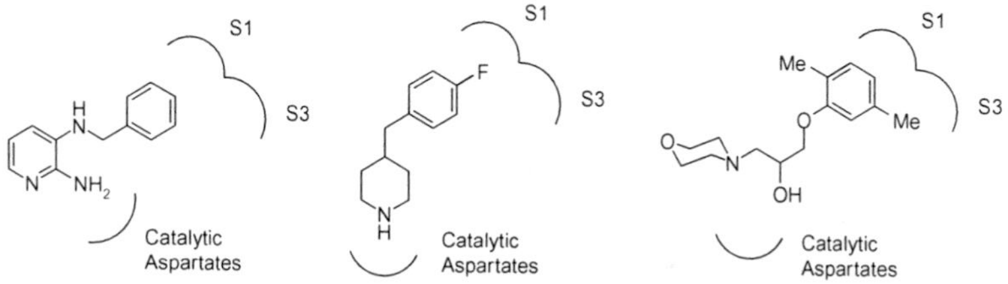

(3)
BACE IC_{50} 310 μM

(4)
BACE 66% inhib. at 1 mM
(estimated IC_{50} 0.5 mM)

(5)
No inhibition detected at 1 mM

Starting from hit (3), the Astex team sought to improve interactions in the S1/S3 region by further extension from the aminopyridine group; this led to (6) with increased activity ($IC_{50} = 40$ μM) [32]. The biaryl group in (6) was changed to indole (7); this introduced an interaction between the indole NH and the Gly230 backbone carbonyl and resulted in an additional increase in activity ($IC_{50} = 9.1$ μM). Proceeding further, structural details suggested that the 6-position of the indole projected under the flap region and toward S2'. Substitution at that position led to even further improvement and resulted in a sub-micromolar active compound (8) ($IC_{50} = 0.69$ μM). Crystallographic analysis of analogues of this compound demonstrated a correct prediction of the binding mode.

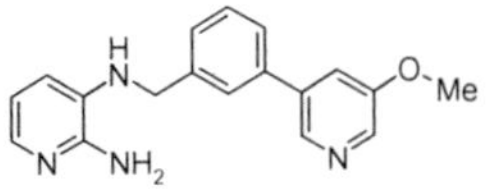

(6)
BACE IC$_{50}$ 40 µM

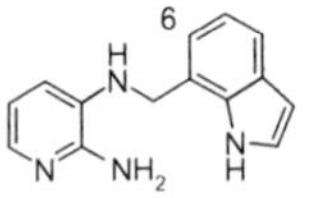

(7)
BACE IC$_{50}$ 9.1 µM

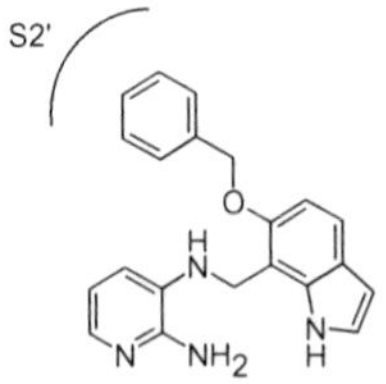

(8)
BACE IC$_{50}$ 0.69 µM

ASTEX/ASTRAZENECA

High-throughput screening by AstraZeneca on many hundreds of thousands of compounds failed to identify useful drug-like hits. Again using fragment-based methods, a screen was carried out on 2,000 low-molecular-weight compounds (MW 150–250 g/mol) using NMR detection [33]. In this approach, pools of 4–6 compounds were screened at individual concentrations of 300 µM in the presence of BACE. Using the water-LOGSY methodology [34], binding of any ligand to the enzyme causes a reversal of its NMR signal and can thus be detected. This approach does not require isotopically labelled enzyme and can detect binding as weak as 1 mM. In this manner, compound (9) was identified as a hit. The binding of the compound (28% inhibition at 1 mM) was determined by surface plasmon resonance spectroscopy; such weak activity is typical for small fragments and illustrates the sensitivity of both the NMR and surface plasmon resonance methods for detecting weak interactions. Following the identification of (9), follow-up screening of analogues led to the identification of (10) with improved affinity (69% inhibition at 1 mM).

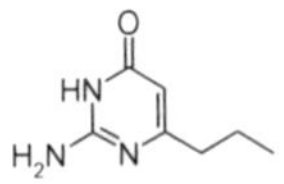

(9)
BACE 28% inhibition at 1 mM

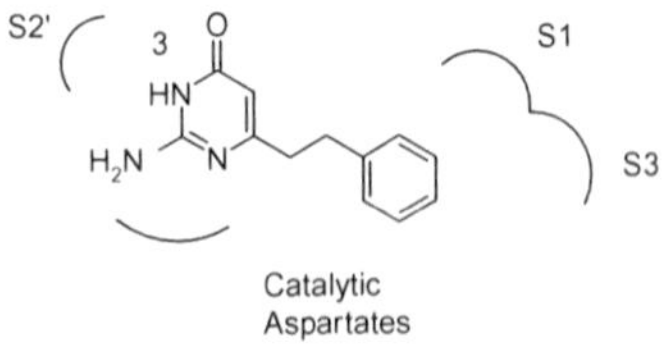

(10)
BACE 69% inhibition at 1 mM

Computational and crystallographic analysis of compounds in this series indicated that the ligand could be extended in both directions. The isocytosine N3 was capped with a methyl group to better fill the S2' region and to eliminate a hydrogen-bond donor (thus improving CNS drug-like properties). This led to (11) with an IC_{50} of 220 µM. Extension further into the S1/S3 region afforded (12) with an IC_{50} of 5.9 µM. Next, it was recognized from related NMR screening hits that the isocytocine ring could be replaced with its reduced analogue. Additionally, the C6 position was methylated to address the potential for oxidation back to the fully aromatic system. When applied to (12), these changes led to (13), which was the first inhibitor with an IC_{50} of < 100 nM [35].

(11)
BACE IC_{50} 220 µM

(12).
BACE IC_{50} 5.9 µM

(13)
BACE IC_{50} 0.08 µM

GSK

GSK undertook a discovery strategy using array chemistry focused around the hydroxyethylamine group (14) that appears frequently among different aspartic protease inhibitors. Final compounds in the designed libraries were selected so that the molecular weight was < 700 g/mol and there were no more than two amide bonds, so as to improve the likelihood of eventually achieving CNS drug-like characteristics [36]. Compound (15) was one of the first active compounds to be identified and had IC_{50} of 5.2 µM. It had nearly equivalent activity at Cat-D (IC_{50} = 7.7 µM), but did show some selectivity over BACE-2 (IC_{50} = 42 µM). The *ortho*-sulfone was replaced with pyrrolidinone so as to reduce the polar surface area and thus increase brain exposure. In addition, it was determined by crystallography that the other *ortho* position was oriented such that any substituents would be directed into the otherwise unoccupied S3 region. Substitution with ethylamine afforded compound (16) and an increase in activity (IC_{50} = 13 nM). This compound had > 100-fold selectivity over BACE-2 and Cat-D and also showed activity in a cellular assay (IC_{50} = 0.31 µM). In an effort to further reduce the molecular weight, the cyclohexyl-containing side was truncated, and it was found that activity was nearly maintained by replacing the terminal peptidic region with the substituted benzyl amines [37].

A wide range of these and related groups maintained activity in the sub-200 nM range. For example, compound (17) had an IC_{50} of 40 nM, showed selectivity over both BACE-2 ($IC_{50} = 4.07 \mu M$) and Cat-D ($IC_{50} = 5.75 \mu M$), and was active in a cellular assay ($IC_{50} = 0.18 \mu M$).

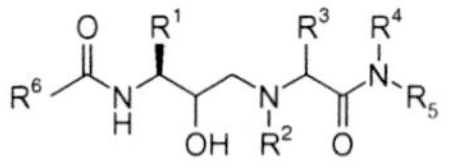

(14)
Library scaffold

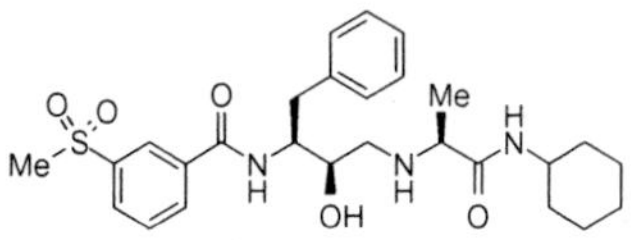

(15)
BACE IC_{50} 5.4 μM
BACE-2 IC_{50} 42 μM
CAT-D IC_{50} 7.7 μM

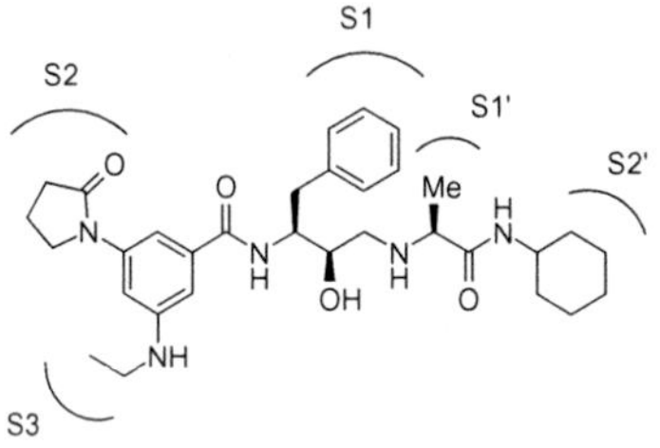

(16)
BACE IC_{50} 0.013 μM
BACE (cellular) IC_{50} 0.31 μM
BACE-2 IC_{50} 1.8 μM
Cat-D IC_{50} 2.7 μM

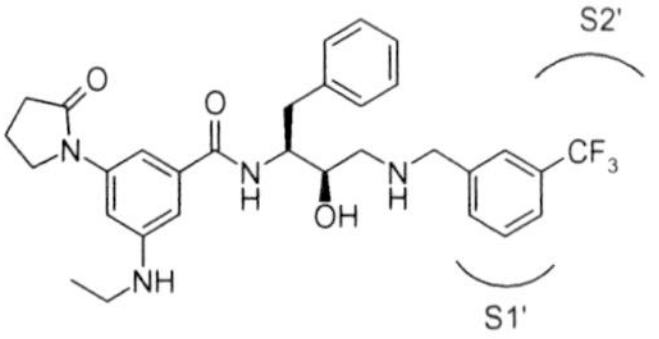

(17)
BACE IC_{50} 0.040 μM
BACE (cellular) IC_{50} 0.31 μM
BACE-2 IC_{50} 4.07 μM
Cat-D IC_{50} 5.75 μM

Changing the S2-binding region from pyrrolidinone to sultam, together with further substitution of the benzamide aryl with *ortho*-fluoro, afforded compound (18). This had even further improved activity ($IC_{50} = 4 nM$) with translation to the cellular assay ($IC_{50} = 27 nM$) and high selectivity over BACE-2 ($IC_{50} = 1.0 \mu M$) and Cat-D ($IC_{50} = 5.9 \mu M$) [38]. The compound was designated as GSK-188909 and further profiling showed it to have favorable pharmacokinetic properties. When administered at 250 mg/kg twice daily, orally for five days to transgenic mice, (18) reduced $A\beta_{40}$ by 18% in the brain. Presumably, the compound is subject to P-glycoprotein-mediated efflux because coadministration of a P-glycoprotein inhibitor, along with a single dose of (18) at 250 mg/kg, led to $A\beta_{40}$ reduction of 68% [39].

(18)
BACE IC$_{50}$ 4 nM
BACE (cellular) IC$_{50}$ 5 nM
BACE-2 IC$_{50}$ 177 nM
Cat-D IC$_{50}$ 2653 nM

(19)
BACE IC$_{50}$ 20 nM
BACE (cellular) IC$_{50}$ 16 nM
BACE-2 IC$_{50}$ 263 nM
Cat-D IC$_{50}$ 7933 nM

In addition to P-glycoprotein mediated efflux susceptibility, compounds such as (18) were subjected to metabolic dealkylation on the aniline nitrogens. To address these problems, as well as to reduce ligand flexibility, a tricyclic heterocycle was designed using crystallographic structural understanding to replace the 3,5-bis-substitued benzamide in (17) [40]. Metabolic oxidation of the terminal benzyl amine region was also a liability and it was found that it could be replaced with small alkyl and cycloalkyl groups [41, 42]. Together, these changes led to the identification of (19). This compound had strong BACE activity in the isolated enzyme (IC$_{50}$ = 20 nM) and cellular (IC$_{50}$ = 16 nM) assays with 10-fold selectivity over BACE-2. Most interestingly, it had improved distributional properties including a blood/brain ratio of 0.37 following i.v. administration to rat, and it had a bioavailability of 17% and 79% in rats and dogs, respectively [43].

HOFFMANN-LA ROCHE

Hoffmann-La Roche employed a fragment-based approach to identify starting points. A pool of low-molecular-weight (100–150 g/mol) screening candidates was identified by computational chemistry and then tested by surface plasmon resonance to afford 48 compounds for soaking into BACE crystals. In this manner, compound (20) was identified [44]. The very weak binding affinity (K_D = 2 mM) demonstrates the exceptional sensitivity of the surface plasmon resonance detection assay and the robust crystallographic methodology. With a molecular weight of just 137 g/mol, compound (20) is among the smallest in size and highest in ligand efficiency (0.37) yet identified. Crystallographic analysis demonstrated that the tyramine nitrogen bound to the catalytic aspartate region through a nondisplaced water molecule. Addition of an ethyl group to the aryl ring led to (21) and further increased activity ($K_D = 660\,\mu M$). This group projected into the S3 binding region and it was proposed that binding might be improved with larger groups. This led to the identification of (22) ($K_D = 350\,\mu M$). Addition of a *para*-methyl group (23) led to further increased activity ($K_D = 60\,\mu M$). Overall, this class of compounds may serve as a starting point to identify others that could be smaller and less polar than the peptide-derived compounds and, hence, not share their typical liabilities for limited brain exposure.

| (20) | (21) | (22) | (23) |
| BACE K_D 2000 μM | BACE K_D 660 μM | BACE K_D 350 μM | BACE K_D 60 μM |

JOHNSON AND JOHNSON

Compound (24), with a K_i of 900 nM, was identified by high throughput screening. Crystallographic analysis demonstrated that the cyclic amidine bound directly to the catalytic aspartates and that the ethyl linker region twisted back toward the aryl region to orient the cyclohexyl group in the S1 pocket. Based on this structural information, the ethyl linker was replaced with an *ortho*-substituted phenyl ring to better predispose the ligand to the preferred binding orientation. This afforded (25) with a K_i of 158 nM [45]. Structural analysis also indicated that the ethyl linker could provide a starting point for additional growth to access the S1' position. Addition of a cyclohexyl group to (24) afforded (26), with an increase in potency to an IC_{50} of 11 nM. Oral administration of (26) at 30 mg/kg to rats reduced $A\beta_{40}$ by 40–70% in plasma after 3 h. However, compound (26) had moderate selectivity against Cat-D ($IC_{50} = 110\,nM$), potent hERG activity ($IC_{50} = 140\,nM$), and potential P-glycoprotein efflux.

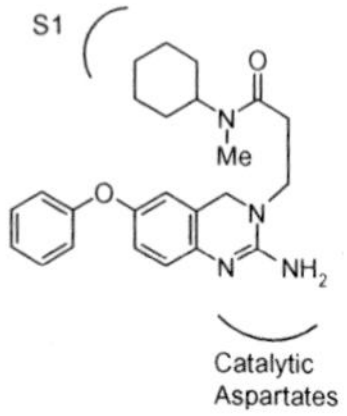

Catalytic
Aspartates

(24)
BACE K_i 900 nM

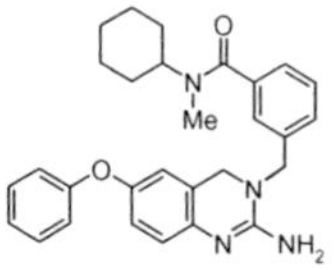

(25)
BACE K_i 158 nM

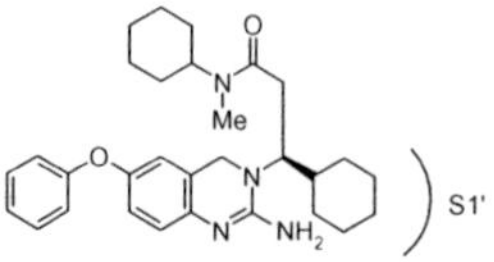

(26)
BACE K_i 11 nM
Renin IC_{50} 2.7 μM
Cat-D IC_{50} 110 nM
hERG IC_{50} 140 nM

MERCK

Isophthalamides serve as cores for a variety of BACE inhibitors from multiple research groups. Among this class, compound (27) was identified by Merck and had potent BACE activity against the Merck group isolated enzyme ($IC_{50} = 11$ nM) and in a cellular assay [46]. In subsequent efforts, the Merck group sought to reduce the P-glycoprotein susceptibility by replacing the hydroxyethylamine transition-state isostere with an imidazolidinone. According to the design, the amide carbonyl oxygen from the imidazolidinone ring could introduce a new hydrogen-bonding interaction with the Thr72 backbone amide, while maintaining all other key interactions [47]. Incorporating these changes, compound (28) showed high potency ($IC_{50} = 2.1$ nM). Reductions in molecular weight were sought within this imidazolidinone series through simplification of the S2/S3-binding isophthalamide region. Compound (29) emerged as the most active example wherein the entire substituted isophthalamide was replaced with a benzyl carbamate ($IC_{50} = 480$ nM).

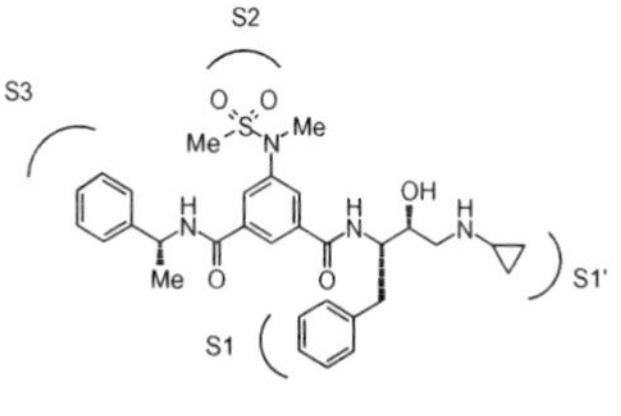

(27)
BACE IC_{50} 11 nM
BACE (cellular) 29 nM

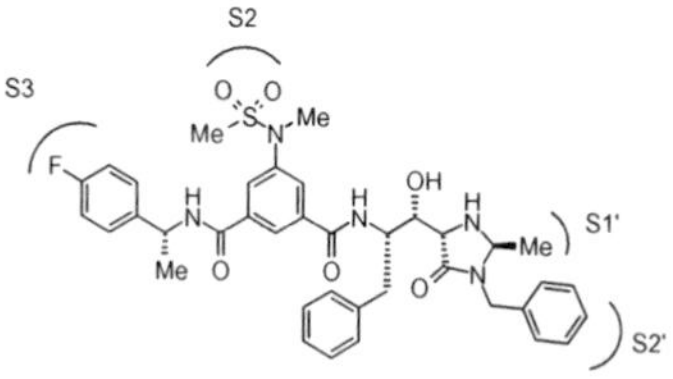

(28)
BACE IC_{50} 2.1 nM

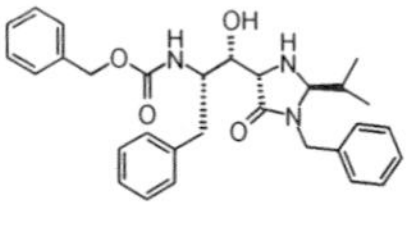

(29)
BACE IC_{50} 480 nM

The isophthalamide chemotype continued to evolve in multiple ways. Merck demonstrated that one of the amides could be replaced with oxadiazole, affording (30) [48, 49]. The second amide could also be replaced to afford compounds such as (31) [48, 49]. Efforts to further improve CNS distribution led Merck to introduce the isonicotinamide as a replacement for the isophthalamide [50]. As exemplified in compound (32), the binding feature comprised a cyclopropyl group similar to (31). Compounds with NH_2 or OH in the Asp32/Asp228 binding region both afforded high activity. Small modifications to (32), changing to isopropyl sulfonamide and adding a fluorine beta to the primary amine to attenuate basicity, afforded (33).

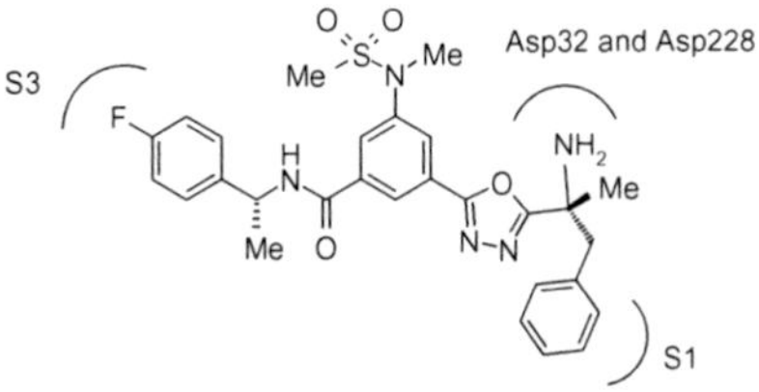

(30)
BACE IC_{50} 12 nM
BACE IC_{50} (cellular) 65 nM
PGP ratio >50

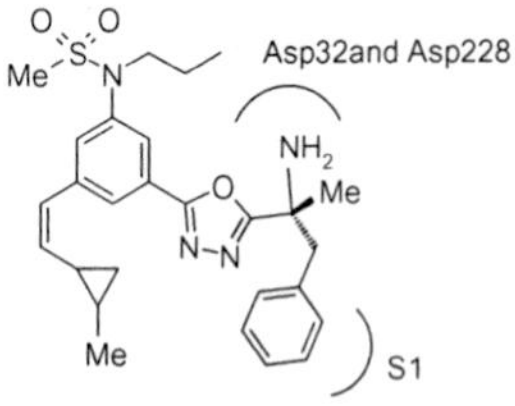

(31)
BACE IC_{50} 59 nM

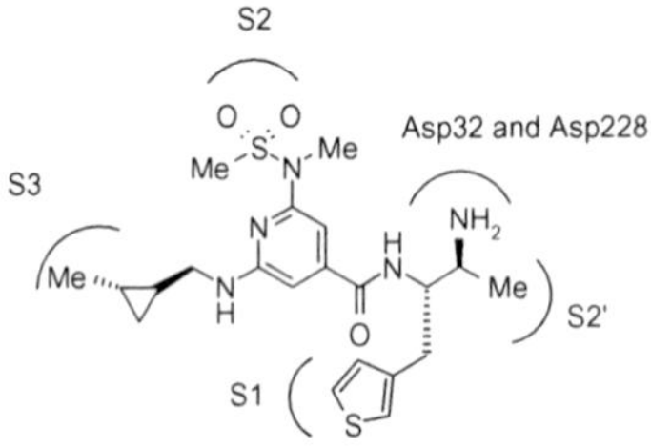

(32)
BACE IC_{50} 32 nM
BACE (cellular) 54 nM

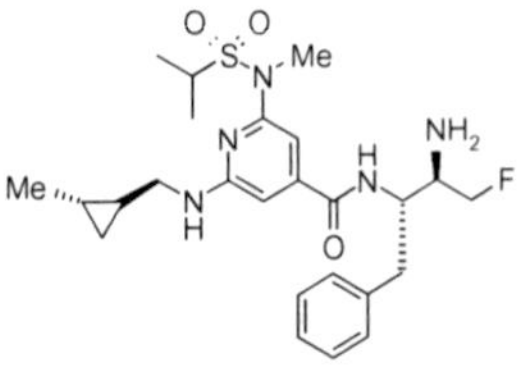

(33)
BACE IC_{50} 2 nM
BACE (cellular) 49 nM

Compound (33) had a greater enzyme potency (IC_{50} = 2 nM) than (32) and comparable activity in the cellular assay (IC_{50} = 49 nM), and it was active in a mouse *in-vivo* assay. Administration of 50 mg/kg, i.v., of (33) to transgenic mice expressing human APP showed a 34% reduction in brain levels of $A\beta_{40}$. Across this general series, reductions in molecular weight were achieved while retaining strong activity. However, most of the compounds also showed a high P-glycoprotein susceptibility and poor CNS distribution.

Combining elements from the cylopropylmethylamino-substituted isoni-cotinamide in (32) with the amine-substituted oxadiazole in (31) afforded (34) [51]. Although cellular potency was reduced ($IC_{50} = 2.5\,\mu M$), (34) had among the lowest P-glycoprotein liability yet seen within this chemical family. Capping of the free NH with a methoxyethyl group led to (35) and had the anticipated effect of further reducing the P-glycoprotein substrate liability, while maintaining high enzymatic potency ($IC_{50} = 6\,nM$). Crystallographic analysis demonstrated that the oxadiazole–aryl bond was twisted out of plane in the bound conformation. Such a twist was expected to introduce an energetic penalty to binding since a more coplanar arrangement would be expected in the free state. Based on this analysis, chlorine was added to form the isonicotinamide (36) enforcing the oxadiazole-twisted geometry. This had the effect of maintaining the high enzymatic potency ($IC_{50} = 1\,nM$) and low P-glycoprotein liability (apical to basolateral efflux ratio of 1.9) while improving activity in the cellular assay ($IC_{50} = 146\,nM$). The addition of the methoxyethyl cap led to (37), which showed even further improved enzymatic ($IC_{50} = 0.4\,nM$) and cellular ($IC_{50} = 40\,nM$) potency and a low P-glycoprotein liability (efflux ratio of 1.9).

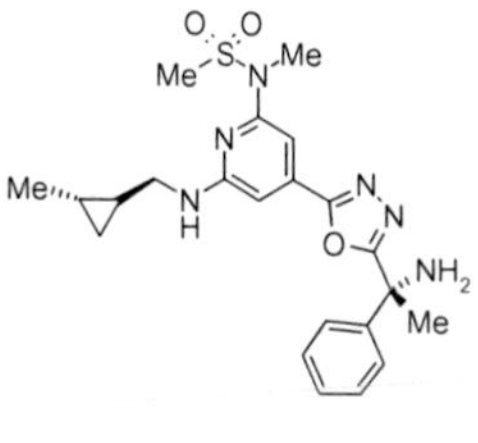

(34)
BACE IC_{50} 25 nM
BACE (cellular) 2500 nM
PGP ratio 3.6

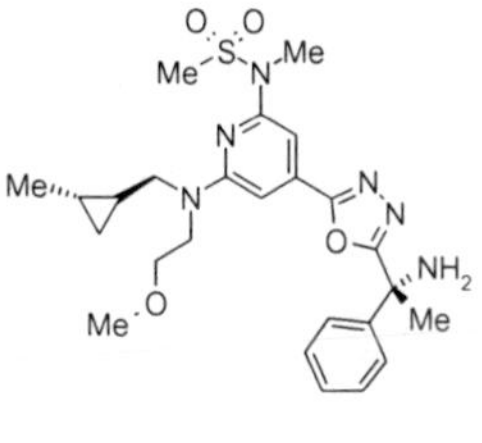

(35)
BACE IC_{50} 6 nM
BACE (cellular) 1300 nM
PGP ratio 1.6

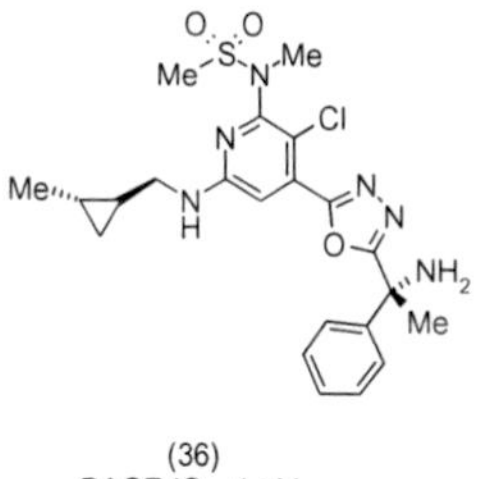

(36)
BACE IC_{50} 1 nM
BACE (cellular) 146 nM
PGP ratio 1.9

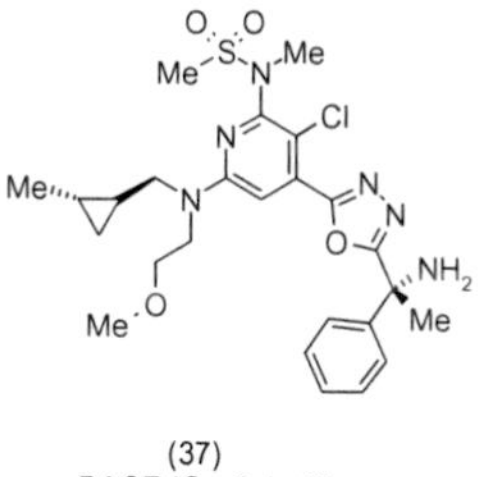

(37)
BACE IC_{50} 0.4 nM
BACE (cellular) 40 nM
BACE-2 IC_{50} 17 nM
Renin IC_{50} >100 μM
Cat-D IC_{50} >100 μM
PGP ratio 1.9

Despite these features, (37) had low brain exposure (brain/plasma ratio 9% in rat), high protein binding (<3% free fraction), high clearance, and poor oral bioavailability in rats, dogs, and monkeys. The high clearance was attributed to extensive oxidation by CYP3A4. Coadministration of (37) together with the CYP3A4 inhibitory compound ritonovir increased bioavailability in rhesus monkey to 83%. When administered in this manner, it provided an opportunity to test BACE inhibitors in primates [52].

Accordingly, compound (37) was administered to monkeys twice daily at 15 mg/kg, p.o., in conjunction with ritonovir at 10 mg/kg, p.o., for three days. Levels of drug and $A\beta_{40}$ in the CSF were continuously monitored. A peak drug concentration of about 35 nM was reached in the CSF with corresponding plasma exposures at about 4.7 µM. During the administration period $A\beta_{40}$ was reduced to about 42%; this serves as the first demonstration of reduction of $A\beta_{40}$ in primate CNS by a BACE inhibitor [51, 52].

Returning to the isophthalamides such as (30), it was known from other researchers, as well as from Merck, that that the S1- and S3-binding regions were very close and formed a relatively continuous pocket that should permit linking between those ligand domains. This was combined with a second insight that installation of a carbonyl function would be well positioned to make a hydrogen bonding contact with the Thr72 flap residue. The resulting compound (38) was potent but, again, had poor CNS penetration due to P-glycoprotein substrate susceptibility [53]. Efforts to improve this was carried out by investigating the S2-binding sulfonamide region. Groups with reduced polarity did improve the efflux liabilities, but also reduced activity. The best compromise was reached with (39), employing a cyanophenyl group in the S2-binding region. This compound had an IC_{50} of 27 nM with good translation to the cellular assay ($IC_{50} = 68$ nM). The compound was not a P-glycoprotein substrate but had poor cellular permeability characteristics.

(38)
BACE IC_{50} 2 nM

(39)
BACE IC_{50} 27 nM
BACE IC_{50} (cellular) 68 nM

Merck discovered (40) by high-throughput, high-concentration screening [54]. From this starting point activity could be improved by more than 10-fold, by elimination of the dimethylamine group, to afford (41). This structural class is unique because it has neither direct nor water-mediated interactions with the catalytic aspartates. SAR studies showed that the aryl methyl group and the sulfonamide were required, but that there was tolerance to modification of the solvent accessible phenyl group. These compounds may serve as a starting point to identify additional classes of BACE inhibitors.

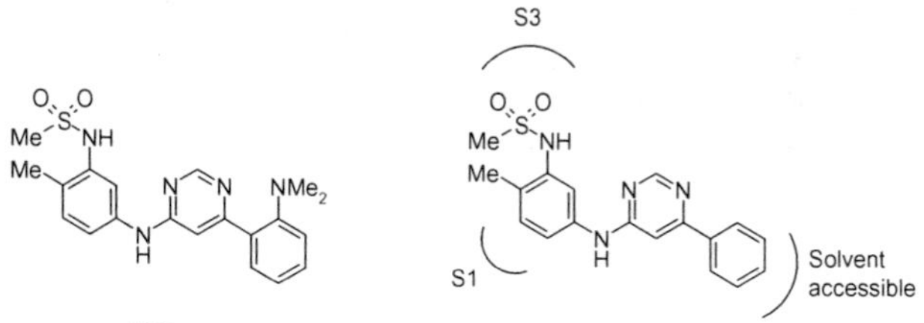

(40)
BACE IC_{50} 317 µM

(41)
BACE IC_{50} 24 µM

Merck discovered another chemotype by screening at 100 µM concentration. Starting hit (42) had an IC_{50} of 22 µM and could be improved by addition of an *ortho*-fluoro group to afford (43) (IC_{50} = 11 µM) [55]. Crystallographic analysis of compounds in this series showed that the piperidine nitrogen interacts indirectly with the catalytic aspartates through water-mediated hydrogen bonds. Potency could be further improved by optimizing contacts in the S3 region; addition of 2-methylphenyl to the benzyl amine afforded (44) with a 10-fold greater activity in the isolated enzyme assay (IC_{50} = 0.11 nM) but with weaker activity in the cellular assay (IC_{50} = 5.2 µM). Administration of (44) at 100 mg/kg, i.p., to transgenic mice expressing human APP resulted in high exposure in plasma (16 µM) and brain (14 µM), and led to a detectable, but statistically non-significant, reduction of brain $A\beta_{40}$ by 8%. This result is consistent with expectations, since when corrected for plasma protein binding, free drug levels in the brain would probably be below the cellular IC_{50} of 5.2 µM. In comparison to virtually all demonstrated BACE inhibitor chemotypes, this series shows good brain exposure for compounds and thus offers promise as an encouraging starting point for further optimisation.

(42)
BACE IC_{50} 22 μM

(43)
BACE IC_{50} 11 μM
BACE (cellular) IC_{50} 32 μM

(44)
BACE IC_{50} 0.11 μM
BACE (cellular) IC_{50} 5.2 μM

A third chemotype reported from Merck was based on an aminothiazole scaffold. The cyclopentyl aminothiazole (45) was identified as a screening hit having an IC_{50} of 217 μM [56]. SAR studies showed that the cyclopentane could be replaced with a 2-methoxy-4-nitrophenyl group and that methoxy substitution was also favorable in the other aryl ring. Together, these changes led to (46) and an improvement in potency ($IC_{50} = 1.1$ μM) in the standard enzymatic assay. Given that virtually all BACE inhibitors suffer from quite a poor CNS penetration, it was of particular interest to find that compounds in this series showed particularly good CNS distribution (brain/plasma ratios as high as 3.9, following i.v. administration). However, compounds such as (46) were inactive in the cellular assay. As is typical across BACE inhibitor research, the standard enzymatic assay is run at pH 4.5, whereas cellular assays are run under near-neutral conditions. Structural analysis indicated that the aminothiazole binds to the catalytic aspartates and would appear to require protonation on the heterocycle. Having a calculated pK_a of 5.5, it was expected that the aminothiazole would be protonated at pH 4.5 and be more readily able to bind to BACE than it would at pH 6.5 when it would not be protonated. According to this rationale, the aminothiazole (calculated pK_a 5.5) was replaced with the more basic aminoimidazole (calculated pK_a 8.1). When tested in the enzymatic assay at pH 4.5 the aminoimidazole (47) maintained nearly equivalent activity to the corresponding aminoimidazole (46), but when tested at pH 6.5 only aminoimidazole (47) had activity. Methylation of one of the imidazole NH groups afforded (48) and increased activity from an IC_{50} of 3.4 to 0.47 μM in the enzymatic assay at pH 6.5. Consistent with expectations, this was the only compound in the group to show activity in the cellular assay. This demonstrates that while measured enzymatic potency may be weaker at pH 6.5 than at pH 4.5, the results are more consistent with findings in the cellular assay. In addition, it is evident that both control of the inhibitor pK_a and gaining an understanding of the biologically relevant pH microenvironment can be important aspects for ligand design.

(45)
BACE IC_{50} 217 µM (pH 4.5)

(46)
Catalytic aspartates
Calc. pKa 5.5
BACE IC_{50} 1.1 µM (pH 4.5)
BACE IC_{50} inactive (pH 6.5)
BACE (cellular) inactive

(47)
Calc. pKa 8.1
BACE IC_{50} 5.7 µM (pH 4.5)
BACE IC_{50} 3.4 µM (pH 6.5)
BACE (cellular) inactive

(48)
Calc. pKa 8.3
BACE IC_{50} not det. (pH 4.5)
BACE IC_{50} 0.47 (pH 6.5)
BACE (cellular) IC_{50} 1.8 µM

NOVARTIS

Novartis initiated a structure-based design program starting with the high-affinity inhibitor OM99-1 (49) [57, 58]. They sought to reduce the molecular weight and polarity to improve CNS drug-like characteristics. In addition to truncating the large peptidic starting compound, they also methylated the amide in the P2 residue corresponding to the Asn in OM99-2. These changes led to the identification of (50), which had an IC_{50} of 2.4 µM and showed activity in a cellular enzyme assay. However, it had poor selectivity over Cat-D. As found by others, structural analysis of this and other related inhibitors indicated that the regions binding into the S1 and S3 pockets were separated by only about 3.8 Å and essentially formed a continuous super-pocket. Cyclization approaches have been applied in various ways to BACE inhibitors [53, 59–63]. In their case, to improve cellular permeability and proteolytic stability, the S1- and S3-binding regions were connected to form a macrocycle with a ring size between 14 and 17 atoms. Crystallography showed that the binding orientations for the macrocycles were largely consistent with predictions from molecular modelling and with the noncyclized analogues. Compound (51), which contained a ring system of 16 atoms, showed an appreciable increase in activity (IC_{50} = 0.15 µM) and improvement in selectivity over Cat-D (IC_{50} = 8.1 µM). However, the BACE cellular potency was not improved as dramatically (IC_{50} = 8.7 µM). In an effort to further improve cellular potency, the hydroxyethylene was replaced with an ethanolamine; both are common in aspartyl protease inhibitors. It was anticipated that the ethanolamine would improve cell penetration properties by eliminating one amide bond. This led to (52) improving in activity (IC_{50} = 22 nM), with consistent translation into the cellular assay (IC_{50} = 45 nM). Moreover, it showed a 53-fold selectivity over Cat-D (IC_{50} = 1.4 µM). However, despite good exposure and pharmacokinetic characteristics in mice, the compound was found to be a

P-glycoprotein substrate and thus likely to suffer from reduced CNS exposure in man. Coadministration of (52) (30 μmol/kg, i.v.) with a P-glycoprotein inhibitor (PSC833, 25 μmol/kg, p.o.) resulted in a 29% reduction in Aβ40 in the brains of APP51 mice.

Further modifications were made to the benzylamine substituent region in (52) to afford (53) to reduce the P-glycoprotein substrate liability. This compound had improved BACE activity (IC_{50} = 2 nM) and strong activity in the BACE cellular assay (IC_{50} = 24 nM); however, Cat-D selectivity was completely lost due to a dramatic increase in Cat-D inhibition (IC_{50} = 1 nM). Compound (53) was administered orally to APP51 mice at 2 × 100 μmol/kg, (without an additional P-glycoprotein inhibitor) but failed to show significant reduction in Aβ40 in brain; however, a significant reduction in CSF levels was observed. While compounds in this series still have challenges, it is notable that the discovery process starting with (49) and leading to (53) enabled reduction in MW from 893 to 558 g/mol and the polar surface area from 416 to 99 Å^2 while retaining similarly potent activity (IC_{50} of around 2 nM in the enzyme assay).

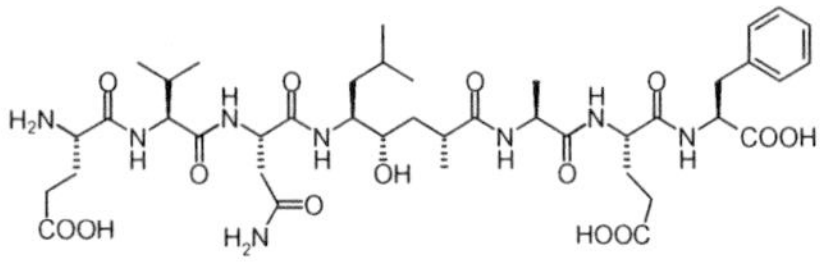

(49)
BACE K_i 1.6 nM

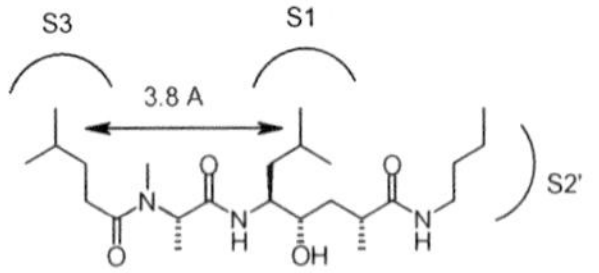

S3   S1
3.8 A
S2'

(50)
BACE IC_{50} 2.4 μM
BACE (cellular) 10% inhibition at 10 μM
Cat-D (cellular) 34% inhibition at 10 μM

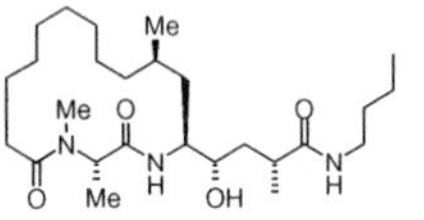

(51)
BACE IC_{50} 0.15 μM
BACE (cellular) 8.7 μM
    (60% inhibition at 10 μM)
Cat-D IC_{50} 8.1 μM

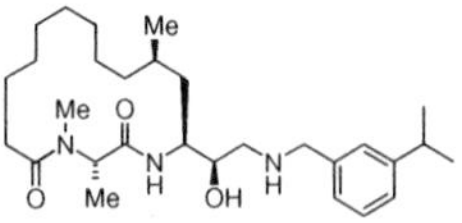

(52)
BACE IC_{50} 0.022 μM
BACE (cellular) 0.045 μM
Cat-D IC_{50} 1.4 μM

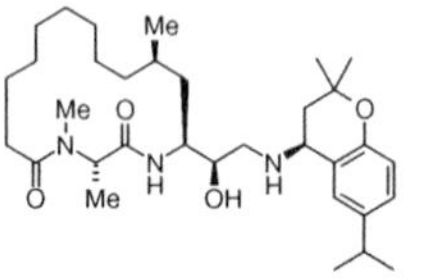

(53)
BACE IC_{50} 0.002 μM
BACE (cellular) 0.024 μM
Cat-D IC_{50} 0.001 μM

PFIZER

As described in sections above, diverse high-affinity inhibitors evolved from the common chemotype based on an isophthalamide. The isophthalamide binds in the S3 region and the hydroxyethylamine isostere interacts with the S1' region. An example is compound (54) from Pfizer with an IC_{50} of 500 nM. Compounds with this chemotype have frequently shown high potency in isolated enzyme assays, but poor cellular potency and CNS distributional characteristics. These liabilities are attributed to their peptidic nature, and extensive efforts have focused on reducing molecular weight, flexibility, and polarity. Other frequent liabilities are poor selectivity with respect to BACE-2 and Cat-D. Pfizer improved the *in-vitro* potency of (54) by 100-fold by modifications to the aryl substituents to afford (55) with an IC_{50} of 5 nM [64]. Further modifications by Pfizer to introduce a pyridine carbamate, capable of binding into the S2 region, afforded (56) and led to an increase in potency ($IC_{50} = 10$ nM) with higher selectivity over Cat-D ($IC_{50} = 444$ nM). Interestingly, this compound had stronger potency in the cellular assay ($IC_{50} = 0.6$ nM) [65].

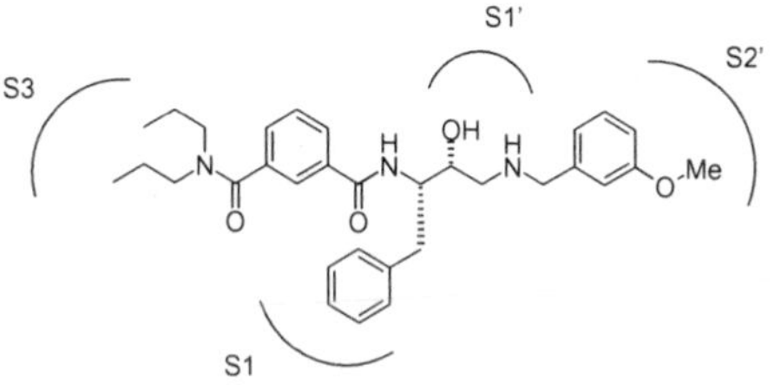

(54)
IC_{50} 500 nM

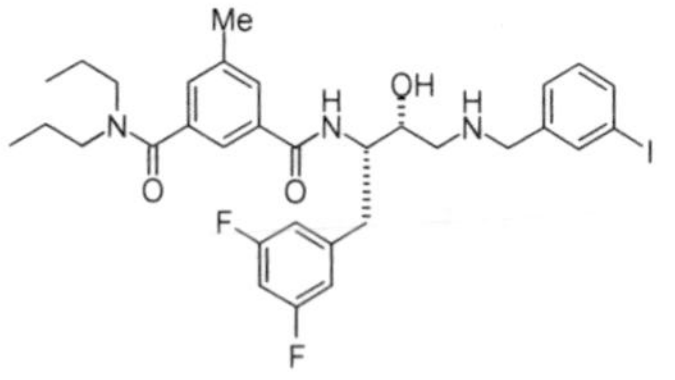

(55)
IC_{50} 5 nM

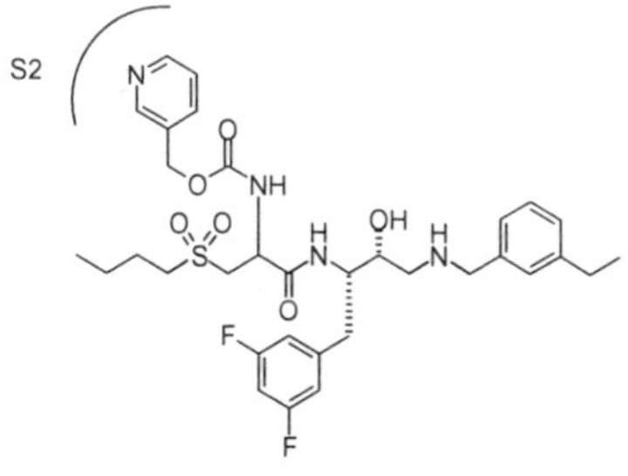

(56)
BACE IC_{50} 10 nM
Cat-D IC_{50} 444 nM
BACE (cellular) IC_{50} 0.60 nM

SCHERING-PLOUGH

Again starting with an isophthalamide hydroxyethylamine isostere similar to that from Merck, Pfizer, and others, Schering-Plough made modifications to the S2' binding region by cyclizing the benzylamino group to afford (57). While this compound had high *in-vitro* potency (IC_{50} = 5 nM), it had no selectivity for cathepsin-E (Cat-E) (IC_{50} = 5 nM), and its potency was 30-fold weaker in the cellular assay (IC_{50} = 150 nM) [66]. Next, the S3-binding dipropylamino region was replaced with methoxymethylpyrrolidine in (58) [67]. This compound had subnanomolar potency in the isolated enzyme assay (IC_{50} = 0.7 nM) and it had increased selectivity over both Cat-D and E. In rodent studies, the compound had a low CNS distribution as indicated from a brain/plasma ratio of 0.1 [67]. In addition, protein binding was high and it was a potent P-glycoprotein substrate (efflux ratio of 174). Nevertheless, oral administration at 10–100 mg/kg to CRND8 transgenic mice resulted in a dose-dependent reduction of plasma levels of $A\beta_{42}$ by as much as 70% relative to control. Subcutaneous administration led to higher exposure and a greater reduction in plasma $A\beta_{42}$, by as much as 88% relative to control, but there was no effect on cortical levels despite the drug reaching concentrations in whole brain 50-fold higher than that of the cellular IC_{50}. Subsequent reports showed that the right-side pyrrolidine region could be further modified to optimize interactions in the S2' pocket by changing it to a piperazinone [68]. This also introduced a new hydrogen-bonding interaction between the piperizinone amide carbonyl and the Thr72 NH from the flap region. The resulting compound (59) had strong *in-vitro* potency (IC_{50} = 3 nM) but worsened CNS distributional characteristics; this compound showed some reduction in plasma $A\beta_{42}$ in the CRND8 transgenic mouse model, but no reduction in brain levels.

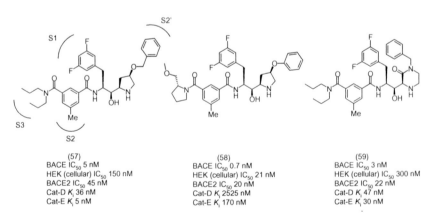

(57)
BACE IC_{50} 5 nM
HEK (cellular) IC_{50} 150 nM
BACE2 IC_{50} 45 nM
Cat-D K_i 36 nM
Cat-E K_i 5 nM

(58)
BACE IC_{50} 0.7 nM
HEK (cellular) IC_{50} 21 nM
BACE2 IC_{50} 20 nM
Cat-D K_i 2525 nM
Cat-E K_i 170 nM

(59)
BACE IC_{50} 3 nM
HEK (cellular) IC_{50} 300 nM
BACE2 IC_{50} 22 nM
Cat-D K_i 47 nM
Cat-E K_i 30 nM

SUNESIS

The Sunesis group, together with Merck, used their tethering technologies [69] to identify starting points for BACE inhibitors. In this approach, cysteine residues are engineered into the expressed-protein at various positions in the vicinity of the active site. The engineered-protein is then incubated with a small screening pool of potential binders that have a free thiol group appended. Because the potential binders are chosen to be small molecules, the binding affinity is not expected to be strong. However, even weak binding for a particular candidate in the active site serves to increase its concentration in the vicinity of the engineered cysteine such that an appropriately binding ligand will form a disulfide tether with the particular engineered protein. The identity of the tethered compound can then be determined by mass spectroscopy.

These approaches were applied to BACE by first preparing 11 different mutant proteins with the cysteine located at different positions near the active site. As expected, this caused a loss of enzymatic activity, but these constructs remained viable for ligand screening. Approximately 15,000 thiol-appended small compounds were tested in groups of 10 pooled together. Among the many hits uncovered, cyclic amines were found as a consistent structural theme despite the fact that they comprised <5% of the total screening set. Compounds (60)–(64) are shown as representative examples. The binding affinity of the non-tethered analogues was weak (no inhibition at 600 µM) and therefore selected compounds were crystallized as their tethered conjugates.

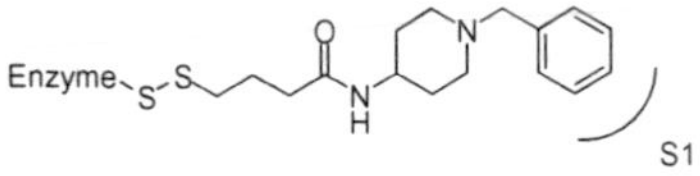

(60)

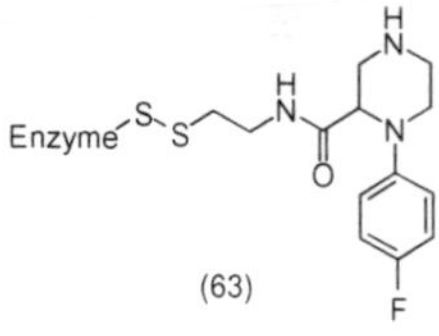

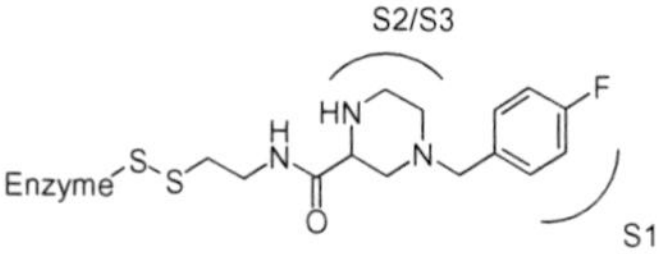

(61)

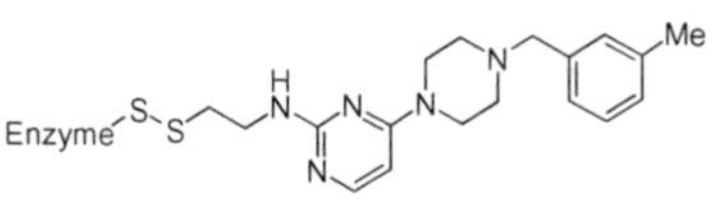

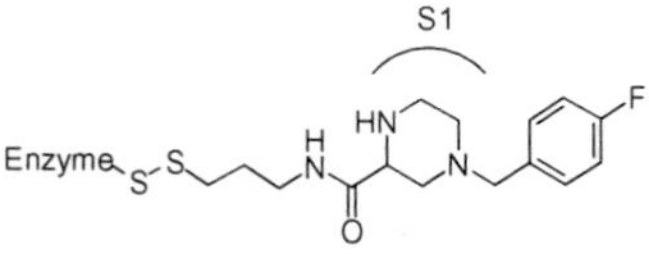

(62)

(63)

(64)

Compound (61) bound with the piperazine in the S2/S3-binding pocket and had no contacts with the catalytic aspartates. Interestingly, compound (64) differed only by having a one-carbon extension in the thiol linker, yet the binding orientation was quite distinct; in this case, the piperazine was oriented in the S1-binding region and it formed a water-mediated contact with Asp32. In the case of compound (60), structural analysis indicated that further growth could be accommodated into the S1 region. Exploiting this hypothesis led to increased binding potency, for example, with compound (65), which had a 1,000-fold stronger binding than the starting point.

The key tests for the tethering methodology are that the non-tethered analogues must bind similarly and that it must be possible to further evolve non-tethered analogues from the initial hit. Replacement of the tethering region of (60) with substituted benzenesulfonamide led to the identification of (66), having an IC_{50} of 74 μM. Crystallographic analysis for this chemotype demonstrated that the binding mode was conserved and the benzenesulfonamide was oriented into the S1/S3-binding pocket.

(65)
1000-fold stronger binding than (60)

(66)
BACE K_i 74 μM

WYETH

Wyeth identified acylguanidine (67) from a high-throughput-screening campaign [70]. This compound had low micromolar activity (IC_{50} = 3.7 μM) in an isolated enzyme assay and was also active in a cellular assay (IC_{50} = 8.9 μM). Crystallographic analysis showed the guanidine interacted with the catalytic aspartates, Asp32 and Asp228. The structure also indicated accessible space to grow into the S3 and S1' regions. In particular, optimization of interactions at S1' could take advantage of differences in this region among other aspartyl proteases (e.g. Cat-D) and thus provide a strategy to increase selectivity [71]. A wide range of types of groups was tolerated with IC_{50} values ranging from 78 to around 300 nM. However, nonpolar aromatic substituents tended to have poorer selectivity relative to Cat-D (0.5–2.5-fold) whereas those with polar substituents had increased selectivity (7.3–85 fold). For example, (68) had an IC_{50} of 78 nM with

7.3 fold selectivity over Cat-D [71]. The aryl substituent binding in the S3 region could also be replaced with adamantane, as in (69). The S3-binding region was explored by replacing the phenyl rings of (67) in a 12×13 combinatorial library. Among the most promising compounds to emerge was (70), with an IC_{50} of 0.7 μM. As expected, the added aryl group interacted with the S3 pocket. The central pyrrole could be replaced by thiophene to afford compounds such as (71) with an IC_{50} as strong as 0.15 μM [72]. Overall, compounds in this class tended to suffer from reduced permeability, thought to result from the guanidine group. Consequently, it was reported that efforts are now directed toward finding suitable surrogates for the guanidine that will achieve a more CNS drug-like character.

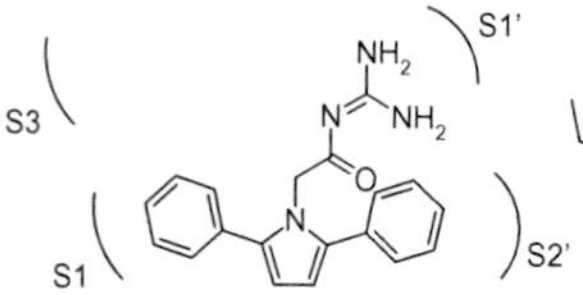

(67)
BACE IC_{50} 3.7 μM
BACE (cellular) IC_{50} 8.9 μM

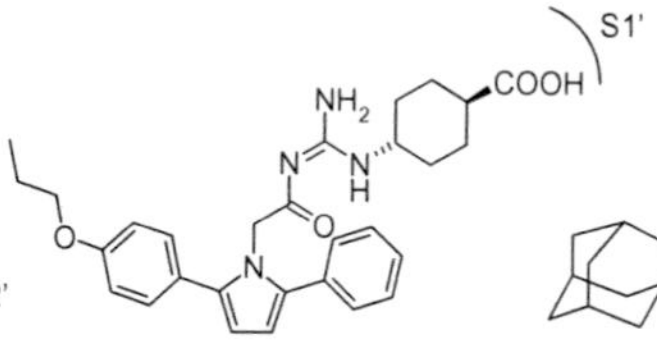

(68)
BACE IC_{50} 0.078 μM
Cat-D IC_{50} 0.57 μM
BACE (cellular) IC_{50} 10.4 μM

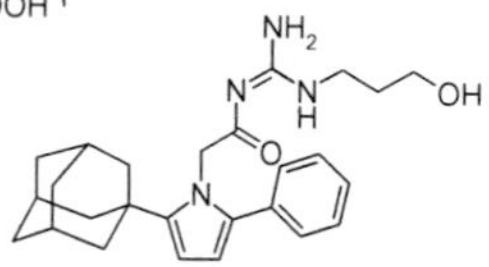

(69)
BACE IC_{50} 0.239 μM
Cat-D IC_{50} 19.9 μM
BACE (cellular) IC_{50} 1.84 μM

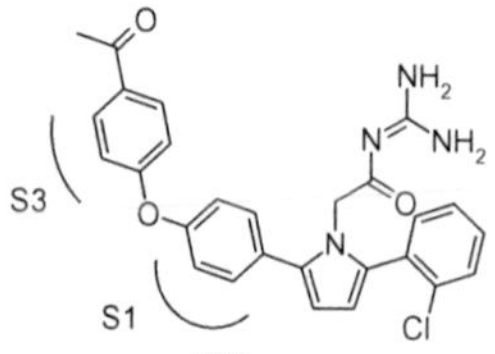

(70)
BACE IC_{50} 0.7 μM
BACE (cellular)ED_{50} 3.8 μM
BACE-2 IC_{50} 2.3 μM
Cat-D IC_{50} 18.9 μM

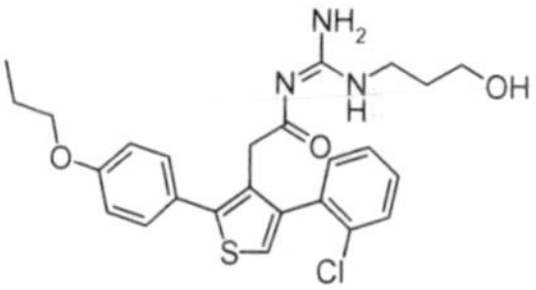

(71)
BACE IC_{50} 0.15 μM
BACE-2 IC_{50} 1.1 μM
Cat-D IC_{50} 3.5 μM

SUMMARY

Since the original identification of BACE in 1999 and until quite recently, BACE was often regarded as a "difficult" drug target, much as renin has proven to be. The reasons for this include the following. First, the long and shallow nature of the substrate binding pocket suggested that it would not be possible to identify small molecule drugs that could have adequate binding affinity. Second, functional groups that typically interact with the active site aspartates are usually highly polarized and, therefore, contribute to reduced CNS localization. Early BACE inhibitors were all designed using knowledge of the peptide substrates and usually contained some variation of a few well-known transition-state isosteres. While these had great impact on fundamental understanding of the enzyme structure and key interaction regions, they were very large, very polar, and had essentially no CNS availability. Continued progress by reducing the peptidic nature of these compounds resulted in incremental advances and has provided compounds that meet, or nearly meet, typical CNS drug-like criteria.

The challenges associated with peptidic starting points inspired innovative new approaches to search for different starting points. Several groups employed high concentration screening (ligand concentration $100\,\mu M$ and higher) to find weak hits after conventional screening (typically at $10\,\mu M$) failed to find more potent ones. Fragment-based methods have also been developed to identify even weaker hits (IC_{50} $1\,mM$ and greater). This was accomplished through the evolution and refinement of several detection methodologies including calorimetry, surface plasmon resonance, NMR, and crystallography. Coupled with detailed structural understanding of ligand–enzyme interactions and focus on maintaining ligand efficiency, these developments have resulted in several examples where potency was improved by 10,000-fold to afford compounds with IC_{50} values $<10\,nM$ and promising drug-like characteristics. Together, all these efforts have afforded a diverse array of chemotypes as BACE inhibitors.

Early work focused on improving BACE potency in isolated enzyme assays. However, most of these compounds showed potency reductions in cellular assays. Continued improvements in drug properties and in understanding of the physiologically relevant conditions have resulted in many compounds that show strong potency in both isolated and cellular assays. Several compounds have shown reduction of Aβ using rodent *in-vivo* models both peripherally and in the brain. Recently, one compound has demonstrated reduction of brain Aβ levels in a non-human primate. Phase I clinical trials were initiated on BACE inhibitor CTS-21166 from CoMentis in July of 2007. This compound derives from the earliest described peptidic inhibitors such as OM99-2 [58] but no details have been reported.

In addition to strategies involving small molecule inhibitors of BACE and γ-secretase to reduce Aβ levels, the application of biological agents has been under investigation since the identification of Aβ. The earliest efforts in this area failed. Despite encouraging results in preclinical models, immunization against Aβ by administration of AN-1792 from Elan led to development of aseptic meningoencephalitis in 6% of the patients receiving the drug. Nevertheless, continued efforts with other biological approaches appear encouraging. Most advanced in clinical trials is bapineuzumab from Elan, which is in Phase III clinical trials. This is a humanized monoclonal antibody against Aβ plaques. A recent monograph is devoted to progress in these areas [73].

Taken together, considerable progress has been made in developing CNS-penetrant agents that reduce Aβ levels and in providing validation that such agents will be therapeutically beneficial for the treatment of Alzheimer's disease.

REFERENCES

[1] Hunt, C.E. and Turner, A.J. (2009) *FEBS J.* **276**, 1845–1859.
[2] Dominguez, D.I. and De Strooper, B. (2002) *Trends Pharm. Sci.* **23**, 324–330.
[3] Selkoe, D.J. and Schenk, D. (2003) *Annu. Rev. Pharm. Toxicol.* **43**, 545–584.
[4] Haass, C. and Selkoe, D.J. (2007) *Nat. Rev. Mol. Cell Biol.* **8**, 101–112.
[5] Lesne, S., Koh, M.T., Kotilinek, L., Kayed, R., Glabe, C.G., Yang, A., Gallagher, M. and Ashe, K.H. (2006) *Nature (London)* **440**, 352–357.
[6] Olson, R.E. and Marcin, L.R. (2007) *Annu. Rep. Med. Chem.* **42**, 27–47.
[7] Selkoe, D.J. and Wolfe, M.S. (2007) *Cell* **131**, 215–221.
[8] Pissarnitski, D. (2007) *Curr. Opin. Drug Discov. Dev.* **10**, 392–402.
[9] Wu, W.-L. and Zhang, L. (2009) *Drug Dev. Res.* **70**, 94–100.
[10] Luo, Y., Bolon, B., Damore, M.A., Fitzpatrick, D., Liu, H., Zhang, J., Yan, Q., Vassar, R. and Citron, M. (2003) *Neurobiol. Dis.* **14**, 81–88.
[11] Cai, H., Wang, Y., McCarthy, D., Wen, H., Borchelt, D.R., Price, D.L. and Wong, P.C. (2001) *Nat. Neurosci.* **4**, 233–234.
[12] Roberds, S.L., Anderson, J., Basi, G., Bienkowski, M.J., Branstetter, D.G., Chen, K.S., Freedman, S.B., Frigon, N.L., Games, D., Hu, K., Johnson-Wood, K., Kappenman, K. E., Kawabe, T.T., Kola, I., Kuehn, R., Lee, M., Liu, W., Motter, R., Nichols, N.F., Power, M., Robertson, D.W., Schenk, D., Schoor, M., Shopp, G.M., Shuck, M.E., Sinha, S., Svensson, K.A., Tatsuno, G., Tintrup, H., Wijsman, J., Wright, S. and McConlogue, L. (2001) *Hum. Mol. Genet.* **10**, 1317–1324.
[13] Luo, Y., Bolon, B., Kahn, S., Bennett, B.D., Babu-Khan, S., Denis, P., Fan, W., Kha, H., Zhang, J., Gong, Y., Martin, L., Louis, J.-C., Yan, Q., Richards, W.G., Citron, M. and Vassar, R. (2001) *Nat. Neurosci.* **4**, 231–232.
[14] Harrison, S.M., Harper, A.J., Hawkins, J., Duddy, G., Grau, E., Pugh, P.L., Winter, P. H., Shilliam, C.S., Hughes, Z.A., Dawson, L.A., Gonzalez, M.I., Upton, N., Pangalos, M.N. and Dingwall, C. (2003) *Mol. Cell. Neurosci.* **24**, 646–655.
[15] Vassar, R., Bennett, B.D., Babu-Khan, S., Kahn, S., Mendiaz, E.A., Denis, P., Teplow, D.B., Ross, S., Amarante, P., Loeloff, R., Luo, Y., Fisher, S., Fuller, J., Edenson, S., Lile, J., Jarosinski, M.A., Biere, A.L., Curran, E., Burgess, T., Louis, J.-C., Collins, F., Treanor, J., Rogers, G. and Citron, M. (1999) *Science (Washington D.C.)* **286**, 735–741.
[16] Sinha, S., Anderson, J.P., Barbour, R., Basi, G.S., Caccavello, R., Davis, D., Doan, M., Dovey, H.F., Frigon, N., Hong, J., Jacobson-Croak, K., Jewett, N., Keim, P., Knops, J., Lieburg, I., Power, M., Tan, H., Tatsuno, G., Tung, J., Schenk, D., Seubert, P., Suomensaari, S.M., Wang, S., Walker, D., Zhao, J., McConlogue, L. and John, V. (1999) *Nature (London)* **402**, 537–540.

[17] Yan, R., Bienkowski, M.J., Shuck, M.E., Miao, H., Tory, M.C., Pauley, A.M., Brashler, J.R., Stratman, N.C., Mathews, W.R., Buhl, A.E., Carter, D.B., Tomaselli, A.G., Parodi, L.A., Heinrikson, R.L. and Gurney, M.E. (1999) *Nature (London)* **402**, 533–537.

[18] Hussain, I., Powell, D., Howlett, D.R., Tew, D.G., Meek, T.D., Chapman, C., Gloger, I.S., Murphy, K.E., Southan, C.D., Ryan, D.M., Smith, T.S., Simmons, D.L., Walsh, F. S., Dingwall, C. and Christie, G. (1999) *Mol. Cell. Neurosci.* **14**, 419–427.

[19] Lin, X., Koelsch, G., Wu, S., Downs, D., Dashti, A. and Tang, J. (2000) *Proc. Natl. Acad. Sci. U.S.A.* **97**, 1456–1460.

[20] Ghosh, A.K., Kumaragurubaran, N. and Tang, J. (2005) *Curr. Top. Med. Chem.* **5**, 1609–1622.

[21] Baxter, E.W. and Reitz, A.B. (2005) *Annu. Rep. Med. Chem.* **40**, 35–48.

[22] Schmidt, B., Baumann, S., Braun, H.A. and Larbig, G. (2006) *Curr. Top. Med. Chem.* **6**, 377–392.

[23] Durham, T.B. and Shepherd, T.A. (2006) *Curr. Opin. Drug Discov. Dev.* **9**, 776–791.

[24] Guo, T. and Hobbs, D.W. (2006) *Curr. Med. Chem.* **13**, 1811–1829.

[25] Hills, I.D. and Vacca, J.P. (2007) *Curr. Opin. Drug Discov. Dev.* **10**, 383–391.

[26] Stachel, S.J. (2009) *Drug Dev. Res.* **70**, 101–110.

[27] Eder, J., Hommel, U., Cumin, F., Martoglio, B. and Gerhartz, B. (2007) *Curr. Pharm. Des.* **13**, 271–285.

[28] Congreve, M., Murray, C.W., Carr, R. and Rees, D.C. (2007) *Annu. Rep. Med. Chem.* **42**, 431–448.

[29] Murray, C.W., Callaghan, O., Chessari, G., Cleasby, A., Congreve, M., Frederickson, M., Hartshorn, M.J., McMenamin, R., Patel, S. and Wallis, N. (2007) *J. Med. Chem.* **50**, 1116–1123.

[30] Abad-Zapatero, C. (2007) *Expert Opin. Drug Discov.* **2**, 469–488.

[31] Hopkins, A.L., Groom, C.R. and Alex, A. (2004) *Drug Discov. Today* **9**, 430–431.

[32] Congreve, M., Aharony, D., Albert, J., Callaghan, O., Campbell, J., Carr, R.A.E., Chessari, G., Cowan, S., Edwards, P.D., Frederickson, M., McMenamin, R., Murray, C.W., Patel, S. and Wallis, N. (2007) *J. Med. Chem.* **50**, 1124–1132.

[33] Geschwindner, S., Olsson, L.-L., Albert, J.S., Deinum, J., Edwards, P.D., de Beer, T. and Folmer, R.H.A. (2007) *J. Med. Chem.* **50**, 5903–5911.

[34] Dalvit, C., Fogliatto, G., Stewart, A., Veronesi, M. and Stockman, B. (2001) *J. Biomol. NMR* **21**, 349–359.

[35] Edwards, P.D., Albert, J.S., Sylvester, M., Aharony, D., Andisik, D., Callaghan, O., Campbell, J.B., Carr, R.A., Chessari, G., Congreve, M., Frederickson, M., Folmer, R. H.A., Geschwindner, S., Koether, G., Kolmodin, K., Krumrine, J., Mauger, R.C., Murray, C.W., Olsson, L.-L., Patel, S., Spear, N. and Tian, G. (2007) *J. Med. Chem.* **50**, 5912–5925.

[36] Clarke, B., Demont, E., Dingwall, C., Dunsdon, R., Faller, A., Hawkins, J., Hussain, I., MacPherson, D., Maile, G., Matico, R., Milner, P., Mosley, J., Naylor, A., O'Brien, A., Redshaw, S., Riddell, D., Rowland, P., Soleil, V., Smith, K.J., Stanway, S., Stemp, G., Sweitzer, S., Theobald, P., Vesey, D., Walter, D.S., Ward, J. and Wayne, G. (2008) *Bioorg. Med. Chem. Lett.* **18**, 1011–1016.

[37] Clarke, B., Demont, E., Dingwall, C., Dunsdon, R., Faller, A., Hawkins, J., Hussain, I., MacPherson, D., Maile, G., Matico, R., Milner, P., Mosley, J., Naylor, A., O'Brien, A., Redshaw, S., Riddell, D., Rowland, P., Soleil, V., Smith, K.J., Stanway, S., Stemp, G.,

Sweitzer, G., Theobald, P., Vesey, D., Walter, D.S., Ward, J. and Wayne, G. (2008) *Bioorg. Med. Chem. Lett.* **18**, 1017–1021.

[38] Beswick, P., Charrier, N., Clarke, B., Demont, E., Dingwall, C., Dunsdon, R., Faller, A., Gleave, R., Hawkins, J., Hussain, I., Johnson, C.N., MacPherson, D., Maile, G., Matico, R., Milner, P., Mosley, J., Naylor, A., O'Brien, A., Redshaw, S., Riddell, D., Rowland, P., Skidmore, J., Soleil, V., Smith, K.J., Stanway, S., Stemp, G., Stuart, A., Sweitzer, S., Theobald, P., Vesey, D., Walter, D.S., Ward, J. and Wayne, G. (2008) *Bioorg. Med. Chem. Lett.* **18**, 1022–1026.

[39] Hussain, I., Hawkins, J., Harrison, D., Hille, C., Wayne, G., Cutler, L., Buck, T., Walter, D., Demont, E., Howes, C., Naylor, A., Jeffrey, P., Gonzalez, M.I., Dingwall, C., Michel, A., Redshaw, S. and Davis, J.B. (2007) *J. Neurochem.* **100**, 802–809.

[40] Charrier, N., Clarke, B., Cutler, L., Demont, E., Dingwall, C., Dunsdon, R., Hawkins, J., Howes, C., Hubbard, J., Hussain, I., Maile, G., Matico, R., Mosley, J., Naylor, A., O'Brien, A., Redshaw, S., Rowland, P., Soleil, V., Smith, K.J., Sweitzer, S., Theobald, P., Vesey, D., Walter, D.S. and Wayne, G. (2009) *Bioorg. Med. Chem. Lett.* **19**, 3664–3668.

[41] Charrier, N., Clarke, B., Demont, E., Dingwall, C., Dunsdon, R., Hawkins, J., Hubbard, J., Hussain, I., Maile, G., Matico, R., Mosley, J., Naylor, A., O'Brien, A., Redshaw, S., Rowland, P., Soleil, V., Smith, K.J., Sweitzer, S., Theobald, P., Vesey, D., Walter, D.S. and Wayne, G. (2009) *Bioorg. Med. Chem. Lett.* **19**, 3669–3673.

[42] Charrier, N., Clarke, B., Cutler, L., Demont, E., Dingwall, C., Dunsdon, R., Hawkins, J., Howes, C., Hubbard, J., Hussain, I., Maile, G., Matico, R., Mosley, J., Naylor, A., O'Brien, A., Redshaw, S., Rowland, P., Soleil, V., Smith, K.J., Sweitzer, S., Theobald, P., Vesey, D., Walter, D.S. and Wayne, G. (2009) *Bioorg. Med. Chem. Lett.* **19**, 3674–3678.

[43] Charrier, N., Clarke, B., Cutler, L., Demont, E., Dingwall, C., Dunsdon, R., East, P., Hawkins, J., Howes, C., Hussain, I., Jeffrey, P., Maile, G., Matico, R., Mosley, J., Naylor, A., O'Brien, A., Redshaw, S., Rowland, P., Soleil, V., Smith, K.J., Sweitzer, S., Theobald, P., Vesey, D., Walter, D.S. and Wayne, G. (2008) *J. Med. Chem.* **51**, 3313–3317.

[44] Kuglstatter, A., Stahl, M., Peters, J.-U., Huber, W., Stihle, M., Schlatter, D., Benz, J., Ruf, A., Roth, D., Enderle, T. and Hennig, M. (2008) *Bioorg. Med. Chem. Lett.* **18**, 1304–1307.

[45] Baxter, E.W., Conway, K.A., Kennis, L., Bischoff, F., Mercken, M.H., De Winter, H. L., Reynolds, C.H., Tounge, B.A., Luo, C., Scott, M.K., Huang, Y., Braeken, M., Pieters, S.M.A., Berthelot, D.J.C., Masure, S., Bruinzeel, W.D., Jordan, A.D., Parker, M.H., Boyd, R.E., Qu, J., Alexander, R.S., Brenneman, D.E. and Reitz, A.B. (2007) *J. Med. Chem.* **50**, 4261–4264.

[46] Stachel, S.J., Coburn, C.A., Steele, T.G., Jones, K.G., Loutzenhiser, E.F., Gregro, A.R., Rajapakse, H.A., Lai, M.-T., Crouthamel, M.-C., Xu, M., Tugusheva, K., Lineberger, J.E., Pietrak, B.L., Espeseth, A.S., Shi, X.-P., Chen-Dodson, E., Holloway, M.K., Munshi, S., Simon, A.J., Kuo, L. and Vacca, J.P. (2004) *J. Med. Chem.* **47**, 6447–6450.

[47] Barrow, J.C., Rittle, K.E., Ngo, P.L., Selnick, H.G., Graham, S.L., Pitzenberger, S.M., McGaughey, G.B., Colussi, D., Lai, M.-T., Huang, Q., Tugusheva, K., Espeseth, A.S., Simon, A.J., Munshi, S.K. and Vacca, J.P. (2007) *ChemMedChem* **2**, 995–999.

[48] McGaughey, G.B., Colussi, D., Graham, S.L., Lai, M.-T., Munshi, S.K., Nantermet, P. G., Pietrak, B., Rajapakse, H.A., Selnick, H.G., Stauffer, S.R. and Holloway, M.K. (2007) *Bioorg. Med. Chem. Lett.* **17**, 1117–1121.

[49] Rajapakse, H.A., Nantermet, P.G., Selnick, H.G., Munshi, S., McGaughey, G.B., Lindsley, S.R., Young, M.B., Lai, M.-T., Espeseth, A.S., Shi, X.-P., Colussi, D., Pietrak, B., Crouthamel, M.-C., Tugusheva, K., Huang, Q., Xu, M., Simon, A.J., Kuo, L., Hazuda, D.J., Graham, S. and Vacca, J.P. (2006) *J. Med. Chem.* **49**, 7270–7273.

[50] Stauffer, S.R., Stanton, M.G., Gregro, A.R., Steinbeiser, M.A., Shaffer, J.R., Nantermet, P.G., Barrow, J.C., Rittle, K.E., Collusi, D., Espeseth, A.S., Lai, M.-T., Pietrak, B.L., Holloway, M.K., McGaughey, G.B., Munshi, S.K., Hochman, J.H., Simon, A.J., Selnick, H.G., Graham, S.L. and Vacca, J.P. (2007) *Bioorg. Med. Chem. Lett.* **17**, 1788–1792.

[51] Nantermet, P.G., Rajapakse, H.A., Stanton, M.G., Stauffer, S.R., Barrow, J.C., Gregro, A.R., Moore, K.P., Steinbeiser, M.A., Swestock, J., Selnick, H.G., Graham, S. L., McGaughey, G.B., Colussi, D., Lai, M.-T., Sankaranarayanan, S., Simon, A.J., Munshi, S., Cook, J.J., Holahan, M.A., Michener, M.S. and Vacca, J.P. (2009) *ChemMedChem* **4**, 37–40.

[52] Sankaranarayanan, S., Holahan, M.A., Colussi, D., Crouthamel, M.-C., Devanarayan, V., Ellis, J., Espeseth, A., Gates, A.T., Graham, S.L., Gregro, A.R., Hazuda, D., Hochman, J.H., Holloway, K., Jin, L., Kahana, J., Lai, M.-T., Lineberger, J., McGaughey, G., Moore, K.P., Nantermet, P., Pietrak, B., Price, E.A., Rajapakse, H., Stauffer, S., Steinbeiser, M.A., Seabrook, G., Selnick, H.G., Shi, X.-P., Stanton, M.G., Swestock, J., Tugusheva, K., Tyler, K.X., Vacca, J.P., Wong, J., Wu, G., Xu, M., Cook, J.J. and Simon, A.J. (2009) *J. Pharmacol. Exp. Ther.* **328**, 131–140.

[53] Lindsley, S.R., Moore, K.P., Rajapakse, H.A., Selnick, H.G., Young, M.B., Zhu, H., Munshi, S., Kuo, L., McGaughey, G.B., Colussi, D., Crouthamel, M.-C., Lai, M.-T., Pietrak, B., Price, E.A., Sankaranarayanan, S., Simon, A.J., Seabrook, G.R., Hazuda, D.J., Pudvah, N.T., Hochman, J.H., Graham, S.L., Vacca, J.P. and Nantermet, P.G. (2007) *Bioorg. Med. Chem. Lett.* **17**, 4057–4061.

[54] Steele, T.G., Hills, I.D., Nomland, A.A., de Leon, P., Allison, T., McGaughey, G., Colussi, D., Tugusheva, K., Haugabook, S.J., Espeseth, A.S., Zuck, P., Graham, S.L. and Stachel, S.J. (2009) *Bioorg. Med. Chem. Lett.* **19**, 17–20.

[55] Barrow, J.C., Stauffer, S.R., Rittle, K.E., Ngo, P.L., Yang, Z., Selnick, H.G., Graham, S.L., Munshi, S., McGaughey, G.B., Holloway, M.K., Simon, A.J., Price, E.A., Sankaranarayanan, S., Colussi, D., Tugusheva, K., Lai, M.-T., Espeseth, A.S., Xu, M., Huang, Q., Wolfe, A., Pietrak, B., Zuck, P., Levorse, D.A., Hazuda, D. and Vacca, J.P. (2008) *J. Med. Chem.* **51**, 6259–6262.

[56] Stachel, S.J., Coburn, C.A., Rush, D., Jones, K.L.G., Zhu, H., Rajapakse, H., Graham, S.L., Simon, A., Katharine Holloway, M., Allison, T.J., Munshi, S.K., Espeseth, A.S., Zuck, P., Colussi, D., Wolfe, A., Pietrak, B.L., Lai, M.-T. and Vacca, J.P. (2009) *Bioorg. Med. Chem. Lett.* **19**, 2977–2980.

[57] Hong, L., Koelsch, G., Lin, X., Wu, S., Terzyan, S., Ghosh, A.K., Zhang, X.C. and Tang, J. (2000) *Science (Washington D.C.)* **290**, 150–153.

[58] Ghosh, A.K., Bilcer, G., Harwood, C., Kawahama, R., Shin, D., Hussain, K.A., Hong, L., Loy, J.A., Nguyen, C., Koelsch, G., Ermolieff, J. and Tang, J. (2001) *J. Med. Chem.* **44**, 2865–2868.

[59] Ghosh, A.K., Devasamudram, T., Hong, L., DeZutter, C., Xu, X., Weerasena, V., Koelsch, G., Bilcer, G. and Tang, J. (2005) *Bioorg. Med. Chem. Lett.* **15**, 15–20.

[60] Stachel, S.J., Coburn, C.A., Sankaranarayanan, S., Price, E.A., Pietrak, B.L., Huang, Q., Lineberger, J., Espeseth, A.S., Jin, L., Ellis, J., Holloway, M.K., Munshi, S., Allison, T.,

Hazuda, J.P., Simon, A.J., Graham, S.L. and Vacca, J.P. (2006) *J. Med. Chem.* **49**, 6147–6150.

[61] Rojo, I., Martin, J.A., Broughton, H., Timm, D., Erickson, J., Yang, H.-C. and McCarthy, J.R. (2006) *Bioorg. Med. Chem. Lett.* **16**, 191–195.

[62] Moore, K.P., Zhu, H., Rajapakse, H.A., McGaughey, G.B., Colussi, D., Price, E.A., Sankaranarayanan, S., Simon, A.J., Pudvah, N.T., Hochman, J.H., Allison, T., Munshi, S.K., Graham, S.L., Vacca, J.P. and Nantermet, P.G. (2007) *Bioorg. Med. Chem. Lett.* **17**, 5831–5835.

[63] Barazza, A., Goetz, M., Cadamuro, S.A., Goettig, P., Willem, M., Steuber, H., Kohler, T., Jestel, A., Reinemer, P., Renner, C., Bode, W. and Moroder, L. (2007) *ChemBioChem* **8**, 2078–2091.

[64] Freskos, J.N., Fobian, Y.M., Benson, T.E., Bienkowski, M.J., Brown, D.L., Emmons, T.L., Heintz, R., Laborde, A., McDonald, J.J., Mischke, B.V., Molyneaux, J.M., Moon, J.B., Mullins, P.B., Prince, D.B., Paddock, D.J., Tomasselli, A.G. and Winterrowd, G. (2007) *Bioorg. Med. Chem. Lett.* **17**, 73–77.

[65] Freskos, J.N., Fobian, Y.M., Benson, T.E., Moon, J.B., Bienkowski, M.J., Brown, D. L., Emmons, T.L., Heintz, R., Laborde, A., McDonald, J.J., Mischke, B.V., Molyneaux, J.M., Mullins, P.B., Prince, D.B., Paddock, D.J., Tomasselli, A.G. and Winterrowd, G. (2007) *Bioorg. Med. Chem. Lett.* **17**, 78–81.

[66] Iserloh, U., Wu, Y., Cumming, J.N., Pan, J., Wang, L.Y., Stamford, A.W., Kennedy, M.E., Kuvelkar, R., Chen, X., Parker, E.M., Strickland, C. and Voigt, J. (2008) *Bioorg. Med. Chem. Lett.* **18**, 414–417.

[67] Iserloh, U., Pan, J., Stamford, A.W., Kennedy, M.E., Zhang, Q., Zhang, L., Parker, E. M., McHugh, N.A., Favreau, L., Strickland, C. and Voigt, J. (2008) *Bioorg. Med. Chem. Lett.* **18**, 418–422.

[68] Cumming, J.N., Le, T.X., Babu, S., Carroll, C., Chen, X., Favreau, L., Gaspari, P., Guo, T., Hobbs, D.W., Huang, Y., Iserloh, U., Kennedy, M.E., Kuvelkar, R., Li, G., Lowrie, J., McHugh, N.A., Ozgur, L., Pan, J., Parker, E.M., Saionz, K., Stamford, A. W., Strickland, C., Tadesse, D., Voigt, J., Wang, L., Wu, Y., Zhang, L. and Zhang, Q. (2008) *Bioorg. Med. Chem. Lett.* **18**, 3236–3241.

[69] Erlanson, D.A., Ballinger, M.D. and Wells, J.A. (2006) *Methods Princ. Med. Chem.* **34**, 285–310.

[70] Cole, D.C., Manas, E.S., Stock, J.R., Condon, J.S., Jennings, L.D., Aulabaugh, A., Chopra, R., Cowling, R., Ellingboe, J.W., Fan, K.Y., Harrison, B.L., Hu, Y., Jacobsen, S., Jin, G., Lin, L., Lovering, F.E., Malamas, M.S., Stahl, M.L., Strand, J., Sukhdeo, M.N., Svenson, K., Turner, M.J., Wagner, E., Wu, J., Zhou, P. and Bard, J. (2006) *J. Med. Chem.* **49**, 6158–6161.

[71] Jennings, L.D., Cole, D.C., Stock, J.R., Sukhdeo, M.N., Ellingboe, J.W., Cowling, R., Jin, G., Manas, E.S., Fan, K.Y., Malamas, M.S., Harrison, B.L., Jacobsen, S., Chopra, R., Lohse, P.A., Moore, W.J., O'Donnell, M.-M., Hu, Y., Robichaud, A.J., Turner, M. J., Wagner, E. and Bard, J. (2008) *Bioorg. Med. Chem. Lett.* **18**, 767–771.

[72] Fobare, W.F., Solvibile, W.R., Robichaud, A.J., Malamas, M.S., Manas, E., Turner, J., Hu, Y., Wagner, E., Chopra, R., Cowling, R., Jin, G. and Bard, J. (2007) *Bioorg. Med. Chem. Lett.* **17**, 5353–5356.

[73] Cribbs, D.H. and Agadjanyan, M.G. (2009) *CNS Neurol. Disord.* **8**, 1–6.

5 Advances in the Design of 5-HT$_6$ Receptor Ligands with Therapeutic Potential

DAVID WITTY[1], MAHMOOD AHMED[2] and TSU TSHEN CHUANG[3]

[1]*Neurosciences Centre of Excellence for Drug Discovery, GlaxoSmithKline, New Frontiers Science Park, Coldharbour Road, Harlow, Essex, CM19 5AW, UK*

[2]*GlaxoSmithKline R&D China, Singapore Research Centre, Biopolis at One-North, 11 Biopolis Way, The Helios, #06-03/04, Singapore 138667*

[3]*GlaxoSmithKline, Sirtris, a GSK Company, 200 Technology Square-Suite 300, Cambridge, MA 01139, USA*

Progress in Medicinal Chemistry – Vol. 48   163
Edited by G. Lawton and D.R. Witty
DOI: 10.1016/S0079-6468(09)04805-X

INTRODUCTION

PHARMACOLOGY OF THE 5-HT$_6$ RECEPTOR

The serotonin (5-hydroxytryptamine, 5-HT) neurotransmitter system in the brain plays a key role in a range of central functions including cognitive, motor, sensory and affective functions, as well as sleep and appetite, and has therefore been the focus of decades of drug-discovery efforts. 5-HT exerts its actions *via* seven plasma membrane receptor subtypes named 5-HT receptors 1–7 (5-HT$_1$–5-HT$_7$), all of which are G-protein coupled receptors (GPCR) except for 5-HT$_3$, a ligand-gated ion channel. For the six GPCRs, ligand binding to 5-HT$_1$ and 5-HT$_5$ receptors leads to reductions in intracellular cyclic adenosine monophosphate, whereas the converse is true for the 5-HT$_4$, 5-HT$_6$ and 5-HT$_7$ receptors: cAMP levels increase. For 5-HT$_2$ receptors, ligand binding activates phospholipase C, affecting an increase in the production of inositol triphosphate and diacylglycerol.

5-HT$_6$ receptors were first identified by molecular cloning in the rat [1, 2], then in man [3] and subsequently in mouse [4]. These receptors are predominantly expressed in the central nervous system, including the olfactory tubercle, nucleus accumbens, striatum, hippocampus and cerebral cortex, with a notably higher level of expression in the striatum of both rat and human brains, though this distinct feature is not reproduced in the mouse brain [1, 5–8]. The initial observation of such a tissue distribution led to the hypothesis that these receptors could be involved in cognitive processes and novelty-seeking behaviour, as well as mood regulation. This was first reinforced by the observation that the knockdown of expression levels of 5-HT$_6$ receptors in the hippocampus induced by antisense oligonucleotide, results in elevated cholinergic-dependant behaviour, a known neurochemical parameter associated with enhanced cognitive functions [9, 10].

When the 5-HT$_6$ receptor was first cloned, the high sequence homology of its transmembrane domains with known 5-HT receptors led to initial pharmacological exploration with the serotonergic ligand radiolabelled LSD, which bound with a K_D in a low nanomolar range. 5-HT was the only biogenic amine that was capable of completely displacing the LSD [1], thus confirming the status of the receptor as a member of the serotonergic family. Pharmacological characterisation shed light on its distinct properties; the receptor shows high affinity for several well-known typical and atypical anti-psychotic compounds [11, 12]. These findings laid the groundwork for subsequent medicinal chemical efforts aimed at the development of a range of 5-HT$_6$ receptor modulators. Both antagonists and agonists have been reported and many studies have been undertaken using animal models to explore the functional role of 5-HT$_6$ [13–15]. Major findings are summarised herein.

Seratonin, 5-HT   LSD

COGNITIVE AND BEHAVIORAL EFFECTS

Several cognitive paradigms have been used to examine the roles of 5-HT$_6$ receptor antagonists. In the Morris water maze test, young and normal rats administered with the antagonists SB-271046 and Ro-04-6790 exhibited an improved ability to retain information, but not the ability to learn [16, 17], though it is noteworthy that such beneficial effects have not always been confirmed by others [18]. In contrast, in aged rats with impaired-cognitive functions, several 5-HT$_6$ receptor antagonists have been demonstrated as able to enhance both learning and retention functions. These compounds include SB-271046, SB-399885 and SB-742457 [19–21]. Whilst the mechanisms underlying these effects are not fully understood, they may be at least partially mediated by the modulatory effects of 5-HT$_6$ receptors on a range of neurotransmitters, including acetylcholine, dopamine and glutamate, as well as the expression of polysialic acid–neural cell adhesion molecules. These ideas have been extensively discussed in a recent review [22].

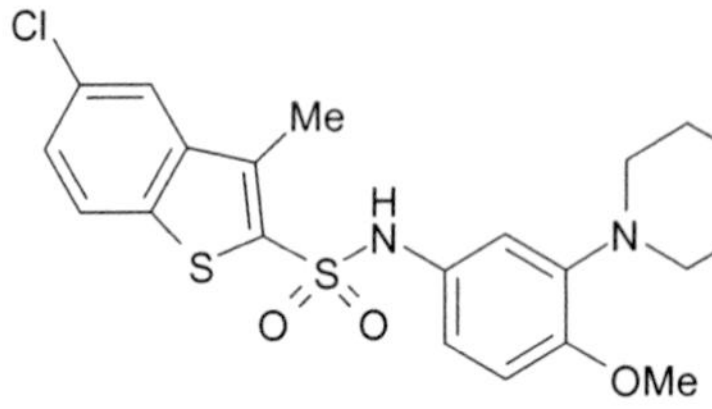

SB-271046

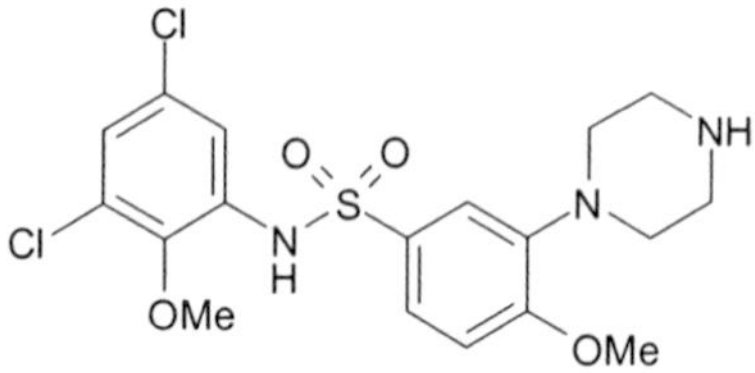

SB-399885

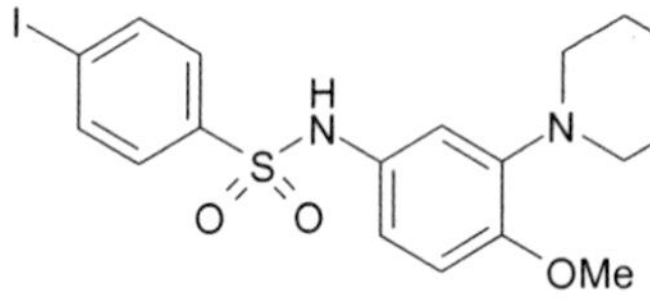

SB-258585

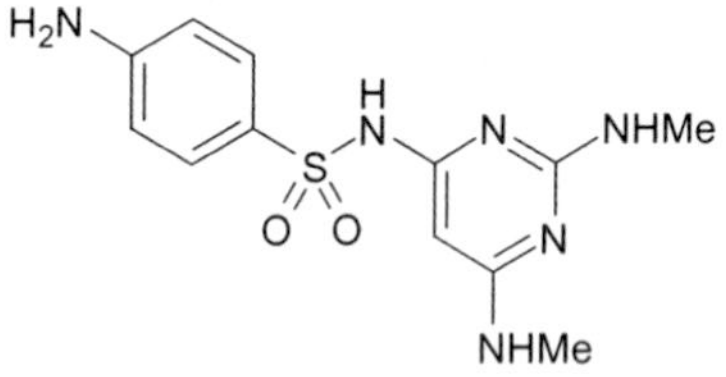

Ro-04-6790

Interestingly, SB-271046 has demonstrated a potential additive or synergistic effect with the cholinesterase inhibitors donepezil and galantamine, in the Morris water maze paradigm, using rats rendered cognitively impaired by different methods [23, 24]. Such an interaction may indicate potential therapeutic benefits by the combined use of a 5-HT$_6$ receptor antagonist and cholinesterase inhibitor for patients with Alzheimer's disease (AD). Other paradigms have also been used to confirm the cognitive enhancing effects observed in the Morris water maze paradigm (which relies mainly on hippocampal functions). These include the novel object recognition (NOR) paradigm, which engages the peri-rhinal and post-rhinal cortex. This is where object processing is considered to be mediated. These cortices have reciprocal innervations with the hippocampus to provide integration of event-related memories [25]. In the NOR paradigm, rats with cognitive impairment, induced by age or pharmacological interventions, responded well to a range of 5-HT$_6$ receptor antagonists, including Ro-04-6790, Ro-43-68554, SB-271046, SB-399885 and SB-742457 [20, 21, 26–28].

The potential therapeutic effects of 5-HT$_6$ receptor antagonists on behavioural symptoms, such as anxiety and depression, have been examined to a lesser extent. SB-399885 and SB-271046 induced positive outcomes in the rat forced-swim-test model of depression, with effects similar in magnitude to that of the anti-depressant imipramine [29, 30]. The study by Hirano *et al.* [30] also demonstrated that the dose of 5-HT$_6$ receptor antagonists correlated with striatal occupancy of the receptors at >60%. At these doses, the compounds were able to induce increased extracellular levels of monoamines in the brain [20, 31]. Interestingly, the cholinesterase inhibitor donepezil, at a dose that could enhance brain cholinergic neurotransmission, failed to exert any anti-depressant effects, indicating that 5-HT$_6$ receptor antagonists may be associated with a wider span of therapeutic benefits in AD by acting on both the cognitive and behavioural functions [30]. There is also a conflicting report indicating that SB-271046 counters the anti-depressant activity of fluoxetine in the tail-suspension anti-depressant model, and that the 5-HT$_6$ receptor agonist 2-ethyl-5-methoxy-*N,N*-dimethyltryptamine was able to induce biochemical and functional effects similar to fluoxetine [32]. Therefore, the nature of the involvement of 5-HT$_6$ receptors in depression remains controversial. The anxiolytic-like activities of SB-399885 were demonstrated in the Vogel conflict drinking test, the elevated plus-maze and the four-plate test, with similar activities to diazepam [29]. The benzodiazepine receptors may be indirectly involved in this effect, possibly due to a functional interaction between 5-HT$_6$ receptors and the gamma-aminobutyric acid/benzodiazepine system [33].

SATIETY, FOOD INTAKE AND BODY WEIGHT

Despite initially negative data on the involvement of 5-HT$_6$ receptors in food intake and body weight [9, 34, 35], Woolley *et al.* [17] reported that receptor knockdown both by anti-sense oligonucleotides and the receptor antagonist Ro-04-6790 reduced food consumption and body weight, suggesting a role for the 5-HT$_6$ receptor in the regulation of feeding. This discrepancy could conceivably be a result of differing degrees of receptor suppression achieved in the different studies. Subsequently, Biovitrum reported the ability of the antagonists BVT.5182 and SB-271046 to reduce food intake and body weight in DIO rats, additionally causing a decrease in visceral adiposity and plasma leptin concentrations [36, 37]. These findings were further supported by other reports using a range of compounds including SUVN-503 and SUVN-504 [38], as well as E-6837 [39].

In summary, the pre-clinical data provided strong rationale for the extensive medicinal chemistry campaigns that ensued in multiple companies to develop brain-penetrant ligands for the 5-HT$_6$ receptors.

MEDICINAL CHEMISTRY OF 5-HT$_6$ RECEPTOR MODULATION

SCOPE

The aim of the medicinal chemistry section of this review is to compare the specific approaches of different institutions to discovering, optimising and exploiting the 5-HT$_6$ receptor ligands. The focus is on scientific publications and patent applications describing new chemical series published since 2005. This review discusses not just 5-HT$_6$ receptor modulation, but also considers approaches targetting dual 5-HT$_6$ and second receptor modulation. Drug-discovery strategies where 5-HT$_6$ receptor binding is a component of a much broader pharmacology will not be reviewed here. For an alternative approach to aspects of this work, a recent review by Liu and Robichaud [40] has analysed certain classes of 5-HT$_6$ receptor *antagonists* by functional group.

The current keen interest in the discovery and application of 5-HT$_6$ receptor modulators is indicated by the fact that some 20 companies and institutions have filed patent applications on 5-HT$_6$ receptor ligands since 2005. Analysis of these patent applications and a welter of published literature reveal a remarkable diversity of approaches with particular groups often focussing on different therapeutic targets, sometimes specifying additional modes of action, including activity at related serotonin receptors and other aminergic receptor targets. Examples of early publications (before 2005) include the SmithKline Beecham compounds SB-271046 and SB-399885, which, as discussed, show activity in cognition models [5, 28], and the Roche pyrimidine Ro-04-6790 [41]. A radio-iodine labelled analogue of SB-271046, SB-258585 has proven to be a valuable ligand for assessing affinity in radioligand-binding studies [8]. The discovery and properties of these early lead compounds have been extensively discussed in the literature and will not be covered in detail here, save to compare properties with later chemical series.

5-HT$_6$ RECEPTOR LIGAND CHARACTERISATION

In general, one of two approaches has been employed to discover and design 5-HT$_6$ receptor antagonists: (i) optimisation of hits following some form of high throughput screening or (ii) elaboration of the natural ligand 5-HT, though conformational constraint and substitution. It is interesting that both approaches have homed in on a consensus pharmacophore. A characterising feature of first generation selective 5-HT$_6$ receptor antagonists, therefore, is the presence of two aromatic groups, one typically an indole,

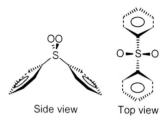

Fig. 5.1   *Preferred geometry of bisarylsulfone.*

linked by a sulfone or sulfonamide with a pendant basic group (typically a piperazine) substituted on the indole. The basic group is usually *para* or *meta* to the sulfonyl attachment point and the pendant aromatic group is often quite lipophilic. This key pharmacophoric unit has been formalised in a computer model [42]. One interpretation of the binding capabilities of this class of antagonist is the fact that a sulfonyl group pre-disposes the two aromatic rings to lie with their centroids pointing close to a line of symmetry between them, with an angle of ~ 100°, between them (Figure 5.1). A carbonyl or oxygen linker, by contrast, typically favours a low-energy confirmation in which the aromatic groups of the benzophenone or bis-aryl ether are twisted relative to each other and with a wider angle between them, such that lines through the two ring centroids do not intercept. The 'bent' disposition of aromatic groups is also potentially a characteristic of several unrelated compounds that show 5-HT$_6$ receptor antagonism, including methiothepin, where the seven-membered ring can also adopt a somewhat 'angular' conformation. For a detailed discussion and pharmacophore overlay, see Kim *et al.* [42], in particular Figure 2 in that publication.

Methiothepin

From these observations, a common lineage can be discovered between many classes of 5-HT$_6$ receptor ligands. Differing approaches have focused on exploiting the key motifs: aryl sulfonyl, core heterocycle, linker, basic sulf group, with the aim of improving specificity and potency by incorporation of differing conformational constraints, varying substituent functionality or inclusion of a heterocycle to achieve distinct molecular properties. In particular, as the 5-HT$_6$ receptor is present in the brain, pharmaceutical groups have often sought to deliver a target compound with optimal *in vitro* and *in vivo* DMPK properties for central penetration, including microsomal stability, an appropriate free fraction, adequate half life *etc*.

A wide variety of molecules fit the basic antagonist pharmacophore, so there has been considerable scope for variation. Within this framework, however, it is apparent that certain chemotypes are associated with differential efficacy in different disease models, though whether this reflects multiple modes of action (i.e. polypharmacology – including targets other than 5-HT$_6$), or different modalities of agonism, antagonism or modulation at the 5-HT$_6$ receptor alone, is not always clear from the published literature. Several groups have now demonstrated that certain classes of tryptamine based 5-HT$_6$ receptor ligands, though often containing a pendant arylsulfonyl moiety, are in fact agonists or partial agonists, rather than full antagonists (see section under Estève).

As the properties of 5-HT$_6$ receptor ligands have become better understood, the potential for dual pharmacology has become apparent. Additional potential indications have been proposed, backed up, in some cases, by animal model studies. More recently, several classes of non-sulfonyl containing 5-HT$_6$ receptor ligands have been published by different research groups and these may offer access to property space inaccessible to sulfones or sulfonamides. The following sections outline these approaches by comparing the compound classes reported by significant players in the field.

GLAXOSMITHKLINE, (GSK, INCORPORATING SMITHKLINE BEECHAM, SB)

Early compound classes

The discovery of tool compounds SB-271046 and SB-399885, with positive effects in rat models of cognition, drove our GSK group to optimise the arylsulfonyl series to find compounds with high *in vivo* efficacy. Since many arylsulfonyl compound classes gave rise to reasonably high levels of binding to the 5-HT$_6$ receptor, and good *in vitro* DMPK properties, we found that a critical strategy in the optimisation of series was the early progression of key compounds into rat *ex vivo* binding studies. This single assay could highlight compounds with good brain exposure following oral dosing, encompassing an in-life measure of both bioavailability and brain penetration. Limited central penetration in early hits was soon seen to be an issue – especially for

compounds bearing a secondary sulfonamide moiety – so we set an objective
to reduce the number of hydrogen-bond donors to improve the brain:blood
ratio. Indeed, the improved brain penetration of SB-399885 relative to SB-
271046 could be ascribed to the masking of the polarity of the sulfonamide
N–H by the adjacent methoxy group, which has the potential to form an
internal hydrogen bond. Our pursuit of this strategy, coupled with
conformational constraint, led to the identification of two classes of tertiary
sulfonamides – indolines (1) pK_i 7.7 and (2) pK_i 8.1 (Figure 5.2), as measured
in a radioligand-binding assay measuring displacement of $[^{125}I]$-SB-258285.
Of these, compound (2) showed a better selectivity profile for 5-HT$_6$ over a
range of alternative aminergic receptors.

While such indolines were susceptible to metabolic oxidation, indole
analogues of (2) proved to be both relatively stabler and more efficacious
antagonists. Several examples were found with excellent affinity for the 5-
HT$_6$ receptor [43]. For example, we prepared a chloroindole SB-699929 that
showed a pK_i of 8.6 for 5-HT$_6$. Indolylsulfonyl piperazines were the subject
of several patent applications independently filed at a similar time by GSK

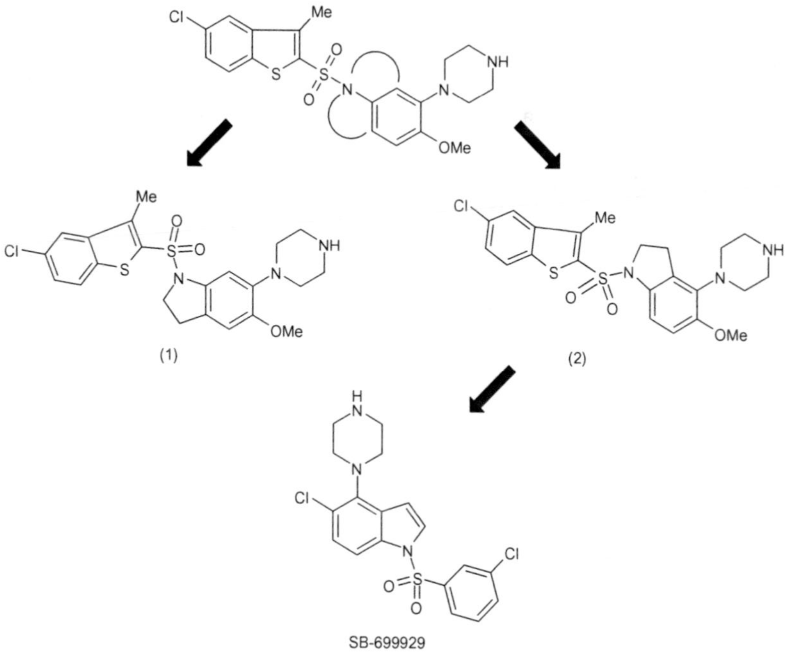

Fig. 5.2 Discovery of indolyl piperazinyl sulfonamides.

[44], American Home Products [45], Biovitrum [46] and Roche [47]. Unfortunately, we also found that such compounds often had a propensity to be substrates for CYP450 enzymes, and many members of this class were also cleared rapidly *in vitro* in liver microsomes. In an effort to reduce the potential for metabolic degradation, which we thought could relate to the electron-rich nature of the indole moiety, we sought compounds containing a range of less electron-rich heterocycles employed in place of indole. We also found that a sulfone was an effective replacement for the sulfonamide moiety, broadening our range of options.

This strategy ultimately led to the discovery of potent and selective 5-HT$_6$ receptor antagonists based on a piperidinylquinolinesulfone template. While the less electron-rich heterocyclic core of the quinoline gave rise to improved P450 inhibitory properties, some members of this class were also found to possess low intrinsic clearance. It was serendipitous that we found the high potency of the 3,8-substitution arrangement – a regioisomeric pattern hitherto scarcely reported in the chemistry of quinolines – which was also associated with a then uniquely high *in vivo* efficacy in the rat [48, 49]. We prepared a range of substituted analogues with many examples giving pK_i values >8. Compound SB-742457 showed particularly high affinity, demonstrated by a pK_i of 9.6, and was an antagonist against the human 5-HT$_6$ receptor, stably expressed in HEK 293 cells. It is currently in Phase II development for the treatment of cognitive deficits associated with mild to moderate AD. The use of this class of compounds in the treatment of pain and of irritable bowel syndrome is the subject of a more recently filed patent application [50].

SB-742457

New classes of 5-HT$_6$ receptor antagonists

Building on this discovery, we sought to exploit the quinoline template to see if high potency was solely the preserve of sulfones or also found with different linker groups. If the sulfonyl side chain is replaced by a sulfonamide then good levels of 5-HT$_6$ receptor affinity are retained. For example both isoindole (3) and anilide (4) amidic side chains are acceptable with pK_i values in the range 7.5–9 for a range of analogues, as measured in a DiscoverX assay of functional antagonism [51]. Although significant 5-HT$_6$ receptor affinity was retained if the sulfonyl group was replaced by a carbonyl or methylene linker [52], indicating the intrinsic affinity of the 8-piperazinylquinoline moiety, such compounds were significantly less effective antagonists than their corresponding sulfones, consistent with pharmacophoric considerations. Interestingly, sulfoxides retained potent affinity for the 5-HT$_6$ receptor but were not pursued because their polar character was likely to limit brain penetration.

Substitution on the terminal piperazine nitrogen led to compounds with higher lipophilicity; these often retained *in vitro* potency with pK_i values above 9, although N-dealkylation could be problematic *in vivo*. Larger alkyl groups led to a reduction in receptor affinity, so intermediate-sized substituents, such as cyclopropylmethyl (5), were found to be the most effective [53]. The terminal position has also proved an ideal site for ^{11}C labelling for use in Positron Emission Tomography studies; compound (6), GSK215083, (pK_i = 9.8) has been used extensively for this purpose [54].

Dual 5-HT$_6$ receptor antagonist/CB$_2$ receptor agonists

The piperazine group of SB-742457 could be replaced with alternative basic side chains with retention of significant affinity for the 5-HT$_6$ receptor [55], for example a dimethylaminopiperidine side chain, as in (7).

Oxygen-linked piperazine replacements also gave effective 5-HT$_6$ receptor antagonists [56], for example (8). While carbon-linked basic side chains generally afforded lower 5-HT$_6$ receptor affinity, interestingly, we found one class of such compounds to show both significant 5-HT$_6$ receptor antagonism and CB$_2$ receptor agonism [57–59]. For example, compound (9) has a pEC$_{50}$ of 7.6 as a CB$_2$ receptor agonist and a pK_i of 7.3 as a 5-HT$_6$ receptor antagonist.

In addition to the piperazinyl quinoline sulfone template, GSK patent applications have described a class of 5-HT$_6$ receptor antagonist based on a benzazepinesulfonamide template [60]. Such compounds possess two hydrogen-bond donors, which limited their brain penetration. We, there-fore, employed a strategy of conformational constraint and hydrogen-bond removal, as in the earlier series. This led to the discovery of novel 7,6,5-tricyclic-heterocycles such as (10) that showed a good 5-HT$_6$ receptor antagonist potency (pK_i = 8.6) and >10-fold selectivity over a range of other aminergic receptors, as well as a good P450 profile and *in vitro* metabolic stability in human liver microsome preparations, despite the unusual 3-chloroindole substitution pattern. In *ex vivo* binding studies, compound (10) showed an ED$_{50}$ of 4 mg/kg following oral dosing, as measured by displacement with the radiolabelled 5-HT$_6$ antagonist SB-258585 [61].

Aycliclic precursor   (10)

(11)   (12)

Sulfonyl indolines

A characteristic of many classes of 5-HT$_6$ receptor antagonists is a tendency to be substrates for P-glycoprotein mediated efflux from the brain and gut, limiting the potential for exposure. Typically, larger and more amphiphilic classes of antagonist are more effective substrates for this transporter. We noticed that certain low molecular weight antagonists were less susceptible to this liability and this prompted a re-examination of indoline-containing sulfone classes, thought originally to be less tractable due to probable poor metabolic stability (in particular a tendency toward CYP450-mediated oxidation to the indole). We found that the combination of a sulfonyl group on the 5-position with a piperidine group on the ring nitrogen afforded a potent stable class of low molecular weight 5-HT$_6$ receptor antagonists, both as indoles, such as (11), and indolines, for example (12) [62].

ROCHE

Early work

Roche were among the first companies to disclose a specific interest in selective 5-HT$_6$ receptor antagonists [41]. Their contribution (along with that of GSK, and others) to the early work in this area has been summarised in a number of reviews [22–31, 63, 64]. These papers and associated patent applications confirm the primary interest of Roche in the use of 5-HT$_6$ receptor antagonists as a treatment for cognitive memory deficits, though food uptake disorders are also discussed in these publications, as are other psychiatric indications.

Early publications indicated exploitation of the arylsulfonyl template and Roche medicinal chemists also made the move from secondary sulfona-mides to tertiary sulfonamides and sulfones, with their concomitant potential for better brain penetration, reflecting a reduction in the number of hydrogen-bond donors, as noted previously. Indolyl piperazinyl sulfone Ro-43-68554 was observed to be effective in a rat object recognition model employing either cholinergic (scopolamine pre-treatment) or seratonergic (tryptophan depletion) induced deficiencies. Reversal of memory deficit was observed with both i.p. and p.o. dosing at between 3 and 10 mg/kg, although the compound did not improve a time-related retention deficit [27].

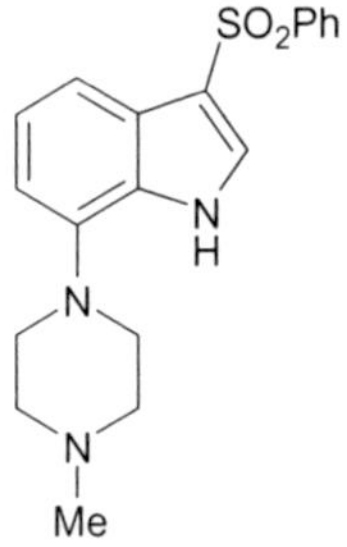

Ro-43-68554

Recent approaches

This review focuses on developments following on from the indole series. Subsequent Roche patent applications describe indole analogues in which the nitrogen heterocycle has been replaced by a benzofuran, a benzopyran or a carbocyclic ring system – presumably in an attempt to find analogues with improved metabolic stability and better potential for brain penetration.

Many Roche patent applications indicate that 5-HT_{2A} affinity is a feature of these chemical series, for example in a radioligand-binding assay, 5HT mimetic (13) exhibits a pK_i of 9.1 at 5-HT_6 and 7.7 at 5-HT_{2A} [65] while the chromane (14) shows a 5-HT_6 pK_i of 9.4 and a 5-HT_{2A} pK_i of 7.3 [66]. The alternatively substituted (15) shows equal affinity for the two receptors with a pK_i both at 5-HT_6 and at 5-HT_{2A} of 7.3 [67]. Further related examples include the much more potent tetrahydronaphthalene class, for example (16), with a pK_i at 5-HT_6 of 9.7 and (17) with a pK_i at 5-HT_{2A} of 9.6 [68].

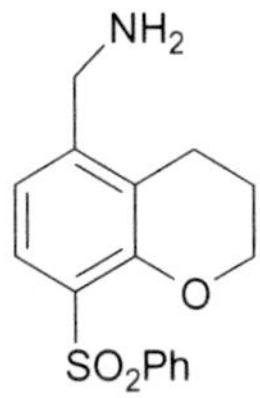

(13) 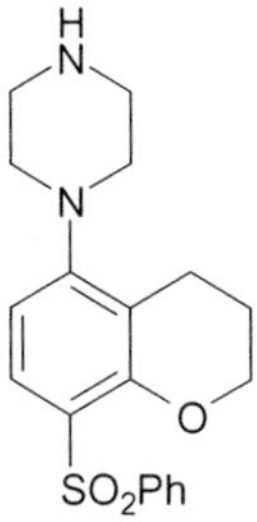   (14) 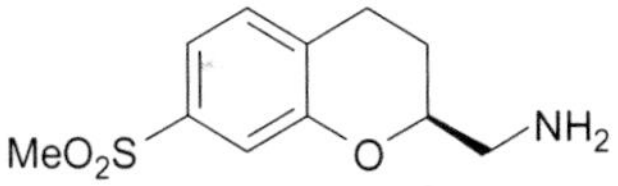   (15)

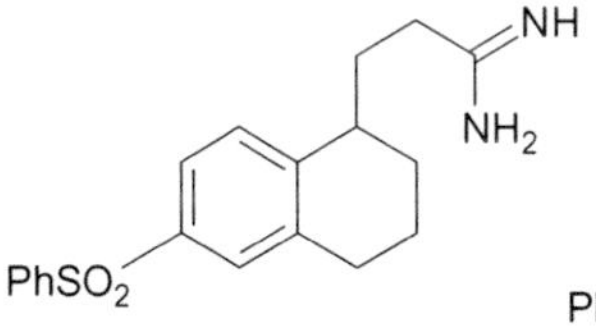

(16) 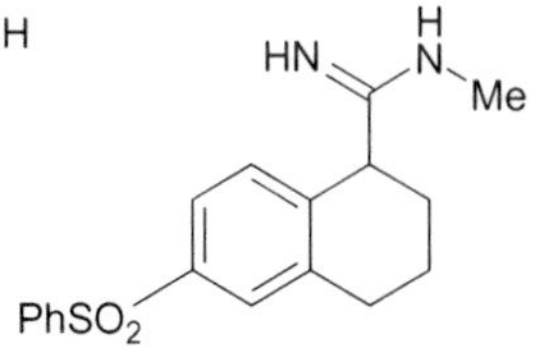   (17)

A patent application on fully unsaturated naphthalene analogues [69] indicates that a 3,8-substitution pattern is favoured (as found with SB-742457) and also discloses ligands with potent affinity for each receptor, for example (18) with a pK_i at 5-HT_6 of 9.4 and (19) with a pK_i at 5-HT_{2A} of 8.3. Unspecified analogues within this latter series were shown to be effective (dosed i.p.) in reversal of scopolamine-induced cognitive deficits in a NOR test using young Wistar rats.

(18)   (19)

A further class of highly potent tetralins containing a sulfonamide linker was disclosed in a more recent patent application [70], which includes, in addition to the usual discussion centred around cognitive enhancement, an

Table 5.1   5-HT$_6$ BINDING AFFINITIES OF 1-SUBSTITUTED 6-AMINOSULFONYL TETRAHYDRONAPHTHALINES

Examples	R	R^1	R^2	Isomer	5-HT$_6$ pK$_i$	5-HT$_{2A}$ pK$_i$
(20)	H	H	NH$_2$	Racemate	9.9	
(21)	F	H	NHMe	Racemate	9.1	
(22)	F	H	NHAc	Racemate	9.9	
(23)	F	H	NHC(O)NH$_2$	Racemate	9.7	
(24)	F	Me	NH$_2$	Racemate	9.6	6.9
(25)	F	H	NH$_2$	R	8.2	
(26)	H	H	NHC(Me)NH$_2$	Racemate	9.0	
(27)	H	H	NHC($=$NH)NH$_2$	Racemate	7.7	
(28)	H	H	NH-1-imidazoline	Racemate	9.5	
(29)	H	H	NMe$_2$	Racemate	9.7	

indication that such compounds may be efficacious in sleep disorders, feeding regulation, panic disorder, attention deficit disorder, withdrawal from drug abuse, disorders associated with spinal trauma and/or head injury (such as hydrocephalus) and functional bowel disorder. Compounds from this series are also reported to be effective in NOR studies. Examples are shown in Table 5.1. The information available in the patent application suggests that higher selectivity over 5-HT_{2A} was available from this series, though it is not clear whether Roche were aiming for dual activity or not.

If the basic group is made into a larger substituent than the imidazoline, for example (28), or a branched or open chain analogue, then 5-HT_6 receptor affinity starts to decrease, for example the glycine derivative (30) with a pK_i of 8. In addition to the compounds shown in Table 5.1, reversed sulfonamides also have high affinity, for example (31) has a pK_i of 8.9 for the 5-HT_6 receptor.

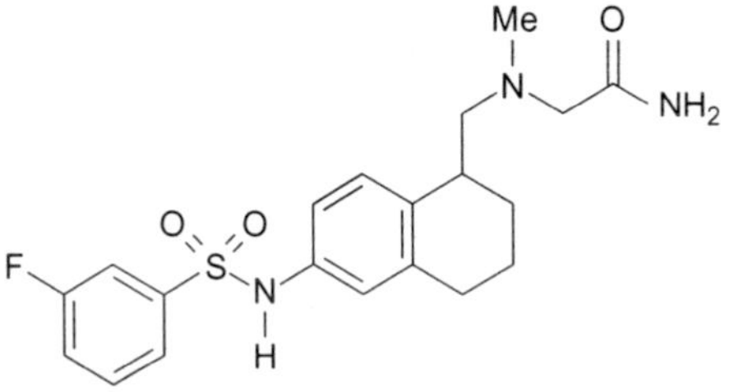

(30)

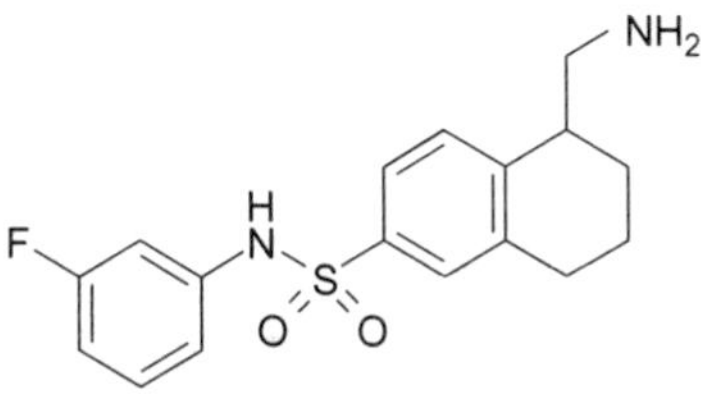

(31)

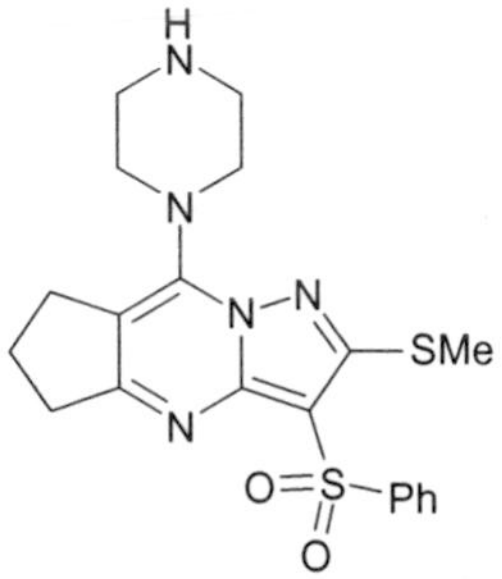

Ro-65-7674

These aromatic templates were further elaborated by Roche medicinal chemists into a potentially less metabolically vulnerable benzoxazine core.

This template is described as having been a developed following the discovery of Ro-65-7674, an example of earlier series of sulfones, based on a more complex 6,7-dihydro-5H-cyclopenta[*d*]pyrazolo[1,5-*a*]pyrimidine lead, which had showed sub-optimal levels of brain penetration despite good *in vitro* potency with a 5-HT$_6$ pK_i of 9.1 and a blood/plasma ratio of 0.1 [71]. Analogues based on the benzoxazine core offer exquisite levels of affinity for the 5-HT$_6$ receptor while moving the cLog *P* closer to a more ideal druggable target range and offering good blood to plasma ratios (Table 5.2) [72].

Halogen substitution on the 2- and/or 3-positions of the phenylsulfonyl group are beneficial for 5-HT$_6$ receptor affinity, for example compound (32). Although piperazine replacement by homopiperazine gave a compound of comparable potency, it offered no advantages, and replacement by other basic side chains tended to lower the potency. It is noticeable that Roche identified that while compounds from this class exhibited a good *in vitro* DMPK profile, including stability in microsomes and a low potential for P450 inhibition, more lipophilic analogues with a cLog *P* > 3 had a

Table 5.2   5-HT$_6$ BINDING AFFINITIES AND SELECTED PROFILES OF 3,4-
DIHYDRO-2H-BENZO[1,4]OXAZINE SULFONAMIDES

Compound	R	Ar	5-HT$_6$ pK_i	hERG inhibition at 1 μM	cLog P	Brain/plasma
(32)	H	2,3-Dichloro-Ph	10.2			
(33)	H	Ph	7.9			
(34)	H	2-F-Ph	8.9	9.9%	2.90	1.8
(35)	H	2-Cl-Ph	8.5	28.2%	3.34	1.6
(36)	Me	Ph	8.6	23.2%	3.25	1.0
(37)	Me	2-F-Ph	9.4	26.5%	3.45	
(38)	Me	2-CN-Ph	8.7	14.2%	2.79	

tendency toward significant hERG blockade, for example compounds (35)–(37).

It is a feature of several 5-HT$_6$ receptor templates that the core can be reoriented with retention of potency. This is demonstrated in a series of benzodioxins, analogous to the benzoxazines, where the point of attachment of the sulfonyl group is not to an amine but the aromatic core, while the basic group is pendant on the aliphatic ring. Though this change favours high 5-HT$_6$ potency, the substitution pattern also results in significant 5-HT$_{2A}$ activity. For example, while (39) has 5-HT$_6$ pK_i of 9.8, compound (40) has a pK_i at 5-HT$_{2A}$ of 7.7. This suggests that the two profiles favour slightly differing substitution patterns – perhaps with differing diastereomeric preferences for the basic side chain [73]. Compounds described in this patent application are reported to show activity in rat models of cognitive deficiency.

(39)   (40)

Non-sulfonyl-based ligands

Roche medicinal chemists were also among the first to successfully move away from the sulfonyl containing template with the disclosure of high levels of 5-HT$_6$ receptor antagonism based on a quinazoline moiety. For example, compound (41) shows a pK_i of 9.2 in a 5-HT$_6$ receptor ligand-binding assay, with a 100-fold or greater selectivity over a range of other aminergic receptors including D$_2$, H$_1$, M$_1$, M$_2$ and 5-HT$_{2C}$. Selectivity is much lower against the 5-HT$_{2A}$ receptor (pK_i = 8.7).

Compound (41) is a member of a well-explored class of 5-HT$_6$ receptor antagonists which contain a 6,6-heterocyclic core (including, for example SB-742457). However, the replacement of the sulfonyl by a methylene group was unique when first reported. It may be supposed that the ring-carbonyl group serves a similar role to one of the sulfonyl oxygen atoms. Patent applications around this motif have identified this series as targeting both the 5-HT$_6$ and 5-HT$_{2A}$ receptors, as for related sulfonyl containing analogues. Two compounds highlighted in the patent applications are (42) and (43) with pK_i values against the 5-HT$_6$ receptor of 9.9 and 8.9, respectively [74], however, the level of 5-HT$_{2A}$ inhibition is not reported. It may be that the function of the 7-substituent is to modulate log P or to block a site of metabolism.

(41)   (42)   (43)

A more recent patent application from Roche also disclosed that a number of 5-HT$_6$ and 5-HT$_{2A}$ receptor antagonists from this series were effective in a maximal electroshock seizure threshold test in reducing seizure threshold, and in several different models of cognitive restoration [75]. Building on these findings, Roche medicinal chemists moved the carbonyl group outside the heterocyclic core and discovered a related class of 5-HT$_6$ receptor antagonists with a pendant amide group, possessing reasonably high levels of 5-HT$_6$ receptor affinity [76]. The patent application cites some 40 examples, all of which show the 1,4- or 3,7-substitution pattern, for example (44) with a 5-HT$_6$ pK_i of 9.9 and (45) with a pK_i of 9.2.

Another Roche patent application [77] describes 3,8-substituted quinolines containing an aromatic side chain linked to the quinoline by an oxygen linker, for example (46). Bis-aryl ethers have been observed by other groups generally to show poor affinity for the 5-HT$_6$ receptor, however, (46) exhibits a pK_i of 9 at 5-HT$_6$. It may be that the 3-methyl group plays a crucial role in favouring the appropriate spatial disposition of aryl groups. No 5-HT$_{2A}$ activity is reported.

(44)   (45)   (46)

The prototypical non-sulfonyl-containing 5-HT$_6$ antagonist is methiothe-pin, and this has also been the subject of derivatisation by Roche medicinal chemists, who have found classes of analogues with potent affinity for both 5-HT$_6$ and 5-HT$_{2A}$ receptors. For example, (47) has a pK_i at the 5-HT$_6$ receptor of 9.6 and (48) is reported to bind to the 5-HT$_{2A}$ receptor with an impressive pK_i of 8.9 [78]. Seven-membered ring tricyclic amines are known to have a fairly rich pharmacology and whether these examples are significantly more selective is not revealed.

(47)   (48)   (49)

New sulfones

Most recently, Roche have returned to a simplified bis-arylsulfone core and describe potent activity in a new class of heterocycle [79], for example (49) with a 5-HT$_6$ pK_i of 10.2. In this patent application, all 140 examples have a *para*-substitution pattern on the benzene ring and the stereochemistry at the pyrrolidine ring is defined.

WYETH

Early research

Along with GSK, Roche and Biovitrum, Wyeth (formerly American Home Products, AHP) were early players in the 5-HT$_6$ story and have been a major force in patenting and publishing in the area both before and since 2005.

They were among a group of four companies to work on arylsulfonyl indolyl piperazines, described in a patent application published in 2002 [45]. Their more recent research builds on this legacy, and a number of the subsequent series are summarised in publications describing the strategy of exploiting conformationally constrained tryptamine derivatives [80, 81], indole sulfonamide ethers [82] and later a related azaindole series, none of which are reviewed here as they have been extensively discussed [83].

Tryptamines

Follow-on publications describe an approach to find activity by replacing the basic side chain of the early hit series by different amino-linked groups. Wyeth has reported an interest in an agonist profile and has extensively investigated a tryptamine template; native ligand analogues are arguably more likely to behave as a serotonin mimetic. Compound (50), which shows a 5-HT$_6$ K_i of 1.5 nM, exemplifies a class of indolylsulfones that retain potency with an ethylamine side chain – functionally a substituted tryptamine. Wyeth measure pK_i values determined by displacement of [^{3}H]-LSD from cloned human 5-HT$_6$ receptors stably expressed in HeLa cells. Recent patent applications include additional indications such as pain, and (undisclosed) 5-HT$_{2A}$ affinity [84].

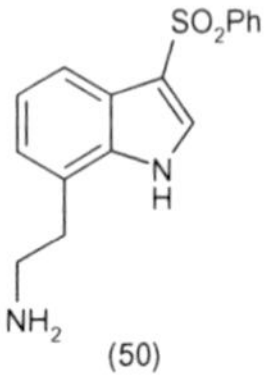

(50) 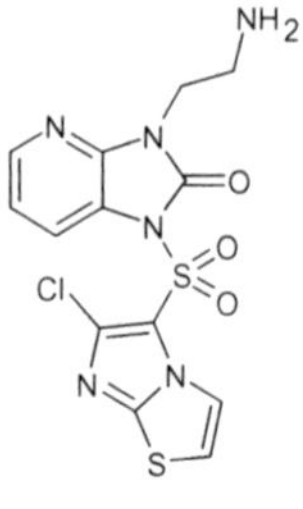   (51) 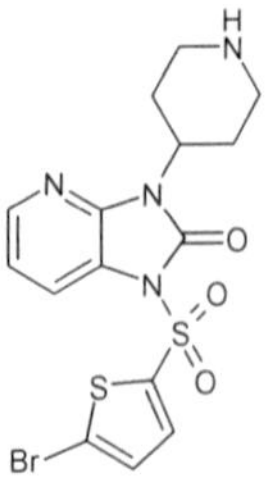   (52)

Imidazolepyrrolidines

The Wyeth group examined a wide range of heterocycles as replacements for the indole core, avoiding its known metabolic liabilities (*vide supra*). These included potentially more stable imidazolopyridines, as exemplified by compounds (51), with a 5-HT$_6$ K_i of 4 nM, and (52), incorporating a 4-piperidine as the basic side chain and with a K_i of 5 nM [85, 86]. Compound (51) was tested in the schedule induced polydipsia model of obsessive-compulsive disorder in rats, and shown to be effective at a dose of 10 mg/kg s.c. [87].

Indazoles

Another approach undertaken by Wyeth (and a number of other groups) was to replace the indole core by an indazole moiety. Using this template, a number of alternate basic side chains were examined in place of ethylamine or piperazine. This approach is exemplified by the 3-aminopropionamide (53) that shows an impressive affinity for the 5-HT$_6$ receptor for a non-piperazine, with a K_i of 0.5 nM, and proved to be a functional antagonist [88]. If the distance between the basic group and the heterocycle is increased further, there is a precipitous loss in affinity, however, the basic side chain could be substituted at the 4-position of the indazole while retaining excellent affinity, for example (54) which has a 5-HT$_6$ K_i of 1 nM [89]. Chemically closer to the piperazine moiety is the aminoazetidine group that could be introduced with retention of activity; for example (55) showed a 5-HT$_6$ K_i of 1.1 nM. It is worth noting that in this case, the indazole geometry is inverted and the point of attachment of the sulfonyl group is in the 3-position, leaving a relatively acidic indazole NH [90]. The principle therapeutic targets of the Wyeth team are similar to Roche and GSK: enhancement of cognition and memory in human diseases, such as AD, and as a treatment for epilepsy.

(53)   (54)   (55)   (56)

Constrained indazoles

A more recent development from this series is a group of ligands describing conformationally constrained amino side chains – often leading to fused-heterocyclic cores. By replacing the amide group of (53) with a fused-heterocyclic ring, another series of fairly potent 5-HT$_6$ receptor ligands was found, exemplified by (56) with a 5-HT$_6$ K_i of 31 nM. It is possible that these compounds resolved some of the hydrolytic issues presented by the pendant amide group and removed the potential for anilinic metabolites [91]. Surprisingly, preferred stereochemistry is not indicated in the patent application, nor is there any evidence of separate enantiomers being profiled. It is interesting to note that this series features a 3-substituted rather than 1-substituted indazole – perhaps an indication of the need to introduce some polarity into an otherwise very lipophilic series.

A further extension of this approach is seen in another patent application in which the indazole heterocycle is fused to form a hetero-tricycle, indicated in examples (57) and (58) with K_is of 1 and 5.1 nM, respectively. The general superiority of the piperazine side chain is apparent in several of these examples [92].

Interestingly, one patent application describes a regioisomeric series, exemplified by (59) and (60), as comparably potent at 5-HT$_6$ (K_i values of 1.6 and 3.0 nM, respectively), indicating the flexibility of substitution around the heterocyclic core [93]. Neither piperazine nor piperidine basic side chains are exemplified in this application.

(57) R = N-methylpiperazine
(58) R = OCH$_2$CH$_2$NHMe

(59) R = NMe$_2$
(60) R = 1-pyrrolidine

The indazole core also features in another series where the basic side chain is a piperidine group substituted at the 3-position of the heterocycle, for example (61) [94]. This time the favoured disposition of basic side chain and sulfonyl substituent is a 3,5-arrangement, which contrasts with findings for

(12), reported by GSK (*vide supra*), in which a 1,5-substitution pattern is claimed. A range of sulfonyl groups is described for analogues of (61), but the naphthalene sulfonyl substituent is always associated with the highest potency. It is interesting that the piperidine group is preferentially *N*-methylated in examples from this case; many other cases describe unsubstituted piperadines thereby avoiding the risk that N-dealkylation could lead to active metabolites.

(61)   (62)

Benzoxazepines

These patent examples illustrate that the naphthalene sulfonyl group features frequently in Wyeth publications. This particular sulfonyl substituent is also present in several of the most potent compounds highlighted in another patent application where the core heterocycle has been exchanged for a benzoxazole. In this case, the group evidently looked at the possibility of alkylating on the piperazine secondary amine, however, none of the examples disclosed is as potent as the parent compound (62), with a K_i of 3.1 nM at 5-HT$_6$; alkylation being associated with a 5–10-fold loss in affinity [95]. The corresponding compounds with a benzothiazole core were also described, but proved to be generally poorer ligands. Various points of attachment around the benzenoid ring were investigated for the sulfonyl group. The 4-position proved to be favoured [96]. These results, and the effects of different aryl substituents, are summarised in Table 5.3.

All of the potent examples described proved to be antagonists at the receptor but with IC$_{50}$ values in a functional assay significantly different from the K_i measures. For example, the lead compound (62) shows an IC$_{50}$ of only 96 nM for antagonism of 5-HT-stimulated cAMP formation. This is comparable to that of significantly lower affinity compounds such as (66) at 117 nM and (70) at 77 nM.

Table 5.3   5-HT$_6$ AFFINITIES OF SELECTED ARYLSULFONYL BENZOXAZEPINES

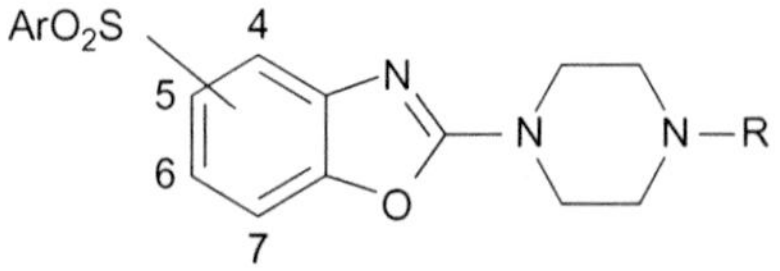

Examples	Position	Ar	R	5-HT$_6$ K$_i$ (nM)
63	4	Ph	H	41
64	4	3-F-Ph	H	54
65	4	4-F-Ph	H	64
66	4	2,5-diCl-Ph	H	13
67	5	1-Naph	H	52
68	6	1-Naph	H	196
69	7	1-Naph	H	7.1
70	4	1-Naph	Me	9.7
71	4	1-Naph	Et	12
72	4	1-Naph	iBu	176

Other heterocyclic cores

An alternative fusion strategy is described in another patent application that builds on an indole template. Compound (73) exemplifies a class of indolyl sulfones where the basic side chain is constrained by a fused tetrahydronaphthalene ring system. High levels of 5-HT$_6$ receptor affinity are reported in a binding assay, for example (73) has a K_i of 1 nM [97].

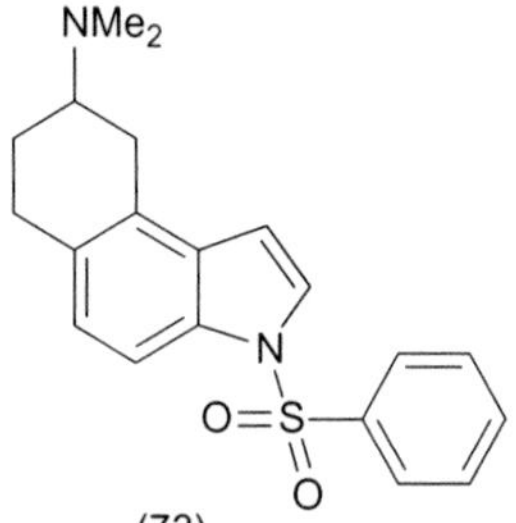

(73)

Like GSK, the Wyeth team also found activity in fused azepines and described a class of tricyclic analogues [98], which are illustrated in Table 5.4. Interestingly, in this series the highest affinity was associated with certain halogenated or alkoxy-substituted arylsulfones, such as (77) or

Table 5.4   5-HT$_6$ BINDING AFFINITIES OF INDOLE-FUSED AZEPINE
SULFONAMIDES

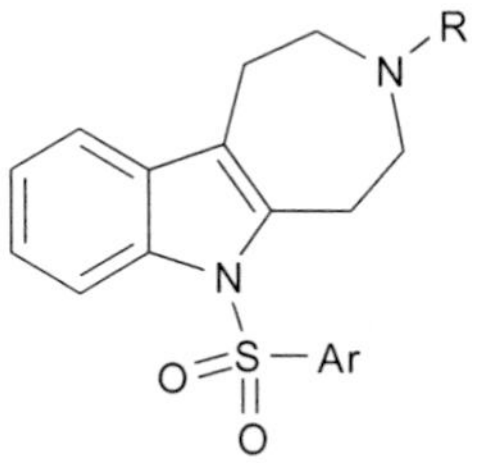

Examples	Ar	R	5-HT$_6$ K$_i$ (nM)
74	Ph	H	193
75	3-F-Ph	H	33
76	4-F-Ph	H	23
77	4-Cl-Ph	H	18
78	2-MeO-Ph	H	19
79	3-MeO-Ph	H	12
80	4-MeO-Ph	H	33
81	3-MeO-Ph	Me	19
82	3-MeO-Ph	Et	50
83	3-MeO-Ph	iPr	85
84	Ph	Et	12

(79). However, this substitution preference did not extend to benzazepines alkylated on the amino nitrogen, for which unsubstituted phenylsulfonyl analogue (84) showed the greatest 5-HT$_6$ receptor affinity.

An adaptation of the 6,6,5-tricyclic heterocyclic core of (73) illustrated above is described in two patent applications in which the phenylsulfone group is replaced by a Roche-like benzylamino substituent, flanked by a carbonyl group, and on an angular 6,6,5-heterocyclic ring system. Both patent applications show 5-HT$_6$ receptor affinities lower by at least an order of magnitude in comparison to typical sulfonyl-based antagonists, irrespective of the relative dispositions of benzyl substituent, heterocyclic core and basic side chain. However, with the appropriate substitution arrangement, high affinity ligands were obtained [99]. For example compounds (85), (86) and (87) have K_i values of 180, 380 and 11 nM, respectively. The second patent application explores a very wide range of angular tricyclic compounds [100]. These patent applications are particularly noteworthy for describing classes of 5-HT$_6$ receptor ligands possessing neither the sulfonyl nor piperazine features found in most series described hitherto.

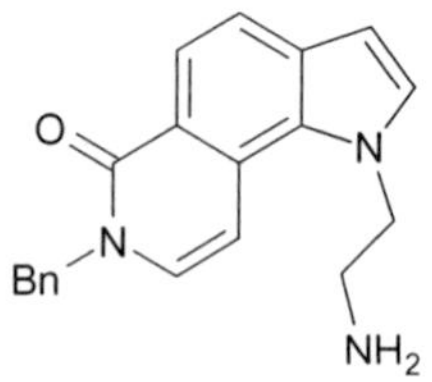

(85) 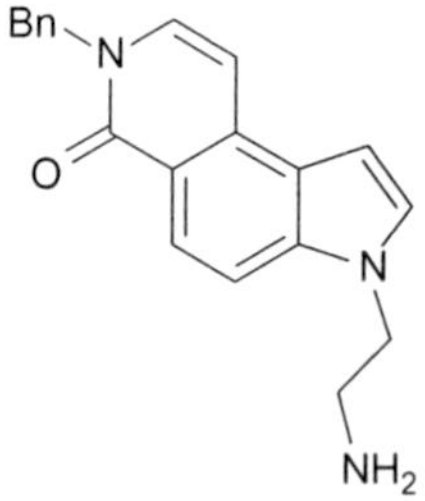   (86) 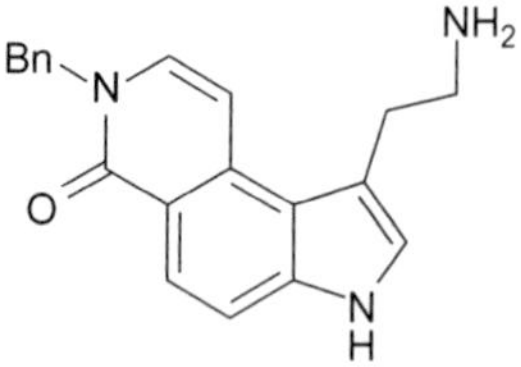   (87)

Efficacy claims

Wyeth underlined their continued interest in cognition targets by filing a patent application directed to a combination of any of the four indazole based lead 5-HT$_6$ receptor antagonists (88) to (91), (K_i values 1.1, 1.5, 1.9 and 0.6 nM, respectively), with another cognitive protective agent such as donepezil [101]. They claim synergistic effects in various model studies.

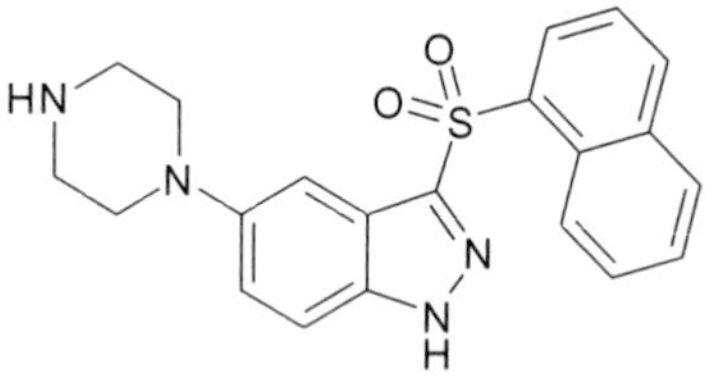

(88) 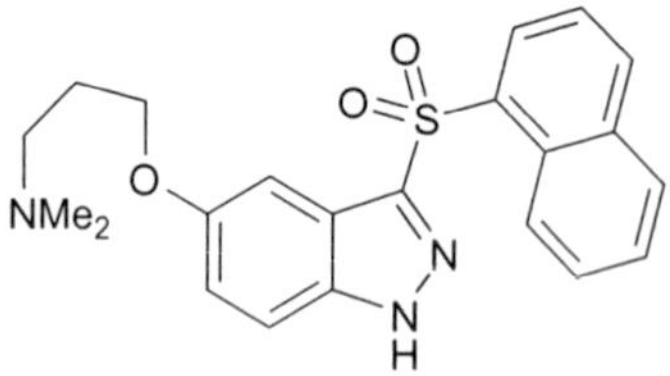   (89)

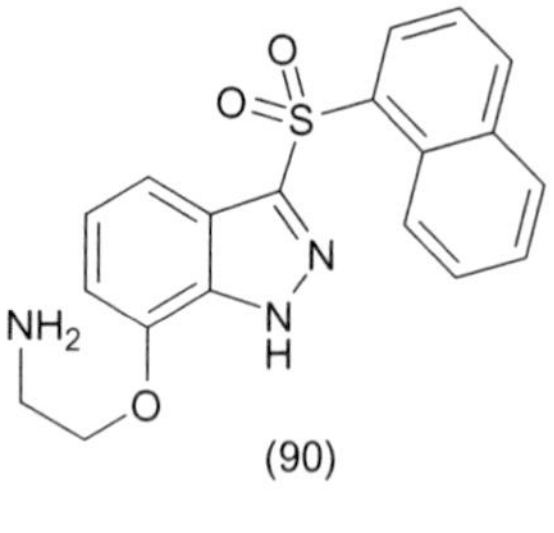

(90) 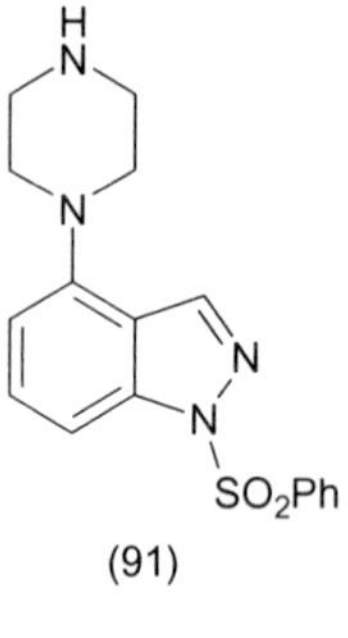   (91)

The efficacy of 5-HT$_6$ receptor antagonists in providing neuroprotection is discussed in a further use patent application from Wyeth, which indicates that compound (92), which shows a 5-HT$_6$ affinity K_i of 5.0 nM, has an

ED_{50} of 48 nM in an assay measuring neuronal survival under conditions of oxygen depletion. This patent application describes a number of techniques for assessing neuroprotective potential, including monitoring neurite outgrowth and the concentration of various brain factors, and gives comparative data for a number of other Wyeth compounds [102].

Alternative compound classes

One patent application from this group describing an indolylsulfonyl class of antagonists is unusual in that it indicates it is possible to achieve 5-HT$_6$ receptor affinity with examples that are neutral or acidic. The compounds claimed are potential metabolites of the tryptamine parent, for example the acetamide (93), with a K_i of 26 nM, and carboxylic acid (94), with a K_i of 248 nM [103].

Wyeth medicinal chemists also sought to exploit a 6,6-heterocyclic core and discovered a potent series of 5-HT$_6$ receptor antagonists incorporating a chromene or chromane moiety. For example (95), which retains the chromene double bond, has a binding K_i of 1 nM at the 5-HT$_6$ receptor [104]. Likewise 5,5-fused systems were examined [105]; among a range of heterocycles fused onto the pyrrole core, the thiophene derivative (96) proved to have the highest affinity for the 5-HT$_6$ receptor. This heterocycle is unusual as it somewhat resembles that of the imidazo[2,1-*b*][1,3]thiazole sulfonyl side chain of (93) (also highlighted in several patent applications from other companies). As a sulfonyl side chain this is readily introduced using a commercially available heterocyclic sulfonyl chloride. In contrast, incorporation of such a substituted thienopyrrole as the 'core' heterocycle requires a much more elaborate synthesis. Their findings suggest that the heterocycle itself provides a significant contribution to binding.

(92)

(93) R = CH$_2$NHAc
(94) R = CO$_2$H

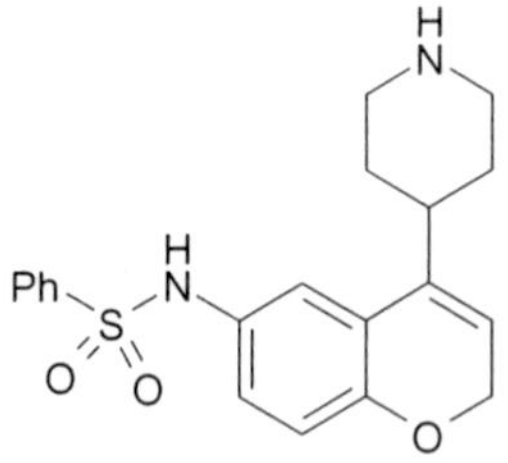

(95) 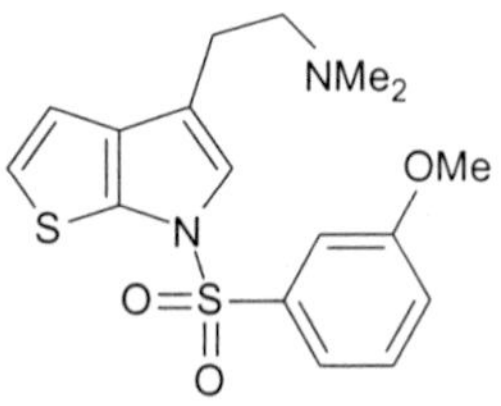   (96)

BIOVITRUM

Early studies

Scientists at Biovitrum have had a focus on 5-HT$_6$ receptor antagonists as a therapeutic target for the treatment of eating disorders, specifically reduction of food intake. Although cognition is discussed in patent applications filed by Biovitrum, the tenor of their publications suggests that effects on eating form their primary indication. They initially reported a link between indolyl piperazine sulfone classes of 5-HT$_6$ receptor antagonist and feeding effects [46] and have sought to build on these findings by optimising analogous series. Like other groups working in the area, their initial investigations focused on finding novelty by moving away from the indolyl piperazine template. Their initial leads evidently also showed significant 5-HT$_{2C}$ activity but subsequent reports do not highlight this aspect, suggesting that more selective compounds were generated by conformationally constraining the basic side chain and utilising heterocyclic replacements for the indole moiety. Like Wyeth, they have homed in on azaindoles, but have also found activity in tricyclic analogues such as (97) and a novel spiro-fused ring system exemplified by (98). This latter compound has a K_i of 140 nM at the 5-HT$_6$ receptor, as measured in a radioligand-binding assay [106]. Unspecified examples from this patent application were reported as showing a positive effect in a rat feeding model over three days.

Benzofuran analogues were also pursued, for example (99). Interestingly, in an echo of Wyeth's benzoxazoles, the team found that in this series the measured 5-HT$_6$ receptor affinity K_i values against the cloned human receptor tended to be an order of magnitude higher than IC$_{50}$

measures of antagonist potency. Thus, for example (99) has an affinity K_i of 2.3 nM but an antagonist potency in a cAMP accumulation assay of 24 nM [107]. Examples from this patent application were also described as efficacious in a feeding assay at doses of between 50 and 200 mg/kg. The patent application specifically cites diet-induced diabetes as a therapeutic target.

This work led to the discovery of benzofuran-linked 5-HT$_6$ receptor antagonists bearing an O-linked basic side chain, including (100), which showed a statistically relevant 4% reduction in rat body weight gain. The study was conducted over a 28 day dosing period, at a relatively low 5 mg/kg oral dose. This suggests that the effect of the antagonist does not reduce over time through toleration [108]. It is interesting to note that this compound is both a secondary sulfonamide and a secondary amine, so the central penetrancy of this compound may be limited; whether the significant effects of (100) may be mediated in part through a peripheral mode of action is therefore debatable.

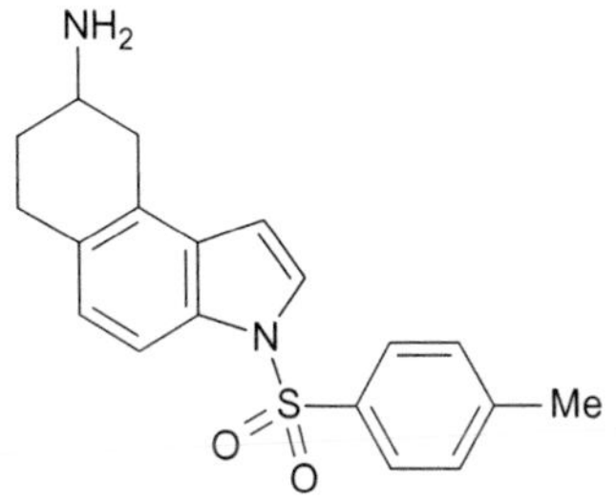

(97) 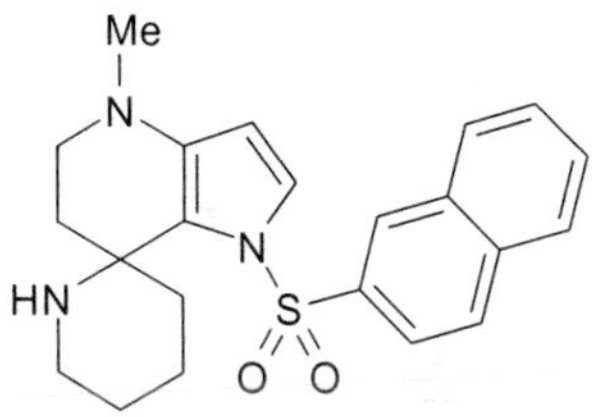   (98)

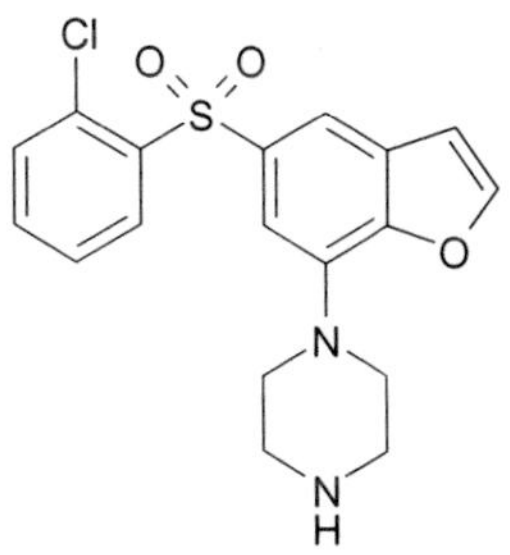

(99) 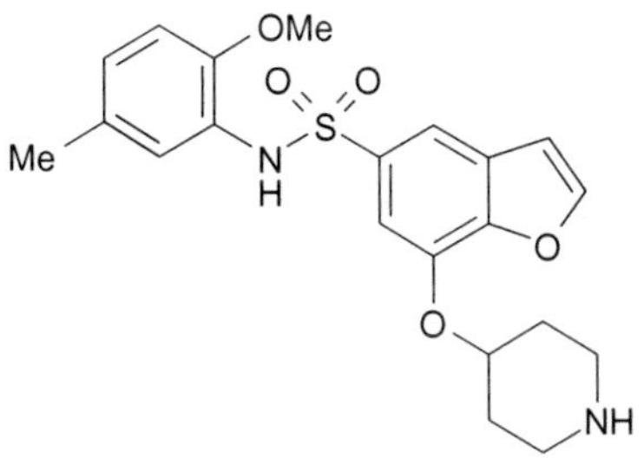   (100)

Recent focus

In more recent patent applications, the Biovitrum group have refocused on additional indolyl sulfonamide derivatives, in particular those bearing constrained or cyclic basic side chains, for example (101), (102) and (103) with 5-HT$_6$ K_i values of 0.6, 1 and 6 nM, respectively. This strategy may suggest that effects on weight gain are particularly associated with an indolylsulfonamide chemotype. A wide variety of sulfonyl side chains are described, including the benzothiazole example of (102), but few show significant advantages in terms of 5-HT$_6$ receptor affinity compared with those reported by other companies [109–111].

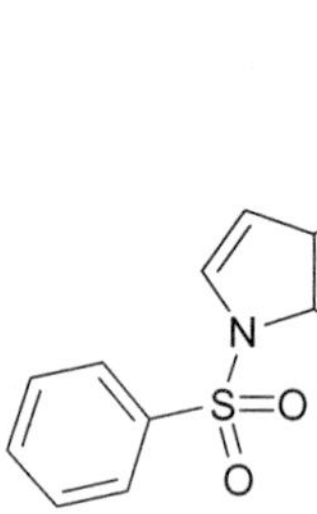

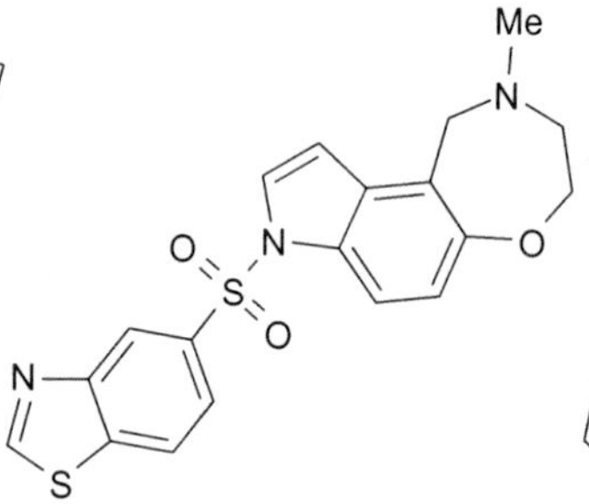

ESTÈVE

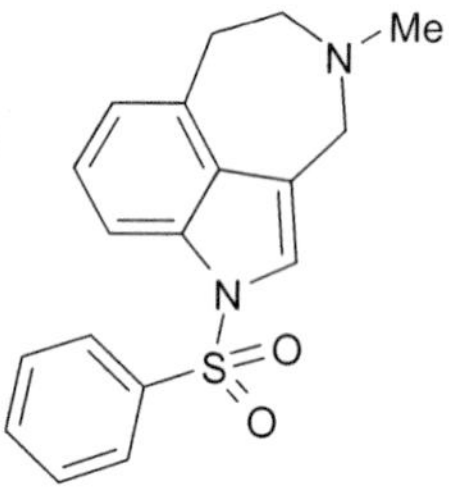

Lead design

Another major player in research into 5-HT$_6$ receptor ligands has been Laboratorios Dr. Estève, which like Biovitrum, has focused on the potential to treat obesity, though interest in cognitive disorders is also apparent from earliest patent filings [112]. Few companies have been more active in publicising their research, with some 17 patent applications appearing since 2005, as well as a number of primary literature publications. The company strategy for finding ligands was based on a computational modelling approach using a high potency lead molecule E-6837 and alignment with earlier external leads such as SB-271046 [113].

Indolyl sulfonamides

Early research appears to have focused on exploiting an indolyl sulfonamide template. Unusually, this group found high 5-HT$_6$ receptor affinity in aliphatic sulfonamides such as (104) with a K_i of 55 nM in a radioligand-binding assay involving displacement of tritiated LSD [114]. Regioisomeric substitution patterns were also examined, for example moving the sulfonamide group to the aromatic ring and the basic side chain to be linked via the indole nitrogen. The most effective inhibitor disclosed, compound (105), possessed a 1,4-substitution pattern and afforded a K_i of 1.9 nM. This patent application also claimed branched-side chains such as shown in (106). Interestingly, the Estève group also found the same favoured imidazothiazole sulfonamide highlighted by a number of teams including Wyeth [115, 116], though this perhaps represents the commercial availability of the sulfonyl chloride rather than specific optimisation around this scaffold.

(104)   (105)   (106)

NPY5 synergy

An unusual aspect discussed in certain Estève patent applications is the advantage of co-dosing a 5-HT$_6$ receptor ligand with a neuropeptide-Y5 antagonist. The group claims to have demonstrated synergy effects in a rat feeding model, although details linking these observations to specific compounds are lacking in the patent applications [117, 118]. This capability extends to another class of non-basic 5-HT$_6$ receptor ligands based on a benzoxazinone scaffold, for example (107) that has a 5-HT$_6$ K_i of 52 nM [119].

(107)   (108)

Indenes and alternative heterocycles

More recent patent filings indicate the group has now dispensed with an indole core and a series of ligands based on an indene template has been described. The most significant example is compound (108) with a K_i of 4.8 nM [120]. In a recent publication [121], Alcalde *et al.* have reviewed the strategy employed by Estève medicinal chemists for optimizing indolyl sulfonamides leading to the discovery of this compound, now identified as E-15136, and reveal (108) to be a full agonist with a pEC_{50} of 9. Their results indicate that many indane-linked compounds are agonists rather than antagonists at the receptor, and suggest that this is the preferred mode of action for compounds targetting eating disorders.

A set of six British patent applications filed during the review period disclose a range of chemotypes having affinity for the 5-HT$_6$ receptor for use in treating food-related disorders. Most intriguing are a class of fluorophenyl substituted imidazoles. Inhibition constants for these compounds are not given; instead the percentage inhibition of binding radiolabelled LSD using a 10 nM solution is reported. Thus, it is apparent that compound (109), identified as E5535, with 100% measured inhibition, showed the highest *in vitro* binding to the 5-HT$_6$ receptor [122]. Related to this series, but possibly less potent, is a class of nitriles exemplified by (110), disclosed as E-5318, with 85% inhibition. Two classes of piperazine derivatives exemplified by (111), (E-4848) and (112) (E-5323), showed 56% and 81% inhibition, respectively. Under the same conditions, a series of oximes exemplified by (113) (E-4251) exhibited 60% inhibition [123–126]. The origins of the compounds in these first five patent applications is unclear, but it is interesting to note that, irrespective of 5-HT$_6$ receptor binding affinity, a wide range have been profiled in feeding studies. It is therefore possible that some of their *in vivo* activity may be due to other mechanisms of action.

In addition to these examples, a class of sulfonyl containing heterocyclic compounds capable of binding to the 5-HT$_6$ receptor was disclosed. For example (114) (E-3894) shows 67% inhibition, though whether these compounds are related to the main classes of 5-HT$_6$ receptor antagonist remains uncertain [127]. Members of all six patent applications are active in a rat feeding paradigm. It is noteworthy that none of these patent applications describe the 5-HT$_6$ receptor ligands as antagonists, leaving open the possibility of different modes of action at the receptor.

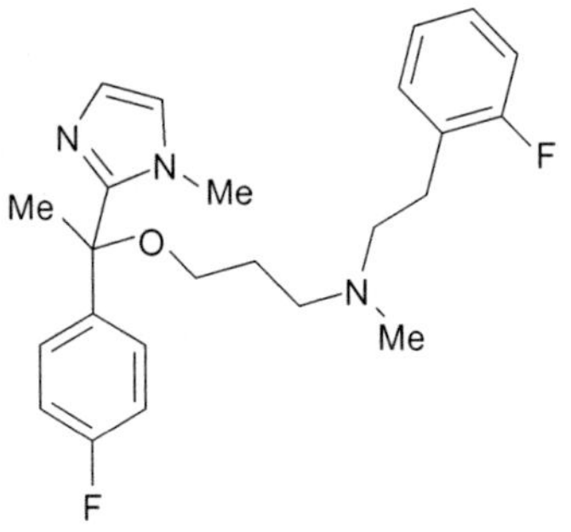

(109) 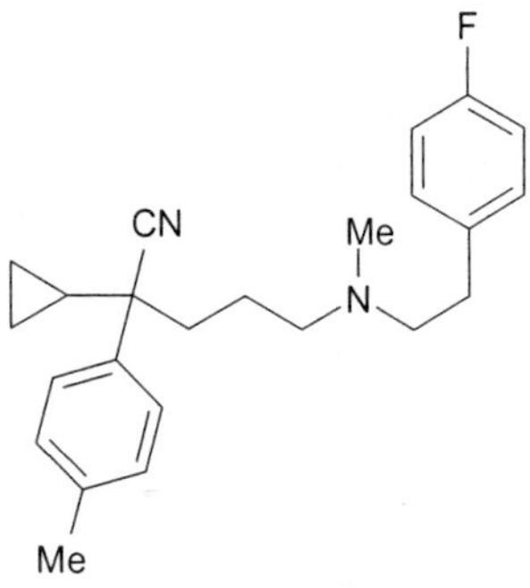   (110)

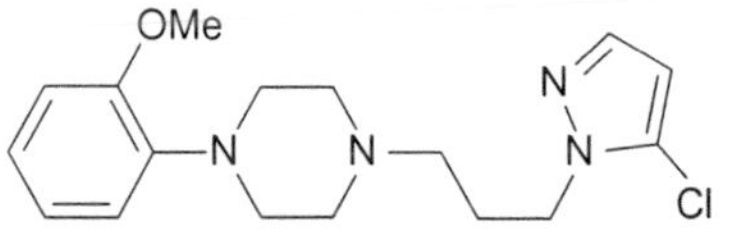

(111) 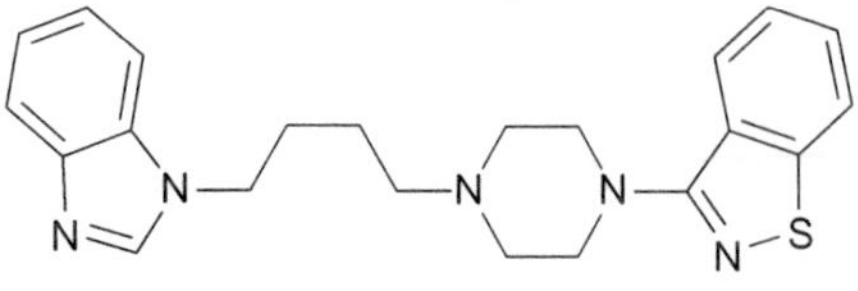   (112)

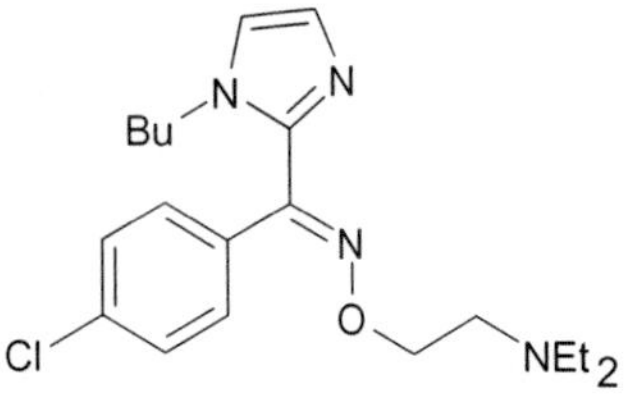

(113) 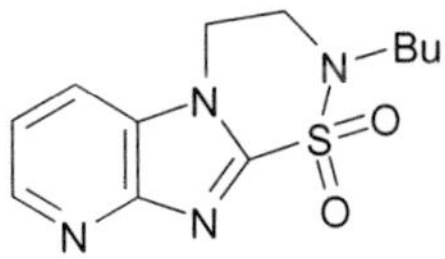   (114)

Recent filings

More recent patent applications from Estève have focused on a conventional 5-HT$_6$ receptor antagonist tryptamine pharmacophore: indolyl sulfonamides. Examples described include chemically interesting substitution patterns such as 3-(2-amino)ethoxyindole-1-arylsulfonamides [128]. The number of compounds exemplified may indicate that they are intended to protect structures of key interest, indeed some examples are within the scope of broader patent applications filed some years earlier. For example, compound (115) is a tryptamine analogue with a K_i of 1.4 nM for binding to the 5-HT$_6$ receptor; one of six close analogues with single figure nanomolar inhibition constants. Interestingly, the most potent compounds discussed have three hydrogen-bond donors and may be zwitterionic [129]. Both eating and cognitive disorders are mentioned. Related to these are examples in which the sulfonamide group contains a heterocycle. For example, the very potent benzothiophenesulfonamide (116), with a K_i of 0.28 nM against the 5-HT$_6$ receptor, is described in a second patent application in which most examples contain a tertiary basic group [130]. A third patent application describes analogues specifically with naphthylsulfonamide side chains such as (117), with a K_i of 2 nM against the 5-HT$_6$ receptor; all compounds described are closely related to E-6837 [131] and include potential metabolites of this key compound.

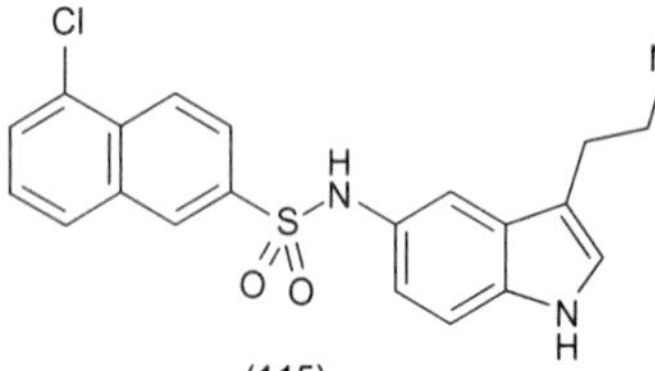

(115)

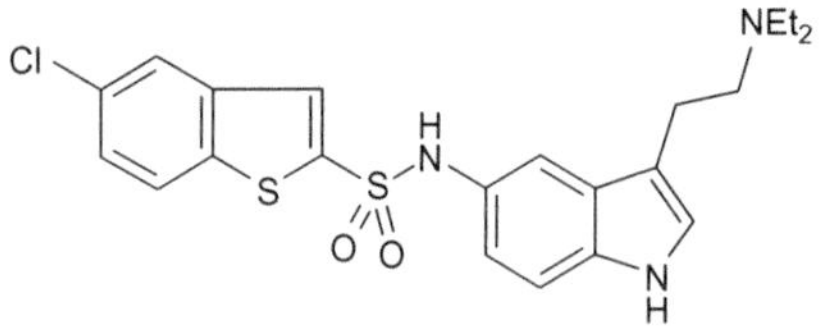

(116) 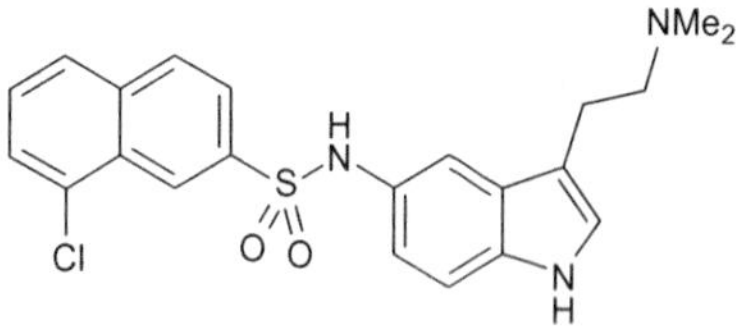   (117)

A tetrahydroisoquinoline piperidine sulfonamide class exemplified by (118) ($K_i = 31$ nM) is, surprisingly, distinct from earlier published series [132], though clearly related. In another case, a guanidine-hydrazone basic side chain is claimed for an indane-based ligand exemplified by (119),

(85% binding inhibition at 10 nM), which is interesting as the functionality may render this class peripherally restricted and perhaps suggests that compounds that are excluded from the CNS may a particular focus for this group [133].

(118)   (119)

Potential development compounds

Compound E-6837 has been a key tool compound for Estève, and has been shown to induce hypophagia and sustained weight loss in diet-induced obese rats, being orally active at 30 mg/kg. It is a 5-HT_6 receptor partial agonist at rat 5-HT_6 ($E_{max} = 67\%$) and a full agonist at the human receptor with a pEC_{50} of 9.2 and a pK_i for binding of 9.1. The compound was able to demonstrate a greater sustained weight loss than silbutramine in the rat feeding model [39]. The development status of this compound is unclear. However, in one of their most recent patent applications the group highlights potential for the use of the 5-HT_6 modulator E-6837 in the treatment of drug-induced weight gain. Figure 5.3 illustrates efficacy in reducing weight gain induced by the anti-diabetic thiazolidinedione rosiglitazone, in a rat model [134].

SUVEN

Although relatively late entrants into the field of 5-HT_6 receptor research with their first publications in the area dating from 2005, Suven have now

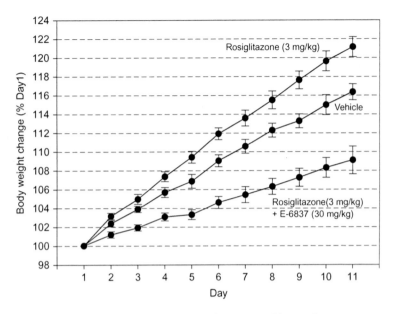

Fig. 5.3   Effect of E-6837 on rat weight gain caused by rosiglitazone.

disclosed a sustained synthetic research effort backed up by a number of patent applications, as well as preliminary clinical data. Early patent applications focused on a typical 5-HT$_6$ pharmacophore: indolylsulfonamides with pendant constrained basic side chains at the 3-position. Example (120) is illustrative, with a K_i of 2.5 nM for displacement of radiolabelled LSD from human 5-HT$_6$ receptors expressed in HEK293 cells. Reduction of food intake is a declared primary therapeutic target. However, the group also claims activity against a wider range of disorders from CNS conditions to pain and cardiovascular activity [135]. These compounds resemble (partial) agonist series described by Estève and others. A related patent application published around the same time utilises the same basic side chain, but describes a constrained sulfonamide in a 6,5,5,6-heterocycle. These compounds have generally much poorer affinity for the receptor, for example (121) with a K_i of 88 nM [136], though this is still comparable with the activity of methiothepin ($IC_{50} = 0.1\,\mu M$) in their assay.

(120)   (121)

A subsequent set of patent applications describes piperazine indolyl sulfonamide compounds in which the piperazine group is connected through the 3-position of the indole, for example (122), which shows a K_i of 9.2 nM, however, the chemical stability of such compounds is not discussed [137]. The primary claim for these compounds may now be psychiatric disorders (e.g. anxiety, obsessive-compulsive disorder *etc.*) though cognition and eating disorders are included. A therapeutic rational for both agonists and antagonists is presented in this Suven patent application, although only antagonist activity is disclosed for compounds described. A number of *in vivo* tests for 5-HT$_6$ receptor mediated actions are also discussed. They demonstrate that 5-HT$_6$ antagonists reverse the suppression of a chewing/yawning/stretching behaviour in the rat, caused by physostigmine. Suven scientists have also employed Morris water maze and passive avoidance studies to gain insights into the efficacy of the compounds as cognitive enhancers and anxiolytics, respectively.

A related series of carbazole derivatives provides an alternative constraint for the basic side chain. The bromo-substituent in the benzenoid ring appears to confer a slight potency advantage, for example (123) has a K_i of 5.9 nM. However, a range of aryl substituents was tolerated in the sulfonamide group [138]. A third patent application from this group describes a distinct class of methylene linked sulfones, joined to the 2-position of the indole ring. While linked pharmacophorically with the earlier examples, this class appears to have somewhat lower affinity for the receptor, for example (124) shows 54% displacement of radioligand at a concentration of 100 nm [139]. The functional activity at the receptor is not described.

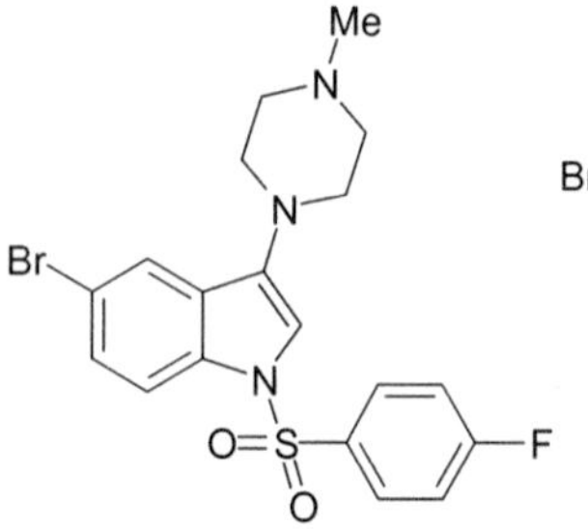

(122) 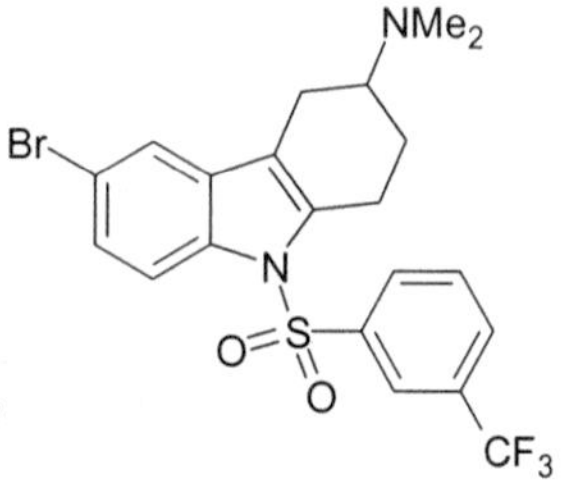   (123)

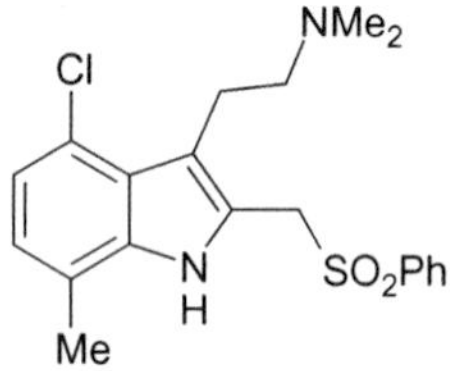

(124)

Recent patent applications

A number of more recent disclosures from Suven indicate a focus on 5-HT$_6$ receptor ligands as anxiolytics and for cognition disorders. The indolyl sulfonamide (125) inhibited 5-HT$_6$ radioligand binding with a K_i of 1.5 nM; it also displayed a positive effect in an object recognition test at 10 mg/kg, and in a Morris water maze assay, reversed scopolamine-induced deficits in spatial learning in rats. While this compound did exhibit brain penetration, it had an oral bioavailability of just 6% [140]. Unusually, (125) bears the basic substituent on the sulfonamide. The distance between the sulfonyl and basic group is evidently not as critical, as the homologue (126), with an extra methylene in the side chain, is comparably potent (K_i = 2.9 nM). An *O*-linked side chain is also described in a related patent application [141], exemplified by compound (127) with a K_i of 1.6 nM. A conventional indolyl sulfonamide substitution pattern is described in a recent patent application [142]. In this case, an aminopiperazine basic substituent furnished the most potent example, compound (128), which showed a K_i of 3.3 nM. The Suven team demonstrated that if the basic side chain is linked by a methylene group, strong affinity is also possible, for example the 4-substituted compound (129) had a K_i of 2.5 nM [143], and the 5-isomer (130) possessed a K_i of 0.94 nM [144].

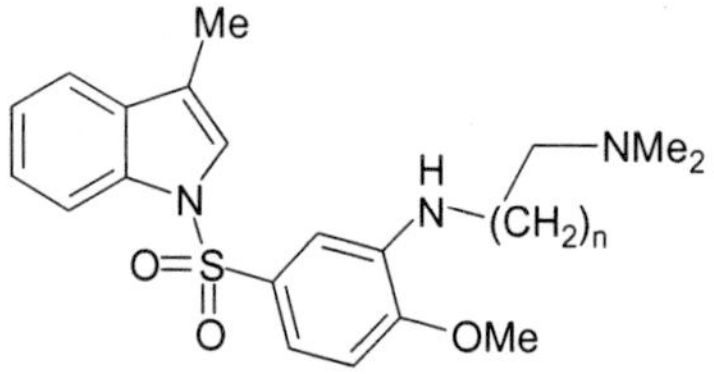

(125) n = 1
(126) n = 2

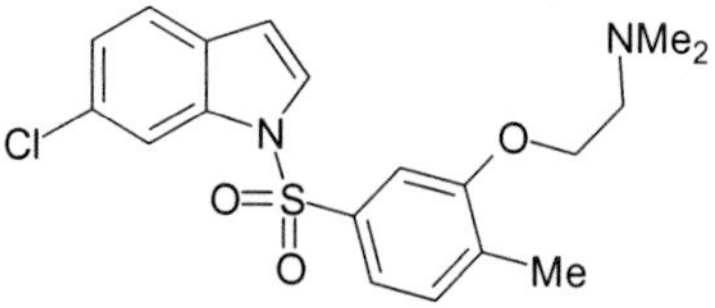

(127)

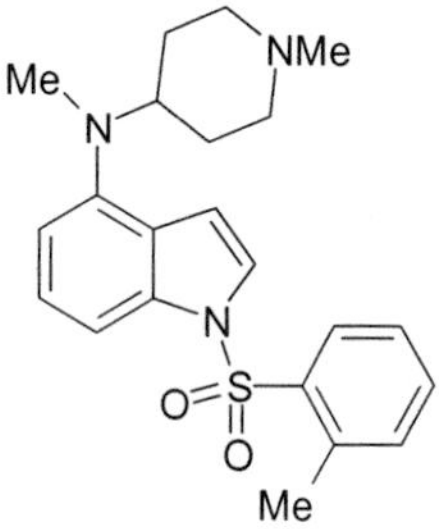

(128)

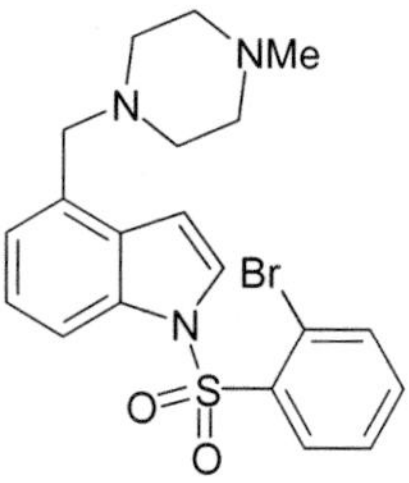

(129)

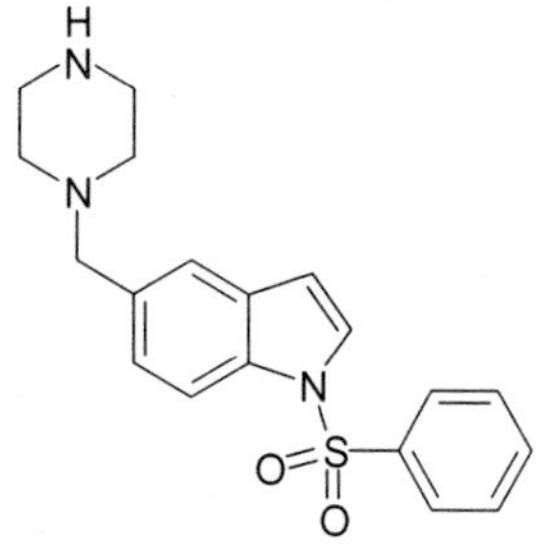

(130)

KRICT

Chemists from the Korean Research Institute of Chemical Technology (KRICT) have described 5-HT$_6$ receptor antagonist ligands in a number of patent applications and papers published since 2006. Their early work is described in a paper, co-authored with groups from several US academic institutions and the Korean Institute of Science and Technology, in which they outline the synthesis and activity of a group of piperazinesulfonyl derivatives somewhat unrelated to most 5-HT$_6$ receptor ligands described hitherto [145]. The most potent of these examples (131) is an antagonist with a modest IC$_{50}$ of 1.5 μM, determined by a binding assay measuring displacement of tritiated LSD from CHO cells stably expressing human 5-HT$_6$. An analysis of their reported SAR suggests that an electron withdrawing group in the 'northern' phenyl group is advantageous, while the substituent on the anilinic ring has only a marginal effect on receptor affinity.

(131)

In their patent applications, KRICT mainly discuss compound classes based around a non-sulfonyl-containing tetrahydroquinolindione series, structurally similar to compounds described by AstraZeneca. It is possible that the dicarbonyl groups on the central heterocycle help both to orient the pendant phenyl group and serve to fulfill one or other of the binding interactions performed by sulfonyl oxygens in other series. It is equally possible that these ligands bind in a different manner; although (132), with a IC$_{50}$ of 15 nM, incorporates a weakly basic anilinic group, non-basic analogues also still show significant affinity for the human 5-HT$_6$ receptor. Compound (132) also has some affinity for the D$_1$ receptor (38.7% inhibition at 10 μM), and this seems to be a characteristic of this class. Examples from this patent application are reported to have a mild anti-psychotic effect in a methamphetamine-induced hyperactivity rat model [146].

The link between the early hits and the standard 5-HT$_6$ receptor pharmacophore is demonstrated by the development of analogues to initial hits bearing a pendant piperazine side chain in place of the 5-position chlorine, for example compound (133) with an IC$_{50}$ at the 5-HT$_6$ receptor of 1.4 nM. In this case, there was significant selectivity for the (S)-isomer (133) over the (R)-isomer (134) (IC$_{50}$ = 9 nM). These, and related examples, show a two to three orders of magnitude selectivity for the 5-HT$_6$ receptor over a range of aminergic receptors including 5-HT$_7$, 5-HT$_{1A}$, 5-HT$_{2A}$, 5-HT$_{2C}$, D$_1$, D$_2$, D$_3$ and D$_4$. The role of the additional benzyl group at N-1 is not clear [147]. This improved class of antagonist is discussed as being of use in diseases associated with feeding, learning and memory dysfunction.

Still more potent analogues were found when the 4-position carbonyl group was replaced by a sulfonyl moiety – whence the link to earlier classes of 5-HT$_6$ receptor antagonists becomes more explicit [148]. Compounds exemplified, such as (135) with a K$_i$ of 0.7 nM, show an affinity comparable to the best of the sulfonyl piperazine series and share the 3-fluoro substituent that was present in potent analogue GSK215083 (6), suggesting a similar binding mode. Studies reported in this patent application indicate that compound (135), with a 5-HT$_6$ IC$_{50}$ of 0.9 nM, was not a sedative, as evidenced by a rotarod study, but had an effect in apomorphine induced disruption of pre-pulse inhibition – a test of learning and memory. Compound (136) maintained an excellent selectivity of > 100-fold over other specified aminergic receptors and was identified as being chosen for further pre-clinical evaluation.

The most recent patent application to emerge from this group identifies the related quinazolinedione core. From over 70 examples, compound (137) is identified as the most potent with an IC$_{50}$ of 0.5 nM against the human 5-HT$_6$ receptor [149].

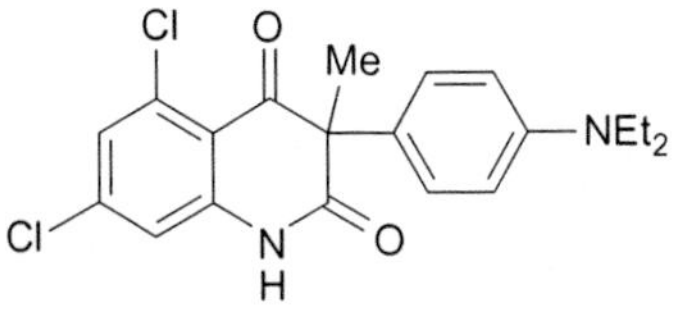

(132)

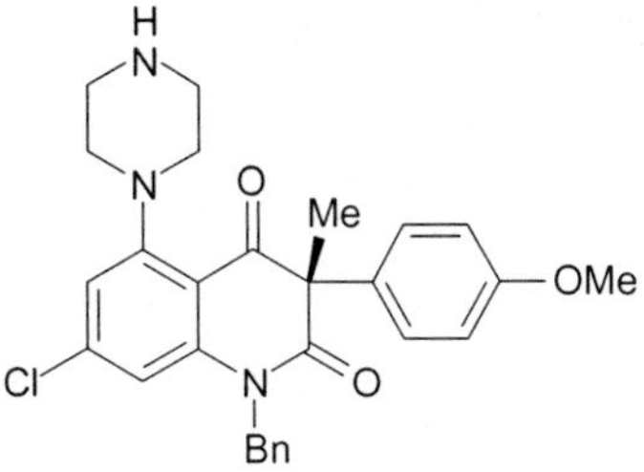

(133) (S)-isomer (shown)
(134) (R)-isomer

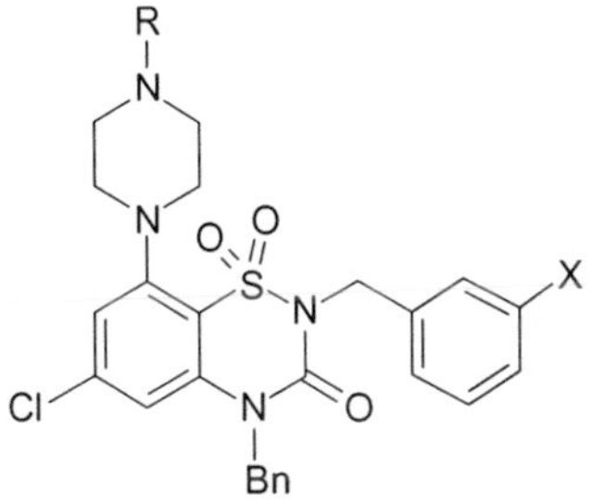

(135) X = F, R = Me
(136) X = H, R = H

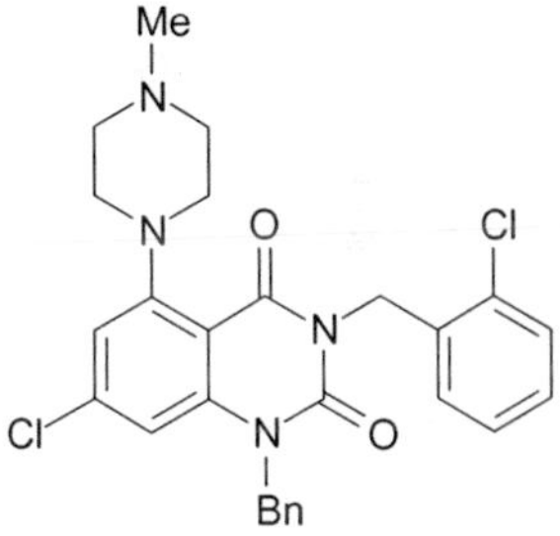

(137)

ASTRAZENECA

AstraZeneca (AZ) researchers have published a number of 5-HT$_6$ receptor ligands in patent applications originating in 2006 and 2007, and more recently for a broader range of therapeutic indications related to 5-HT$_6$ modulation. Their early patent applications are based around classical sulfonamide templates related to SB-271046, for example (138), with a K_i of 1.7 nM in a radioligand-binding assay, is clearly a conformationally constrained analogue. A number of variations to the base-substituted heterocycle are described [150]. The reverse sulfonamides were also found to be active [151]. In addition to tetrahydronaphthalenes, chromane derivatives, such as (139), were comparable in their affinity for the receptor.

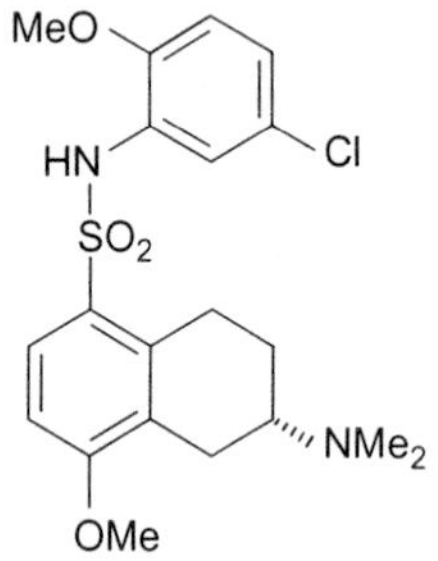

(138) 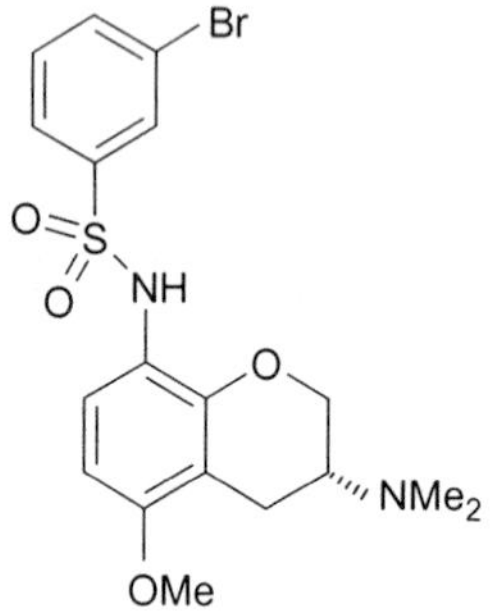   (139)

The next group of patent applications from AZ shows another interesting parallel with work undertaken elsewhere as they focus on 6,7-membered azo-fused ring systems; this time principally benzoxazepines rather than the benzazepines which were pursued by GSK. Both benzoxazepine sulfonamides, such as (140) with a 5-HT$_6$ K_i of 31 nM, and reverse sulfonamides such as (141), with a K_i of 82 nM, are described [152, 153].

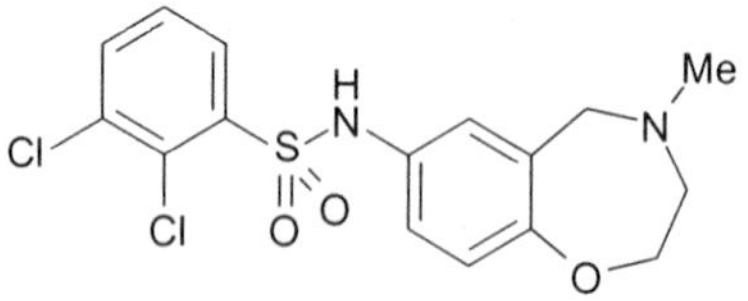

(140) 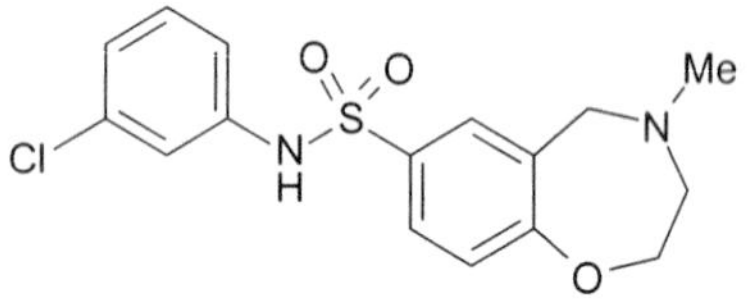   (141)

AZ patent applications specify a range of potential indications for 5-HT$_6$ ligands, including anxiety, depression and convulsive disorders, obsessive-compulsive disorder, migraine, memory dysfunction, sleep, feeding, drug withdrawal, schizophrenia, attention deficit hyperactivity disorder (ADHD), dementia, pain and impaired neuronal growth. However, they do not specify whether AZ compounds act as agonists or antagonists of the 5-HT$_6$ receptor.

A third group of patent publications centres around quinazoline templates. Two series are disclosed, one exemplified by (142) with a K_i of 200 nM, and a more optimised class incorporating features such as a fluoroethyl group, perhaps in an attempt to moderate metabolic liabilities, for example (143) with a K_i of 1.4 nM [154–156]. No functional data is presented in these claims and the lead compounds in each patent application are not readily identified.

In a structurally unrelated patent application from the AZ group, the core heterocycle of (138) is reversed and a class of tetrahydronaphthaline piperazines substituted by aryl sulfonamides is described. This patent application highlights compound (144), with a 2,8-substitution pattern, as having a K_i of 0.8 nM [157]. It is interesting to note that the AZ team appear to have changed the focus of their research in their recent disclosures, with newer compounds based around a more classical 5-HT$_6$ pharmacophore incorporating features described in several of the series already discussed, including the piperazinyl quinoline moiety and naphthyl sulfonyl substituent.

(142)   (143)   Me   (144)

ABBOTT

The Abbott Group have also entered the 5-HT$_6$ arena relatively recently, with seven patent applications published during 2007. However, their applications incorporate very large numbers of examples, reflecting a sustained research effort. Their first publication [155] describes a class of compounds that have affinity for both the 5-HT$_6$ and D$_3$ receptors. The Abbott Group notes that this characteristic makes them suitable for targetting disorders associated with both of these receptors, for example AD-associated cognitive dysfunction, depression, schizophrenia and Parkinson's disease (PD). Their preferred compounds conform to the classical 5-HT$_6$ receptor antagonist pharmacophore: a phenyl core with a basic group and arylsulfonamide unit attached in either a *meta*- or *para*-position with respect to one another. Over 3000 examples are given, with compounds described exhibiting good affinities for the 5-HT$_6$ receptor ($K_i < 50$ nM) and some compounds achieving similar levels of affinity at the D$_3$ receptor. For example compounds (145) and (146) are both specified as having K_i values at 5-HT$_6$ and D$_2$ of < 10 nM while being > 150-fold selective for D$_3$ over D$_2$.

A second patent application published during the same year [158] discloses more than 4,500 examples, again with the claimed templates having a classical 5-HT$_6$ receptor antagonist pharmacophore. The affinity range for both 5-HT$_6$ and D$_3$ receptors is similar to that specified in the first patent application, but there is particular mention that treatment using a molecule with dual 5-HT$_6$/D$_3$ receptor antagonism can lead to increased levels of acetylcholine in the medial prefrontal cortex and hippocampus, when compared to that with a selective D$_3$ ligand alone. Typical examples include (147), (148) and (149), which contain a pyridyl core.

(145) R = Me
(146) R = Cl

(147) R = H
(148) R = Pr

(149)

A third patent application contains some 300 examples around a class of compounds comprising a 6,5-bicyclic core, with azetidine as the basic group, and an arylsulfone side chain. These are reported as potent and selective 5-HT$_6$ receptor ligands [159] targeted at treating a range of neurological and psychiatric disorders including cognitive dysfunction in AD and schizophrenia, PD, and eating and sleeping disorders. All the compounds of the patent application are reported to exhibit fairly good affinity for the 5-HT$_6$ receptor ($K_i < 250$ nM), with a number of examples having a K_i of < 5 nM in a ligand-binding assay. A typical example from this latter class (150) is also reported to be at least 200-fold selective over H$_1$, D$_2$ and α_1-adrenergic receptors.

The Abbott group has published two patent applications that claim compounds in which there is a quinoline heterocycle at the core of the structures [160, 161]. In the former case, their arylsulfonyl heterocycles are related to templates discussed in earlier sections, with the difference that they describe a preference for a carbon-linked basic side chain, instead of the piperazine moiety utilised by Roche and GSK. The preferred compounds (> 100 examples given) incorporate azetidine, pyrrolidine, piperidine or homopiperidine basic groups, for example (151) with a 5-HT$_6$ K_i of < 100 nM. In the latter patent application, the basic group is a fused bicyclic or tricyclic system, with a preference for the arylsulfonyl unit to be attached at the 3-positon of the quinoline core, for example (152) with a 5-HT$_6$ K_i of <10 nM. Interestingly, while both carbonyl and methylene groups are claimed as linkers to connect the quinoline core and aryl side chain, of the 200 or so examples given in this patent application, all feature a 3-arylsulfonylquinoline.

In the most recent disclosure from Abbott, the structures claimed in the patent application appear closely related to SB-271046. This series is also described as comprising potent and selective 5-HT$_6$ ligands. There are over 800 examples specified with a claimed 5-HT$_6$ K_i of <250 nM. The generic structure comprises a phenyl core, with piperazine performing the function of the basic group, and an arylsulfonamide moiety in the *meta*-position relative to this substituent [162]. Notably, there is preference for incorporating a difluoromethylene unit in the arylsulfonamide side chain, for example (153) with a K_i of <10 nM. It is possible that this feature increases the lipophilicity of the molecule by masking the polarity of the sulfonamide N–H thereby improving central penetrancy, in a similar manner to that posulated for SB-399885 (*vide supra*). It is notable that throughout their publications, Abbott refers to their 5-HT$_6$ receptor ligands as modulators, leaving open the possibility that some classes may act through agonist, antagonist or other modulatory mechanisms.

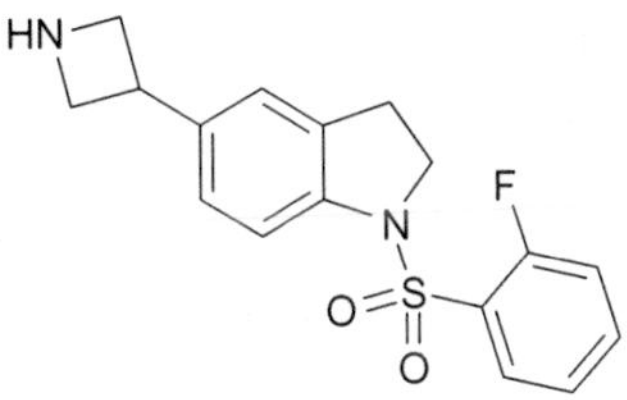

(150)

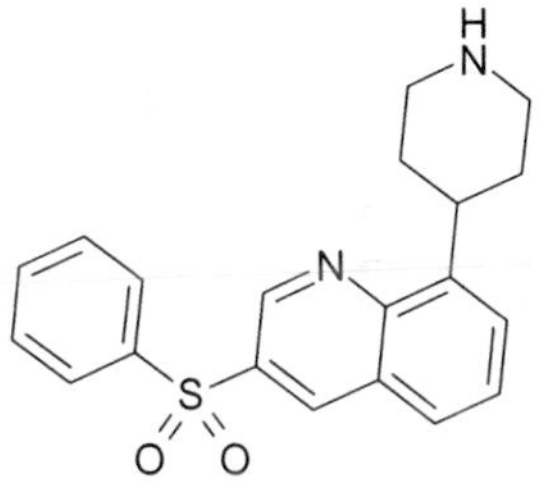

(151)

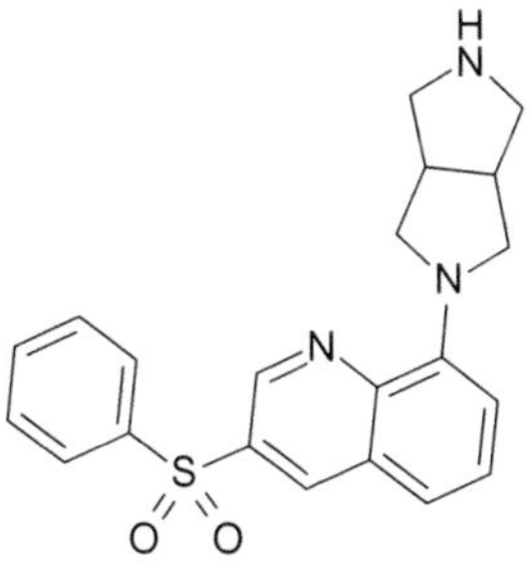

(152)

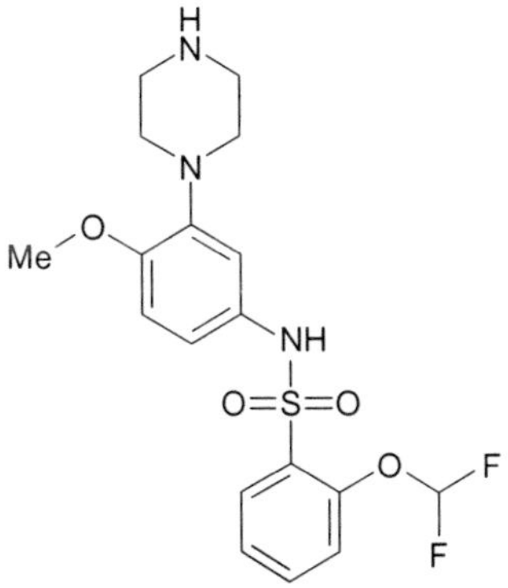

(153)

MEMORY PHARMACEUTICALS

This group first made known an interest in 5-HT$_6$ receptor agonists with a patent application in 2007 in which they describe a class of 3-piperidinyl indazole sulfonamide derivatives, closely related to Wyeth compounds, but with a distinct substitution pattern on the indazole ring, for example compound (154). Although the specific activity of all the 150 examples described is not disclosed, all are reported to have K_i values in a radioligand-binding assay of < 100 nM and a pA_2 value > 6 [163]. Interestingly, despite the agonist focus, their target indication is described as being mood and/or cognitive disorders, though the patent application also describes potential for treatment of gastrointestinal and polyglutamine-repeat disorders. The benzoxazinesulfonyl side chain appears to be particularly important: this group features in a second class of compounds with a classical indolyl or indazolyl piperazine substitution pattern. The 1,6-arrangement between sulfonyl and basic side chain maintains a similar relative special disposition of the two groups, for example in compound (155). Similar potency claims are made for preferred compounds as in the first claimed indazole class [164]. A benzoxazine sulfonamide features in a third indolylsulfonamide patent application in which the more frequently encountered 1,4-substitution pattern is represented, for example compound (156) [165].

The most recent publication from this group reverts to a substitution pattern closer to that of (154) but with an azaindole core, and this time the preferred sulfonyl substituents include the 3-pyridyl group [166]. The tetrahydropyridine substituent of (157) is similar to that highlighted by Estève on an indole template, for example compound (104). However, whereas Estève are targetting agonists, the Memory group appear to be focusing on antagonist molecules in this patent application.

(154)   (155)

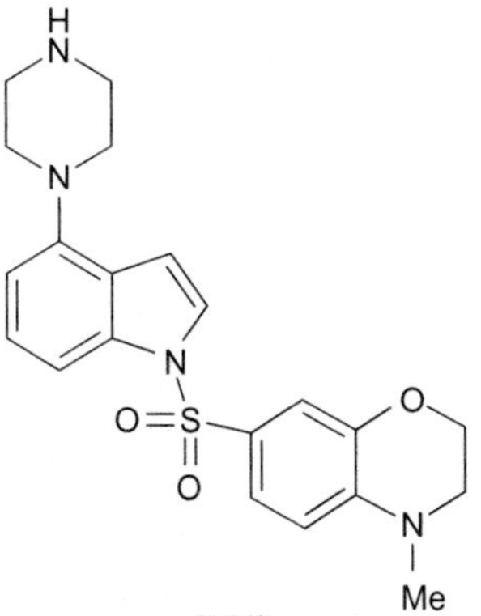

(156) 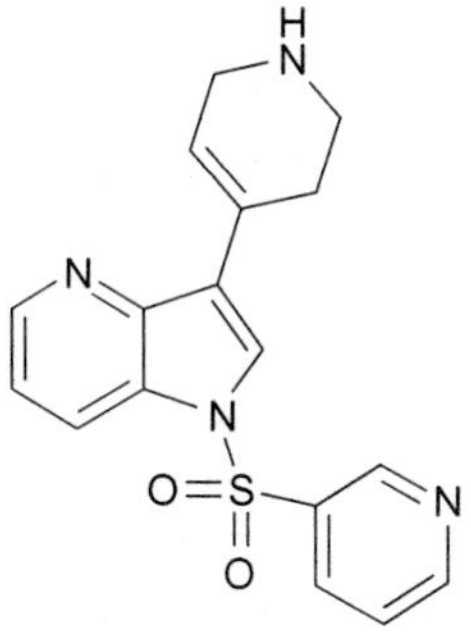   (157)

MERCK

Merck have only recently become patentees in the field of 5-HT_6 receptor modulators. Interestingly, while they started with the classic indolyl phenylsulfonamide template, they have reached out beyond the usual basic binding site to find additional affinity with a class of substituted benzyl groups. Highlighted, for example, is compound (158) with an IC_{50} of 0.2 nM measured in a radioligand-binding assay. Such compounds are described as 5-HT_6 receptor antagonists [167].

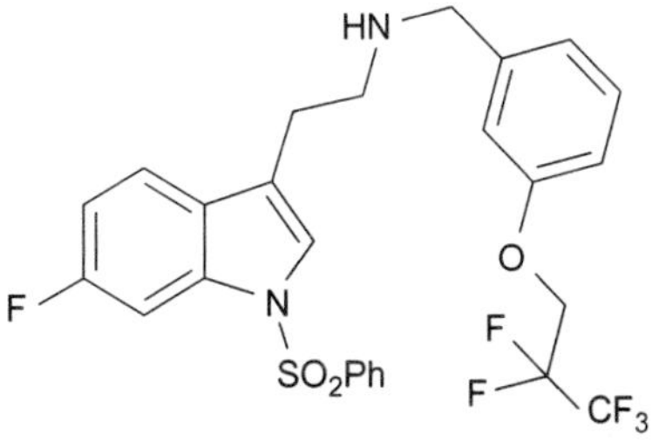

(158) 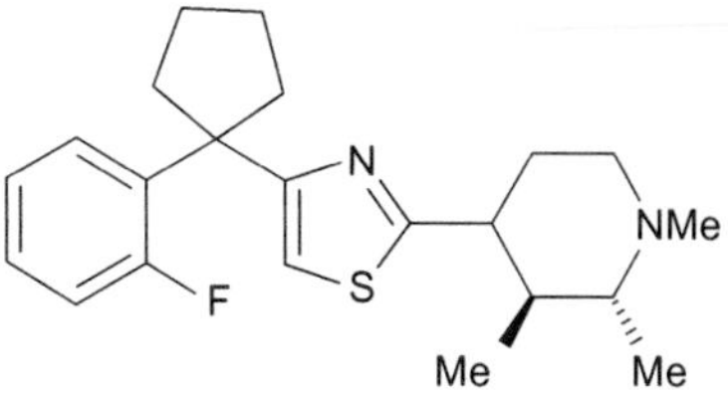   (159)

PSYCHENOMICS

The group from Psychenomics Inc. has published a single patent application describing a class of dual 5-HT_6 receptor and D_3 (or dopamine transporter – DAT) modulators – similar to Abbott. The common feature shared by their hits is a 1,1-cyclopentyl group separating a 2-fluorophenyl group from a thiazole bearing a basic substituent. Compound (159) is illustrative with the 5-HT$_6$ receptor affinity shown by an IC$_{50}$ value of 30 nM and the DAT affinity IC$_{50}$ of 40 nM against the human protein [168].

The patent application does not distinguish whether the compounds act through an agonist or antagonist mechanism; however, their major therapeutic target is bipolar disorder. It may be supposed that the cyclopentane group serves to hold the two aromatic groups in the 'butterfly wings' conformation preferred by bisarylsulfonyl 5-HT$_6$ receptor ligands (see Figure 5.1).

SOLVAY

Solvay has filed two patent applications on the use of 5-HT$_6$ receptor ligands with a focus on preventing relapse into addiction [169, 170]. Their first use patent application relates to compounds from other companies but the second patent application describes compounds, such as (160) (pK_i = 8.8), which bears some familiar 5-HT$_6$ receptor modulator features. This example includes a thiazoloimidazole sulfonamide, together with a unique basic side chain. The compounds are claimed as antagonists of the 5-HT$_6$ receptor.

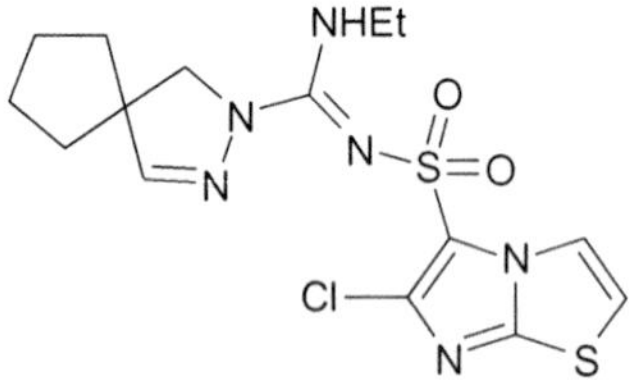

(160)

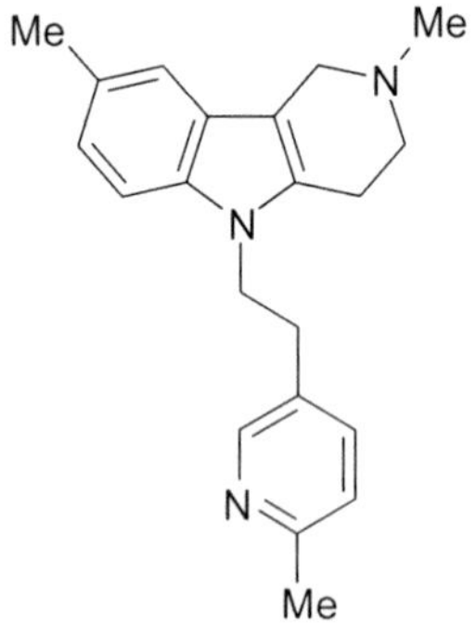

Dimebon

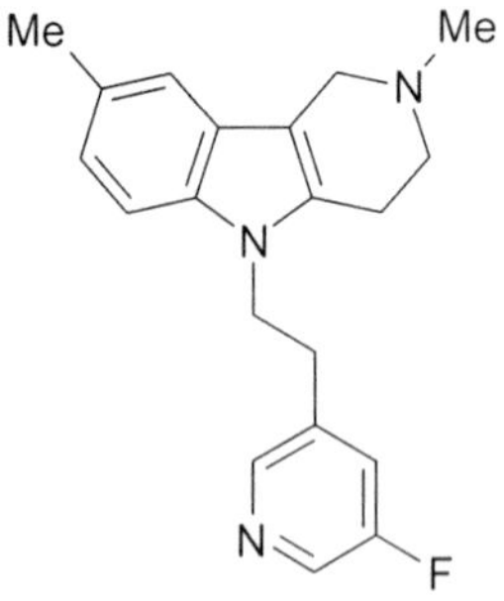

(161)

D2E TECHNOLOGIES, LLC

This group have recently published a patent application covering a class of fluorinated compounds, related to the known *N*-nitrosodimethylamine antagonist dimebon, which also act as 5-HT$_6$ antagonists. These are carboline derivatives, such as D2E 1 (161) [171], which shows complete inhibition of the binding of radiolabelled LSD at 10 μM. The target of their application appears to be to find a dual mode of action compound for the treatment of cognitive disorders. The 5-HT$_6$ receptor binding affinity of Dimebon is not reported.

EGNMR

The Hungarian group, Egis Gyogyszergyar, has filed a patent application on a class of 5-HT$_6$ ligand with a benzofuran core group. At least two examples show an affinity pK_i of <30 nM for 5-HT$_6$, including examples (162) and (163) [172].

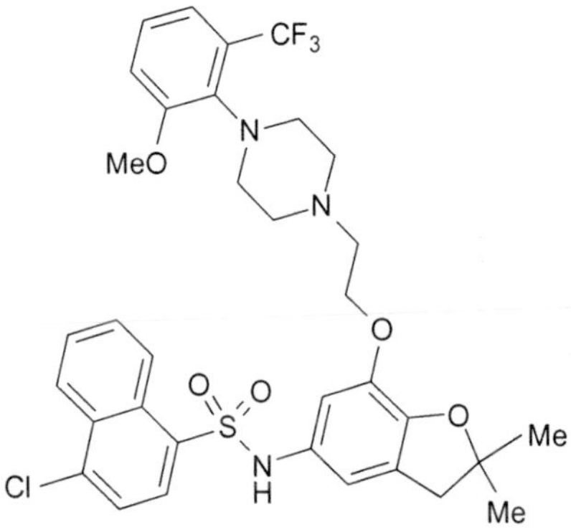

(162)

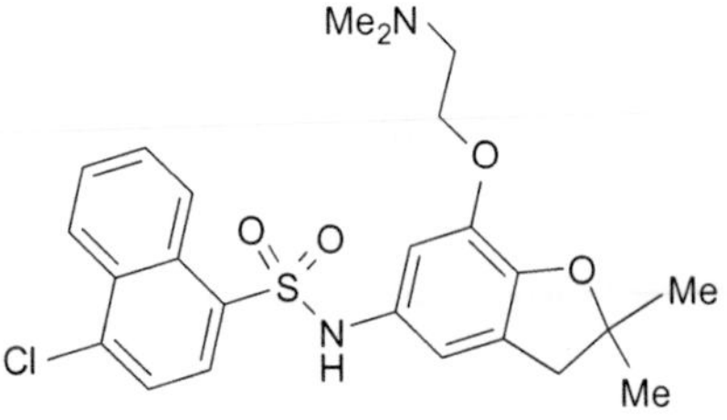

(163)

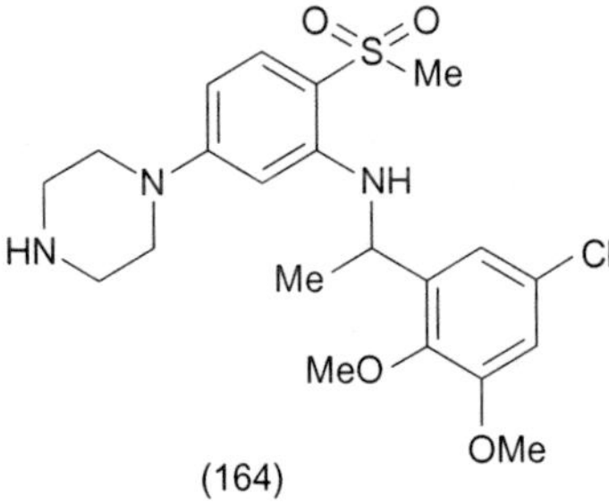

(164)

AVINEURO

The Avineuro Pharmaceutical company has reported that it developed a brain-penetrant and highly selective 5-HT$_6$ receptor antagonist [173] with a binding IC$_{50}$ of 0.84 nM and a functional IC$_{50}$ of 38 nM. The undisclosed compound is also reported to have clean P450, *in vivo* and *in vitro* PK profiles in pre-clinical species, and to restore scopolamine-induced cognitive dysfunction in a rat model.

SERVIER

An interest in the development of 5-HT$_6$ receptor ligands has been indicated by Servier in a recent paper studying the coupling of 5-HT$_6$ to Gαs, and analysing the properties of a number of reported 5-HT$_6$ receptor ligands [174]. These include WAY208,466 and WAY181,187, which they determine to be partial agonists while determining SB-399885 as a potent antagonist. No indication of specific compounds from this company has yet appeared.

EPIX

Epix has published only one patent application outlining interest in 5-HT$_6$ receptor modulation [175]. The claim is specifically focused on the combination of a class of 5-HT$_6$ receptor ligands with acetylcholinesterase inhibitors. The patent application gives a single example of a bis-aminosulfone antagonist, (164), unrelated to most series described hitherto, in effect a vinylogous sulfonamide. The 5-HT$_6$ binding affinity is disclosed as having a K_i of 4 nM, however, efficacy in a cognition study (novel object discrimination at 3 mg/kg) is claimed together with synergistic activity in the same study in combination with donepezil (0.1 mg/kg) requiring a dose of only 0.3 mg/kg of the 5-HT$_6$ receptor ligand to be effective. The patent application does not disclose the functional activity of the compound.

ACADEMIC STUDIES

In addition to the wider range of chemotypes synthesised by pharmaceutical companies, several academic groups have shown an interest in designing 5-HT$_6$ receptor ligands. A team comprising scientists from Virginia Commonwealth University and Case Western Reserve University has described a class of naphthyl piperazine derivatives as novel templates for 5-HT$_6$ receptor ligands [176]. However, such compounds have been addressed in more depth by patent applications from pharmaceutical groups. In a follow-up paper, a wider group of tryptaminergic ligands are described, some bearing sulfonyl groups on their indole nitrogen rings, and several containing fused examples [177]. They conclude that N-substituted and benzene-sulfonyl substituted tryptamines bind to the 5-HT$_6$ receptor in a dissimilar manner, based on the tolerance of different indole substituents. They also offer empirical evidence for the preference for the aromatic groups of an arylsulfonyl indole to prefer a non-coplanar conformation.

CLINICAL ACTIVITIES

No fewer than 60 ligands for 5-HT$_6$ receptors have been specifically reported over the last decade, most notably by Roche, GlaxoSmithKline, Wyeth, Biovitrum, Estève, Saegis/Eli Lilly, and a number of other groups, as well as hundreds more in the patent literature. Several of these have reportedly entered clinical trials for AD, obesity and schizophrenia. For AD, a number of 5-HT$_6$ receptor antagonists have successfully undergone Phase I clinical studies in healthy volunteers with good safety and tolerability profiles after single or repeated doses, albeit in small trial populations to date. Information in the public domain indicates that the most advanced compound is SB-742457 being developed by GlaxoSmithKline. Two Phase II trials have recently been completed in subjects with mild-to-moderate AD. The first was reportedly a dose-ranging trial comparing SB-742457 with placebo, and the second was an exploratory efficacy study with both SB-742457 and donepezil arms. Overall, these studies have demonstrated that SB-742457 was generally safe and well tolerated in AD patients. Regarding therapeutic efficacy, SB-742457 produced an improvement in both cognitive and global function in AD as assessed by ADAS-cog (AD assessment scale-cognitive subscale) and CIBIC+ (clinician's interview-based impression of change-plus caregiver input), respectively. The effect size measured at the dose of 35 mg SB-742457 was similar to that of donepezil [178]. These preliminary data represent the first demonstration of symptomatic benefit in AD for a 5-HT$_6$ receptor antagonist. Manuscripts reporting the results of both studies are currently being prepared for submission to a peer-reviewed journal.

Compounds from a number of Wyeth series are being progressed through clinical trials. Agonist WAY-181187 (SAX-187,12) [81] has entered Phase I studies. SAM-531, also from Wyeth, is entering a 24-week, placebo-controlled Phase II clinical trial in outpatients with moderate to severe AD, comprising three doses of the compound and donepezil as the active reference group [179].

Synosia has assumed the clinical development of two Roche compounds SYN-114 and SYN-120, and both are in Phase I trials for cognitive disorders including AD [180].

SUVN-502 from Suven has successfully completed a Phase I clinical trial for the symptomatic treatment of AD, schizophrenia and other disorders of memory and cognition such as ADHD. Suven plans to initiate the POC studies (Phase IIa) during the last quarter of 2009 [181].

In 2005, Saegis completed a small Phase II trial with SGS-518/LY-483518 involving schizophrenia patients stable on anti-psychotic medication. The drug was safe and well tolerated with preliminary evidence suggesting improvement in cognition using the Brief Assessment of Cognition in Schizophrenia scale [182].

For obesity, two compounds have successfully completed Phase I clinical trials with no further information in the public domain regarding their progression. These are BVT.74316 [183] of Biovitrum and PRX-07034 of EPIX [184], the latter being developed for AD and obesity. The Phase Ib trial results for PRX-07034 indicated that a greater proportion of subjects on the drug experienced weight loss during the one month dosing period compared with subjects on placebo, suggesting pharmacologic activity for obesity.

CONCLUSION

This is an exciting time in the field of 5-HT$_6$ receptor modulation. The first generation of ligands is now in later phase clinical studies; preliminary Phase II results have reported efficacy in humans for at least one of the potential indications – cognition. At the same time, our basic understanding of the role of the receptor in human physiology has improved and the number of potential therapeutic areas that could be impacted by 5-HT$_6$ receptor agonism or antagonism continues to increase, especially in the fields of neurology and psychiatry. Compound classes with synergistic co-activities as agonist or antagonists at a number of other receptors including CB_2, D_2, D_3, NK_1, 5-HT$_{2A}$ and 5-HT$_{2C}$ have been demonstrated. There is now a clear potential, within the next decade, for new medications incorporating 5-HT$_6$ receptor modulation as a key mode of action to become registered drugs.

REFERENCES

[1]   Monsma, F.J., Jr., Shen, Y., Ward, R.P., Hamblin, M.W. and Sibley, D.R. (1993) *Mol. Pharmacol.* **43**, 320–327.

[2]   Ruat, M., Traiffort, E., Arrang, J.-M., Tardivel-Lacombe, J., Diaz, J., Leurs, R. and Schwartz, J.C. (1993) *Biochem. Biophys. Res. Commun.* **193**, 268–276.

[3]   Kohen, R., Metclaf, M.A., Khan, N., Druck, T., Huebner, K., Lachowicz, J.E., Meltzer, H.Y., Sibley, D.R., Roth, B.L. and Hamblin, M.W. (1996) *J. Neurochem.* **66**, 47–56.
[4]   Kohen, R., Fashingbauer, L.A., Heidmann, D.E.A., Guthrie, C.R. and Hamblin, M.W. (2001) *Mol. Brain Res.* **90**, 110–117.
[5]   Hirst, W.D., Abrahamsen, B., Blaney, F.E., Calver, A.R., Aloj, L., Price, G.W. and Medhurst, A.D. (2003) *Mol. Pharmacol.* **64**, 1295–1308.
[6]   East, S.Z., Burnet, P.W.J., Leslie, R.A., Roberts, J.C. and Harrison, P.J. (2002) *Synapse* **45**, 191–199.
[7]   Gerard, C., Martres, M.P., Lefevre, K., Miquel, M.C., Verge, D., Lanumey, L., Doucet, E., Hamon, M. and El Mestikawy, S. (1997) *Brain Res.* **746**, 207–219.
[8]   East, S.Z., Burnet, P.W.J., Leslie, R.A., Roberts, J.C. and Harrison, P.J. (2002) *Synapse* **45**, 191–199.
[9]   Bourson, A., Borroni, E., Austin, R.H., Monsma, F.J., Jr. and Sleight, A.J. (1995) *J. Pharmacol. Exp. Ther.* **274**, 173–180.
[10] Sleight, A.J., Monsma, F.J., Jr., Borroni, E., Austin, R.H. and Bourson, A. (1996) *Behav. Brain Res.* **73**, 245–248.
[11] Roth, B.L., Craigo, S.C., Choudhary, M.S., Uluer, A., Monsma, F.J., Jr., Shen, Y., Meltzer, H.Y. and Sibley, D.R. (1994) *J. Pharmacol. Exp. Ther.* **268**, 1403–1410.
[12] Biess, F.G., Monsma, F.J., Carolo, C., Meyer, V., Rudler, A., Zwingelstein, C. and Sleight, A.J. (1997) *Neuropharmacology* **36**, 713–720.
[13] Bentley, J.C., Bourson, A., Boess, F.G., Fone, K.C.F., Marsden, C.A., Petit, N. and Sleight, A.J. (1999) *Br. J. Pharmacol.* **126**, 1537–1542.
[14] Francis, P.T., Palmer, A.M., Snape, M. and Wilcock, G.K. (1999) *J. Neurol. Neurosurg. Psychiatry* **66**, 137–147.
[15] Perry, E., Walker, M., Grace, J. and Perry, R. (1999) *Trends Neurosci.* **22**, 273–280.
[16] Rogers, D.C. and Hagan, J.J. (2001) *Psychopharmacology (Berlin)* **158**, 114–119.
[17] Woolley, M.L., Bentley, J.C., Sleight, AJ., Marsden, C.A. and Fone, K.C. (2001) *Neuropharmacology* **41**, 210–219.
[18] Lindner, M.D., Hodges, D.B., Jr., Hogan, J.B., Orie, A.F., Corsa, J.A., Barten, D.M., Polson, C., Robertson, B.J., Guss, V.L., Gillman, K.W., Starrett, J.E., Jr. and Gribkoff, V.K. (2003) *J. Pharmacol. Exp. Ther.* **307**, 682–691.
[19] Foley, A.G., Murphy, K.J., Hirst, W.D., Gallagher, H.C., Hagan, J.J., Upton, N., Walsh, F.S. and Regan, C.M. (2004) *Neuropsychopharmacology* **29**, 93–100.
[20] Hirst, W.D., Stean, T.O., Rogers, D.C., Sunter, D., Pugh, P., Moss, S.F., Bromidge, S.M., Riley, G., Smith, D.R., Bartlett, S., Heidbreder, C.A., Atkins, A.R., Lacroix, L.P., Dawson, L.A., Foley, A.G., Regan, C.M. and Upton, N. (2006) *Eur. J. Pharmacol.* **553**, 109–119.
[21] Chuang, A.T.T., Foley, A., Pugh, P.L., Sunter, D., Xin, T., Regan, C., Dawson, L.A., Medhurst, A.D. and Upton, N. (2006) *Alzheimers Demen.* **2**, S631–S632 (abstract).
[22] Upton, N., Chuang, T.T., Hunter, A.J. and Virley, D.J. (2008) *Neurotherapeutics* **5**, 458–469.
[23] Callahan, P.M., Ilch, C.P., Rowe, W.B. and Tehim, A. (2004) *Soc. Neurosci. Abstr.* 776.19 (abstract).
[24] Marcos, B., Chuang, T.T., Gil-Bea, F.J. and Ramirez, M.J. (2008) *Br. J. Pharmacol.* **155**, 434–440.
[25] Bussey, T.J., Muir, J.L. and Aggleton, J.P. (1999) *J. Neurosci.* **19**, 495–502.

[26] Woolley, M.L., Marsden, C.A., Sleight, A.J. and Fone, K.C. (2003) *Psychopharmacology (Berlin)* **170**, 358–367.
[27] Lieben, C.K., Blokland, A., Sik, A., Sung, E., van Nieuwenhuizen, P. and Schreiber, R. (2005) *Neuropsychopharmacology* **30**, 2169–2179.
[28] King, M.V., Sleight, A.J., Woolley, M.L., Topham, I.A., Marsden, C.A. and Fone, K.C. (2004) *Neuropharmacology* **47**, 195–204.
[29] Wesołowska, A. and Nikiforuk, A. (2007) *Neuropharmacology* **52**, 1274–1283.
[30] Hirano, K., Piers, T.M., Searle, K.L., Miller, N.D., Rutter, A.R. and Chapman, P.F. (2009) *Life Sci.* **84**, 558–562.
[31] Mitchell, E.S. and Neumaier, J.F. (2005) *Pharmacol. Ther.* **108**, 320–333.
[32] Svenningsson, P., Tzavara, E.T., Qi, H., Carruthers, R., Witkin, J.M., Nomikos, G.G. and Greengard, P. (2007) *J. Neurosci.* **27**, 4201–4209.
[33] Wesołowska, A. (2008) *Eur. J. Pharmacol.* **580**, 355–360.
[34] Yoshioka, M., Matsumoto, M., Togashi, H., Mori, K. and Saito, H. (1998) *Life Sci.* **62**, 1473–1477.
[35] Hamon, M., Doucet, E., Lefèvre, K., Miquel, M.C., Lanfumey, L., Insausti, R., Frechilla, D., Del Rio, J. and Vergé, D. (1999) *Neuropsychopharmacology* **21**(Suppl. 2), 68S–76S.
[36] Svartengren, J., Axelsson-Lendin, P., Edling, N., Fhölenhag, K., Isacson, R., Hillegaart, V. and Grönberg, A. (2004) Washington, DC: Society for Neuroscience P75.8.
[37] Svartengren, J., Öhman, B., Edling, N., Svensson, M., Fhölenhag, K., Axelsson-Lendin, P., Klingström, G. and Larsson, C. (2003) *Int. J. Obes.* **27**(T1), 1–94.
[38] Pendharkar, V.V., Vishwakarma, L.S., Patel, A.S., Shirsath, V.S., Kambhampati, S.R. and Nirogi, R.V.S. (2005) *Soc. Neurosci. Abstr.* 533.7 (abstract).
[39] Fisas, A., Codony, X., Romero, G., Dordal, A., Giraldo, J., Mercè, R., Holenz, J., Heal, D., Buschmann, H. and Pauwels, P.J. (2006) *Br. J. Pharmacol.* **148**, 973–983.
[40] Liu, K.G. and Robichaud, A.J. (2009) *Drug Dev. Res.* **70**, 145–168.
[41] Sleight, A.J., Boess, F.G., Bös, M., Levet-Trafit, B., Riemer, C. and Bourson, A. (1998) *Br. J. Pharmacol.* **123**, 556–562.
[42] Kim, H.-J., Doddareddy, M.R., Choo, H., Cho, Y.S., No, K.T., Park, W.-K. and Pae, A.-N. (2008) *J. Chem. Inf. Model.* **48**, 197–206.
[43] Ahmed, M., Briggs, M.A., Bromidge, S.M., Buck, T., Campbell, L., Deeks, N.J., Garner, A., Gordon, L., Hamprecht, D.W., Holland, V., Johnson, C.N., Medhurst, A.D., Mitchell, D.J., Moss, S.F., Powles, J., Seal, J.T., Stean, T.O., Stemp, G., Thompson, M., Trail, B., Upton, N., Winborn, K. and Witty, D.R. (2005) *Bioorg. Med. Chem. Lett.* **15**, 4867–4871.
[44] Bromidge, S.M. (2002) *PCT Int. Appl.*, WO 2002041889; *Chem. Abstr.* 136:401783.
[45] Kelly, M.G. and Cole, D.C. (2002) *PCT Int. Appl.*, WO 2002036562; *Chem. Abstr.* 136:369738.
[46] Jossan, S., Nilsson, B.M., Sakariassen, K.S. and Svartengren, J. (2002), *PCT Int. Appl.*, WO 2002008178; *Chem. Abstr.* 136:129087.
[47] Briggs, A.J., Clark, R.D., Harris III., R.N., Repke, D.B. and Wren, D.L. (2002) *PCT Int. Appl.*, WO 2002102774; *Chem. Abstr.* 138:55865.
[48] Witty, D.R. (2006) SCIPharm, Edinburgh.
[49] Ahmed, M., Johnson, C.N., Jones, M.C., MacDonald, G.J., Moss, S.F., Thompson, M., Wade, C.E. and Witty, D. (2003) *PCT Int. Appl.*, WO 2003080580; *Chem. Abstr.* 139:292163.

[50] Bruton, G., Orlek, B.S., Stemp, G. (2008) *PCT Int. Appl.*, WO 2008113818; *Chem. Abstr.* 149:394689.
[51] Johnson, C.N., Moss, S.F., Tait, M.M. and Witty, D.R. (2005) *PCT Int. Appl.*, WO 2005021530; *Chem. Abstr.* 142:298134.
[52] Johnson, C.N., Moss, S.F. and Witty, D.R. (2005) *PCT Int. Appl.*, WO 2005030724; *Chem. Abstr.* 142:373864.
[53] Johnson, C.N. and Witty, D.R. (2005) *PCT Int. Appl.*, WO 2005026125; *Chem. Abstr.* 142:336394.
[54] Gee, A., Martarello, L., Johnson, C.N. and Witty, D.R. (2006) *PCT Int. Appl.*, WO 2006053785; *Chem. Abstr.* 145:8187.
[55] Johnson, C.N., Stemp, G., Thompson, M. and Witty, D. (2005) *PCT Int. Appl.*, WO 2005113539; *Chem. Abstr.* 144:22823.
[56] Ahmed, M.A., Johnson, C.N., Miller, N.D., Trani, G. and Witty, D. (2005) *PCT Int. Appl.*, WO2005095346; *Chem. Abstr.* 143:386931.
[57] Ahmed, M., Giblin, G.M.P., Johnson, C.N., Livermore, D.G.H., Miller, N.D., Moss, S.F. and Witty, D.R. (2007) *PCT Int. Appl.*, WO 2007039219; *Chem. Abstr.* 146:401840.
[58] Ahmed, M., Brown, A., Giblin, G.M.P., Haslam, C., Johnson, C.N., Johnson, M.R., Livermore, D.G.H., Miller, N.D., Moss, S.F. and Witty, D.R. (2008) Second RSC/SCI Symposium: G-protein Coupled Receptors in Medicinal Chemistry, Gothenberg, Sweden.
[59] Ahmed, M., Brown, A., Giblin, G.M.P., Goldsmith, P., Haslam, C., Johnson, C.N., Johnson, M.R., Livermore, D.G.H., Miller, N.D., Moss, S.F. and Witty, D.R. (2009) XV RSC/SCI Medicinal Chemistry Symposium, Cambridge, UK.
[60] Bromidge, S.M., Johnson, C.N., Moss, S.F., Rahman, S.S. and Witty, D.R. (2003) *PCT Int. Appl.*, WO 2003068751; *Chem. Abstr.* 139:197390.
[61] Trani, G., Baddeley, S.M., Briggs, M.A., Chuang, T.T., Deeks, N.J., Johnson, C.N., Khazragi, A.A., Mead, T.L., Medhurst, A.D., Milner, P.H., Quinn, L.P., Ray, A.M., Rivers, D.A., Stean, T.O., Stemp, G., Trail, B.K. and Witty, D.R. (2008) *Bioorg. Med. Chem. Lett.* **18**, 5698–5700.
[62] Ahmed, M., Johnson, C.N., Miller, N.D., Milner, P.N., Rivers, D.A. and Witty, D.R. (2006) *PCT Int. Appl.*, WO 2006038006; Chem. Abstr. 144:390735.
[63] Geldenhuys, W.J. and Van der Schyf, C.J. (2008) *Curr. Top. Med. Chem.* **8**, 1035–1048.
[64] Holenz, J., Pauwels, P.J., Díaz, J.L., Merce, R., Codony, X. and Buschmann, H. (2006) *Drug Discov. Today* **11**, 283–299.
[65] Krauss, N.E. and Zhao, S.-H. (2006) *PCT Int. Appl.*, WO 2006066746; *Chem Abstr.* 145, 103551.
[66] Krauss, N.A. and Zhao, S.-H. (2006) *PCT Int. Appl.*, WO 2006066756; *Chem. Abstr.* 145:103552.
[67] Berger, J., Caroon, J.M., Lopez-Tapia, F.J., Lowrie, L.E., Nitzan, D. and Zhao, S.-H. (2007) US Patent Application US20070099908; *Chem. Abstr.* 146:481925.
[68] Greenhouse, R., Harris, R.N., Jaime-Figueroa, S., Kress, J.M., Repke, D.B. and Stabler, R.S. (2006) *PCT Int. Appl.*, WO 2006066790; *Chem. Abstr.* 145:103432.
[69] Harris, R.N. (2007) US Patent Application US 20070293526; *Chem. Abstr.* 148:54628.
[70] Harris, R.N., Kress, J.M., Repke, D.B. and Stabler, R.S. (2007) *PCT Int. Appl.*, WO 2007147762; *Chem. Abstr.* 148:100391.
[71] Boes, M., Riemer, C. and Stadler, H. (1999) European Patent Application EP941994; Stadler, H., Boes, M., Borroni, E., Bourson, A., Martin, J., Poli, S., Riemer, C. and Sleight, A.J. (1999) *37th IUPAC Congress*, Berlin, Abstract MM-7.

[72] Zhao, S.-H., Berger, J., Clark, R.D., Sethofer, S.G., Krauss, N.E., Brothers, J.M., Martin, R.S., Misner, D.L., Schwab, D. and Alexandrova, L. (2007) *Bioorg. Med. Chem. Lett.* **17**, 3504–3507.

[73] Berger, J. and Zhao, S.-H. (2005) *PCT Int. Appl.*, WO2005105776; *Chem Abstr.* 143:460162.

[74] Sui, M. and Zhao, S.-H. (2005) *PCT Int. Appl.*, WO 2005067933; *Chem Abstr.* 143:172899.

[75] Bonhaux, D.W. and Martin, R.S. (2006) US Patent Application US 20060069094; *Chem. Abstr.* 144:324862.

[76] Bamberg, J.T., O'Yang, C., Sui, M. and Zhao, S.-H. (2008) *PCT Int. Appl.*, WO 2008055808; *Chem. Abstr.* 148:538073.

[77] Harris, R.N., Kress, J.M., Repke, D.B. and Stabler, R.S. (2007) US Patent Application US 20070027161; *Chem. Abstr.* 146:206338.

[78] Harris, R.N., Kress, J.M., Repke, D.B. and Stabler, R.S. (2006) *PCT Int. Appl.*, WO 2006061126; *Chem. Abstr.* 145:62932.

[79] Lopez-Tapia, F.J., Lowrie L.E. Jr. and Nitzan, D. (2008) *PCT Int. Appl.*, WO 2008055847; *Chem. Abstr.* 148:561718.

[80] Cole, D.C., Lennox, W.J., Stock, J.R., Ellingboe, J.W., Mazandarani, H., Smith, D.L., Zhang, G., Tawa, G.J. and Schechter, L.E. (2005) *Bioorg. Med. Chem. Lett.* **15**, 4780–4785.

[81] Cole, D.C., Ellingboe, J.W., Lennox, W.J., Mazandarani, H., Smith, D.L., Stock, J.R., Zhang, G., Zhou, P. and Schechter, L.E. (2005) *Bioorg. Med. Chem. Lett.* **15**, 379–383.

[82] Zhou, P., Yan, Y., Bernotas, R., Harrison, B.L., Huryn, D., Robichaud, A.J., Zhang, G.M., Smith, D.L. and Schechter, L.E. (2005) *Bioorg. Med. Chem. Lett.* **15**, 1393–1396.

[83] Elokdah, H., Li, D., McFarlane, G., Bernotas, R.C., Robichaud, A.J., Magolda, R.L., Zhang, G.M., Smith, D. and Schechter, L.E. (2007) *Bioorg. Med. Chem.* **15**, 6208–6226.

[84] McDevitt, R.E., Li, Y., Robichaud, A.J., Heffernann, G.D., Coghlan, R.D. and Bernotas, R.C. (2007) *PCT Int. Appl.*, WO 2007084841; *Chem. Abstr.* 147:211728.

[85] Cole, D.C. US Patent Application (2005) US 20050020575; *Chem. Abstr.* 142:176840.

[86] Cole, D.C. and Bernotas, R.C. (2005), US Patent Application US 20050020596; *Chem. Abstr.* 142:176841.

[87] Cole, D.C., Stock, J.R., Lennox, W.J., Bernotas, R.C., Ellingboe, J.W., Boikess, S., Coupet, J., Smith, D.L., Leurg, L., Zhang, G.-M., Feng, X., Kelly, M.F., Galante, R., Huang, P., Dawson, L.A., Marquis, K., Rosenzweig-Lipson, S., Beyer, C.E. and Schechter, L.E. (2007) *J. Med. Chem.* **50**, 5535–5538.

[88] Liu, K., Lo, J.R., Robichaud, A.J. and Elokdah, H.M. (2007) US Patent Application US 20070281922; *Chem. Abstr.* 148, 33727.

[89] Zhou, P. and Kelly, M.G. (2005) US Patent Application US 2005009819; *Chem. Abstr.* 142:134588.

[90] Haydar, S.N. and Andrae, P. (2008), US Patent Application US 2008293688; *Chem. Abstr.* 150:746.

[91] Greenfield, A.A., Grosanu, C., Elokdah, H.M. and Robichaud, A.J. (2007) US Patent Application US 20070244179; *Chem. Abstr.* 147, 462303.

[92] Greenfield, A.A., Grosanu, C., Elokdah, H.M. and Robichaud, A.J. (2007) US Patent Application US20070244106; *Chem. Abstr.* 147, 462302.

[93] Greenfield, A.A., Grosanu, C., Elokdah, H.M. and Robichaud, A.J. (2007) US Patent Application US 20070244105; *Chem. Abstr.* 147:462301.

[94] Lo, J.R., Liu, K., Elokdah, H.M. and Robichaud, A.J. (2007) US Patent Application US 200702387; *Chem. Abstr.* 149:127393.

[95] Liu, K., Robichaud, A.J. and Elokdah, H.M. (2007) US Patent Application US 20070281945; *Chem. Abstr.* 148:33772.

[96] Liu, K.G., Lo, J.R., Comery, T.A., Zhang, G.M., Zhang, J.Y., Kowal, D.M., Smith, D.l., Di, L., Kerns, E.H., Schechter, L.E. and Robichaud, A.J. (2009) *Bioorg. Med. Chem. Lett.* **19**, 3214–3216.

[97] Cole, D.C. and Mahwah, M.A. (2005) US Patent Application US 20050101596; *Chem. Abstr.* 142:463598.

[98] Liu, K.G., Lo, J.R., Comery, T.A., Zhang, G.M., Zhang, J.Y., Kowal, D.M., Smith, D.L., Di, L., Kerns, E.H., Schechter, L.E. and Robichaud, A.J. (2008) *Bioorg. Med. Chem. Lett.* **18**, 3929–3931.

[99] Kelly, M.G., Lenleek, S.E. and Palmer, Y.L. (2007) US Patent Application US 2007099912; *Chem. Abstr.* 146, 482048.

[100] Zhou, P. and Li, Y. (2007) US Patent Application US 2007099911; *Chem. Abstr.* 146:482047.

[101] Comery, T.A. and Schechter, L.E. (2007) US Patent Application US 20070167431; *Chem. Abstr.* 147:158506.

[102] Schechter, L.E., Pong, K. and Zaleska, M.M. (2006) US Patent Application US 20060128744; *Chem. Abstr.* 145:21210.

[103] Wang, Y., Shaw, C.-C., Bernotas, R.C., Wang, C.-C.P., Cai, P. and Wang, Z. (2006) *PCT Int. Appl.*, WO 2006002125; *Chem. Abstr.* 144:108204.

[104] Greenblatt, L.P. (2005) *PCT Int. Appl.*, WO 2005037830; *Chem. Abstr.* 142:430148.

[105] Cole, D.C. (2005) *PCT Int. Appl.*, WO 2005012311; *Chem. Abstr.* 14:219264.

[106] Johansson, G., Brandt, P., Dykes, G.J. and Nilsson, B.M. (2005) *PCT Int. Appl.*, WO 2005037834; *Chem. Abstr.* 142:430255.

[107] Dykes, G. (2006) *PCT Int. Appl.*, WO 2006062481; *Chem. Abstr.* 145:46093.

[108] Caldirola, P., Johansson, G. and Sutin, L. (2006) *PCT Int. Appl.*, WO 2006134150; *Chem. Abstr.* 146:75330.

[109] Johansson, G., Angbrant, J. and Ringom, R. (2008) *PCT Int. Appl.*, WO 2008003703; *Chem. Abstr.* 148:144646.

[110] Berts, W., Brandt, P., Hammer, K., Henriksson, S., Lindqvist, B., Ringberg, E. and Ringom, R. (2008) *PCT Int. Appl.*, WO 2008054288; *Chem. Abstr.* 148:517753.

[111] Angbrant, J., Brandt, P., Ringom, R. and Lindqvist, B. (2008) *PCT Int. Appl.*, WO2008110598; *Chem. Abstr.* 149:378560.

[112] Merce-Vidal, R., Andaluz-Mataro, B. and Frigola-Constansa, J. (2003) *PCT Int. Appl.*, WO 2003042175; *Chem. Abstr.* 138:401602.

[113] Tasler, S., Kraus, J., Wuzik, A., Müller, O., Aschenbrenner, A., Cubero, E., Pascual, R., Quintana-Ruiz, J.-R., Dordal, A., Mercè, R. and Codony, X. (2007) *Bioorg. Med. Chem. Lett.* **17**, 6224–6229.

[114] Merce, V.R., Codony Soler, X. and Dordal Zueras, A. (2005) *PCT Int. Appl.*, WO 2005013974; *Chem. Abstr.* 142:219147.

[115] Merce, V.R., Codony Soler, X. and Dordal Zueras, A. (2005) *PCT Int. Appl.*, WO 2005013979; *Chem. Abstr.* 142:219149.

[116] Eoman, M.V. (2006) European Patent Application EP1632491; *Chem. Abstr.* 144:274133.

[117] Torrens Jover, A., Mas Prió, J. and Dordal Zueras, A. (2005) *PCT Int. Appl.*, WO 2005014045; *Chem. Abstr.* 142:240323.

[118] Torrens Jover, A., Mas Prió, J., Dordal Zueras, A., Codony Soler, X., Merce Vidal, R., Aurelio Castrillo Perez, J., Frigola Constansa, J. and Buschmann H.-H. (2005) *PCT Int. Appl.*, WO 2005014000; *Chem. Abstr.* 142:240322.

[119] Torrens-Jover, A. and Mas-Prió, J. (2005) *PCT Int. Appl.*, WO 2005014589; *Chem. Abstr.* 142:240439.

[120] Frigola-Constansa, J., Merce-Vidal, R., Holenz, J., Alcalde-Pais, M.E., Mesquida-Estevez, M.N., Lozez-Perez, S. (2007) *PCT Int. Appl.*, WO 2007054257; *Chem. Abstr.* 146:521555.

[121] Alcalde, E., Mesquida, N., López-Pérez, S., Frigola, J. and Mercè, R. (2009) *J. Med. Chem.* **52**, 675–697.

[122] Buschmann, H.H., Ruiz, J.-R.Q., Ajona, J.C. and Soler, X.C. (2007) UK Pat. Appl. GB 2435826; *Chem. Abstr.* 147:365161.

[123] Buschmann, H.H., Ruiz, J.-R.Q., Ajona, J.C. and Soler, X.C. (2007) UK Pat. Appl. GB 2435824; *Chem. Abstr.* 147:365275.

[124] Buschmann, H.H., Ruiz, J.-R.Q., Vidal, R.M. and Soler, X.C. (2007) UK Patent Application GB 2435828; *Chem. Abstr.* 147:365524.

[125] Buschmann, H.H., Quintana Ruiz, J.-R., Merce Vidal, R. and Codony Soler, X. (2007) Brit. UK Patent Application GB 2435827; *Chem. Abstr.* 147:344113.

[126] Buschmann, H.H., Quintana Ruiz, J.-R., Merce Vidal, R.M. and Codony Soler X., (2007) Brit. UK Patent Application GB 4235830; *Chem. Abstr.* 147:336389.

[127] Buschmann, H.H., Quintana Ruiz, J.-R., Merce Vidal, R. and Corbera Arjona, J., (2007) Brit. UK Patent Application GB 2435825; *Chem. Abstr.* 147:365498.

[128] Beller, M., Alex, K. and Diaz Fernandez, J.L. (2008) European Patent Application EP 1947085; *Chem. Abstr.* 149:176177.

[129] Diaz-Fernandez, J.-L. (2008) *PCT Int. Appl.*, WO 2008092666; Chem. Abstr. 149, 224091; (2008) European Patent Application EP 1953141; *Chem Abstr.* 149:224090.

[130] Diaz-Fernandez, J.-L. (2008) *PCT Int. Appl.*, WO 2008092665; Chem. Abstr. 149, 224092; (2008) European Patent Application EP 1953153; *Chem. Abstr.* 149:224089.

[131] Merce-Vidal, R. (2009) *PCT Int. Appl.*, WO 2009036955; Chem Abstr. 150, 329618; (2009) European Patent Application EP 2036888; *Chem. Abstr.* 150:329616.

[132] Diaz-Fernandez, J.L., Merce-Vidal, R. and Novak, L. (2009) European Patent Application EP 2016943; *Chem. Abstr.* 150:144517.

[133] Alcalde-Pais, M. De Las E., Mesquida-Estevez, M. De Les N., Lopez-Perez, S., Frigola-Constansa, J., Holenz, J. and Merce-Vidal, R. (2008) US Patent Application Publication US 2008027073; Chem. Abstr. 148:191738.

[134] Fisas-Escasany, M.A. and Buschmann, H.H. (2009) *PCT Int. Appl.*, WO 2009013010; *Chem. Abstr.* 150:160175.

[135] Jasti, V., Ramakrishna, V.S.N., Kambhampati, R.S., Shirsath, V.S. and Vishwottam, N.K. (2005) *PCT Int. Appl.*, WO 2005066157; *Chem. Abstr.* 143:153282.

[136] Jasti, V., Ramakrishna, V.S.N., Kambhampati, R.S., Shirsath, V.S. and Vishwottam, N.K. (2005) *PCT Int. Appl.*, WO 2005066184; *Chem. Abstr.* 143:153284.

[137] Ramakrishna, V.S.N., Shirasath, V.S., Kambhampati, R.S., Deshpande, A.D., Daulatbad, A.V., Vishwakarma, S. and Jasti, V. (2007) *PCT Int. Appl.*, WO 2007138611; *Chem. Abstr.* 148:33622.

[138] Ramakrishna, V.S.N., Kambhampati, R.S., Shirsath, V.S., Konda, J.B., Vishwakarma, S. and Jasti, V. (2007) *PCT Int. Appl.*, WO 2007046111 *Chem Abstr.* 146:462135.

[139] Ramakrishna, V.S.N., Kambhampati, R.S., Shirsath, V.S., Deshpande, A.D., Vishwa-karma, S. and Jasti, V. (2007) *PCT Int. Appl.*, WO 2007046112; *Chem. Abstr.* 146:462134.

[140] Ramakrishna, N.V.S., Kambhampati, R.S., Shinde, A.K., Kandikere, N.V. and Jasti, V. (2009) *PCT Int. Appl.*, WO 2009053997; *Chem. Abstr.* 150:472558.

[141] Nirogi, R., Kambhampati, R.S., Shinde, A.K., Daulatabad, A.V., Dwarampudi, A.R., Kandikere, N.V., Vishwakarma, S. and Jasti, V. (2008) *PCT Int. Appl.*, WO 2008136017; *Chem. Abstr.* 149:556440.

[142] Ramakrishna, N.V.S., Shinde, A.K., Kambhampati, R.S. and Jasti, V. (2009) *PCT Int. Appl.*, WO 2009034581; *Chem. Abstr.* 150:329617.

[143] Ramakrishna, N.V.S., Kambhampati, R.S., Deshpande, A.D. and Jasti, V. (2009) *PCT Int. Appl.*, WO 2008084491; *Chem. Abstr.* 149:176376.

[144] Ramakrishna, N.V.S., Kambhampati, R.S., Konda, J., Kothmirkar, P. and Jasti, V. (2008) *PCT Int. Appl.*, WO 2008084492; *Chem. Abstr.* 149:176374.

[145] Jeon, S.A., Choo, H., Park, W.-K., Rhim, H., Ko, S.Y., Cho, Y.S., Koh, H.Y. and Pae, A.N. (2007) *Bull. Korean Chem. Soc.* **28**, 285–291.

[146] Seong, C., Jung, Y., Choi, J., Park, W., Cho, H., Kong, J., Jung, D., Kang, S., Song S. and Kwark, K. (2006) US Patent Application US 20060084676; *Chem. Abstr.* 144:412383.

[147] Seong, C., Park, N., Choi, J., Park, W., Kong, J., Cho, H., Kang, S. and Song, S. (2007) *PCT Int. Appl.*, WO2007032572; *Chem. Abstr.* 146:358879.

[148] Seong, C., Choi, J., Park, C.M., Park, W., Kong, J., Kang, S. and Park, C. (2007) *PCT Int. Appl.*, WO 2007108569; *Chem. Abstr.* 147:406846.

[149] Seong, C., Park, N., Choi, J., Park, C.M., Park, W. and Kong, J. (2008) *PCT Int. Appl.*, WO 2008004716; *Chem. Abstr.* 148:121736.

[150] Chu, C., Lister, A., Nordvall, G., Petersson, C., Rotticci, D. and Sohn, D. (2006) *PCT Int. Appl.*, WO 2006126938; *Chem. Abstr.* 146:27727.

[151] Chu, C., Lister, A., Nordvall, G., Petersson, C., Rotticci, D. and Sohn, D. (2006) *PCT Int. Appl.*, WO 2006126939; *Chem. Abstr.* 146:27726.

[152] Nordvall, G., Petersson, C. and Sehgelmeble, F. (2007) *PCT Int. Appl.*, WO 2007004959; *Chem Abstr.* 146:142690.

[153] Nordvall, G. and Sehgelmeble, F. (2007) *PCT Int. Appl.*, WO 2007004960; *Chem. Abstr.* 146:142692.

[154] Nordvall, G. and Yngve, U. (2007) *PCT Int. Appl.*, WO2007108744; *Chem. Abstr.* 147:386023.

[155] Grandel, R., Braje, W.M., Haupt, A., Turner, S.C., Lange, U., Drescher, K., Unger, L. and Plata, D. (2007) *PCT Int. Appl.*, WO 2007118899; *Chem. Abstr.* 147:486320.

[156] Nordvall, G. and Yngve, U. (2007) *PCT Int. Appl.*, WO 2007108743; *Chem. Abstr.* 147:386021.

[157] Alcaraz, L., Nordvall, G., Rotticci, D. and Sohn, D. (2007) *PCT Int. Appl.*, WO 2007108742; *Chem. Abstr.* 147:386022.

[158] Grandel, R., Braje, W.M., Haupt, A., Turner, S.C., Lange, U., Drescher, K. and Unger, L. (2007) *PCT Int. Appl.*, WO 2007118900; *Chem. Abstr.* 147:486459.

[159] Schultz, T., Braje, W., Turner, S.C., Haupt, A., Lange, U., Drescher, K., Wicke, K., Unger, L., Mezler, M. and Mayrer, M. (2008) *PCT Int. Appl.*, WO 2008116833; *Chem. Abstr.* 149:402186.

[160] Turner, S.C., Haupt, A., Braje, W., Lange, U., Drescher, K., Wicke, K., Unger, L., Mezler, M., Wernet, W. and Mayrer, M. (2008) *PCT Int. Appl.*, WO 2008116831; *Chem. Abstr.* 149:425817.

[161] Turner, S.C., Braje, W., Haupt, A., Lange, U., Drescher, K., Wicke, K., Unger, L., Mezler, M., Wernet, W. and Mayrer, M. (2009) *PCT Int. Appl.*, WO 2009019286; *Chem. Abstr.* 150:214191.

224 ADVANCES IN THE DESIGN OF 5-HT$_6$ RECEPTOR LIGANDS

[162] Braje, W., Turner, S.C., Haupt, A., Lange, U., Drescher, K., Wicke, K., Unger, L., Mezler, M., Wernet, W., Mayrer, M., Jongen-Relo, A.L., Bespalov, A. and Zhang, M. (2009) *PCT Int. Appl.*, WO 2009056632; *Chem. Abstr.* 150:515206.

[163] Dunn, R., Nguyen, T.M., Xie, W. and Tehim, A. (2007) *PCT Int. Appl.*, WO 2007098418; *Chem. Abstr.* 147:322994.

[164] Dunn, R., Nguyen, T.M., Xie, W. and Tehim, A. (2007) *PCT Int. Appl.*, WO 2008101247; *Chem. Abstr.* 149:288783.

[165] Dunn, R., Nguyen, T.M., Xie, W. and Tehim, A. (2008) *PCT Int. Appl.*, WO 2008147812; *Chem. Abstr.* 150:20153.

[166] Dunn, R., Xie, W. and Tehim, A. (2009) *PCT Int. Appl.*, WO 2009023844; *Chem. Abstr.* 150:237586.

[167] Pasternak, A. and Szymonifka, M.J. (2009) *PCT Int. Appl.*, WO 2009073118; *Chem. Abstr.* 151:33409.

[168] Hauske, J.R. (2006) *PCT Int. Appl.*, WO 2006091703; *Chem. Abstr.* 145:278328.

[169] De Bruin, N.M.W.J., Van Loevezijn, A., Wijnen, J., Herremans, A.H.J., Kruse and C. G. (2008) *PCT Int. Appl.*, WO 2008087123; *Chem. Abstr.* 149:168025; (2008) US Patent Application Publication US 2008171779; *Chem. Abstr.* 149:144007.

[170] Iwema, B.W.I., Keizer, H.G., Van der Neut, M.A.W., Kruse, C.G., Loevezijn, V.A. and Zorgdrager, J. (2008) *PCT Int. Appl.*, WO 2008034863; *Chem. Abstr.* 148:403215.

[171] Sablin, S., Bachurin, S., Beznosko, B., Sokolov, V. and Aksinenko, A. (2009) *PCT Int. Appl.*, WO 2009038764; *Chem. Abstr.* 150:352122.

[172] Rivo, E., Prauda, I., Reiter, J., Barkoczy, J., Gacsalyi, I., Kompagne, H., Sziray, N., Pallagi, K., Egyed, A., Hegedues, E., Levay, G. and Harsing, L.G. (2008) *PCT Int. Appl.*, WO 2008146063; *Chem. Abstr.* 150:19990.

[173] Tkachenko, S. (2009) 9th Interantional Conference on Alzheimer's and Parkinson's Disease, Prague, Abstract 475.

[174] Dupuis, D.S., Mannoury la Cour, C., Chaput, C., Verriè, L., Lavielle, G. and Millan, M.J. (2008) *Eur. J. Pharmacol.* **588**, 170–177.

[175] Gannon, K. and Shacham, S. (2008) *PCT Int. Appl.*, WO 2008002539; *Chem. Abstr.* 148:106221.

[176] Lee, M., Rangisetty, J.B., Pullagurla, M.R., Dukat, M., Setola, V., Roth, B.L. and Glennon, R.A. (2005) *Bioorg. Med. Chem. Lett.* **15**, 1707–1711.

[177] Kolanos, R., Dukat, M., Roth, B.L. and Glennon, R.A. (2006) *Bioorg. Med. Chem. Lett.* **16**, 5832–5835.

[178] GlaxoSmithKline (2007). Available at: http://www.gsk.com/investors/presentations/2007/neurosciences-seminar-ec07/jackie-hunter.pdf.

[179] UK Clinical trials database (2009). Available at: http://clinicaltrials.gov/ct2/show/NCT00895895?term = sam-531&rank = 1.

[180] Synosia (2009). Available at: http://www.synosia.com.

[181] Suven (2009). Available at: http://www.suven.com/news.htm.

[182] Saegis (2009). Available at: http://www.saegispharma.com.

[183] Biovitrum (2009). Available at: http://www.biovitrum.com.

[184] Epix (2009). Available at: http://www.epixpharma.com/products/prx-07034.asp.

Subject Index

Cumulative Index of Authors for Volumes 1–48

The volume number, (year of publication) and page number are given in that order.

Cumulative Index of Subjects for Volumes 1-48

The volume number, (year of publication) and page number are given in that order.